FOUR ANTARCTIC YEARS IN THE SOUTH ORKNEY ISLANDS

José Manuel Moneta

1 : *José Manuel Moneta in the South Orkney Islands*

Four Antarctic Years in the South Orkney Islands

An Annotated Translation of *Cuatro Años en las Orcadas del Sur*

José Manuel Moneta

Assistant in the 1923 Expedition, Assistant Leader in 1925,
Leader of the 1927 and 1929 Expeditions
☼ 9 July 1900 + 30 March 1973

Translators Kathleen Skilton and Kenn Back

Editor Robert Keith Headland

Bernard Quaritch Ltd
2017

Original published by
EDICIONES PEUSER,
Buenos Aires, Argentina

Editions
1st, December 1939 (308 pages)
2nd, September 1940 (308 pages)
3rd, September 1944 (299 pages)
4th, January 1946 (302 pages)
5th, April 1948 (342 pages)
6th, May 1949 (342 pages)
7th, June 1951 (342 pages)
8th, February 1954 (344 pages)
9th, February 1954 (344 pages)
10th, June 1958 (352 pages)
11th, July 1958 (352 pages)
12th, October 1963 (354 pages)

Annotated English Edition
2017 (440 pages)

Published by
Bernard Quaritch Ltd
40 South Audley Street
London W1K 2PR
United Kingdom

rarebooks@quaritch.com
www.quaritch.com

ISBN 978-0-9955192-0-6

Designed by *et al design consultants*
Printed and bound in Great Britain
by Blissetts, London

Foreword from the Author's Family

Several years ago we were contacted by Bob Headland with the idea of translating *Cuatro Años en las Orcadas del Sur* into English. Many Argentines have read this book in its original Spanish version which has been published in twelve editions during a quarter of a century (1939–1963).

We were very supportive of the translation project from its inception. We were also much impressed by the work of the translators, Kathleen Skilton and Kenn Back, who very quickly grasped the spirit of the work. The continuous development and polishing of the manuscript resulted in a very accurate English version.

We thank Bob Headland for updating and completing scientific and historical facts. The annotations and other information included in this edition are a practical and enriching addition to the book. Although most Argentines will be familiar with the localisms, place names and cultural references these will no doubt be beneficial for foreign readers.

Modern technology has allowed the original photographs from the 1920s, a fascinating part of the book, to be reproduced with much better quality than in any previous edition.

Argentine stations have contributed much to Antarctic knowledge and no other country has had a longer continuously operating presence in the area. Our father was an important part of this accomplishment. The Antarctic was his lifelong passion and he kept his special interest in the South Orkney Islands continuously, visiting them to commemorate the fiftieth anniversary of his first expedition just a month prior to his death.

We are deeply touched to see this important historical document, and key part of our father's legacy, translated. We hope the English version will contribute to the knowledge and understanding of the Heroic Age of Antarctic exploration of which our father was a part.

Dora Moneta de Landívar
Martha Moneta de López Naguil
Buenos Aires, 2017

Map 1: Geographical location of the South Orkney Islands

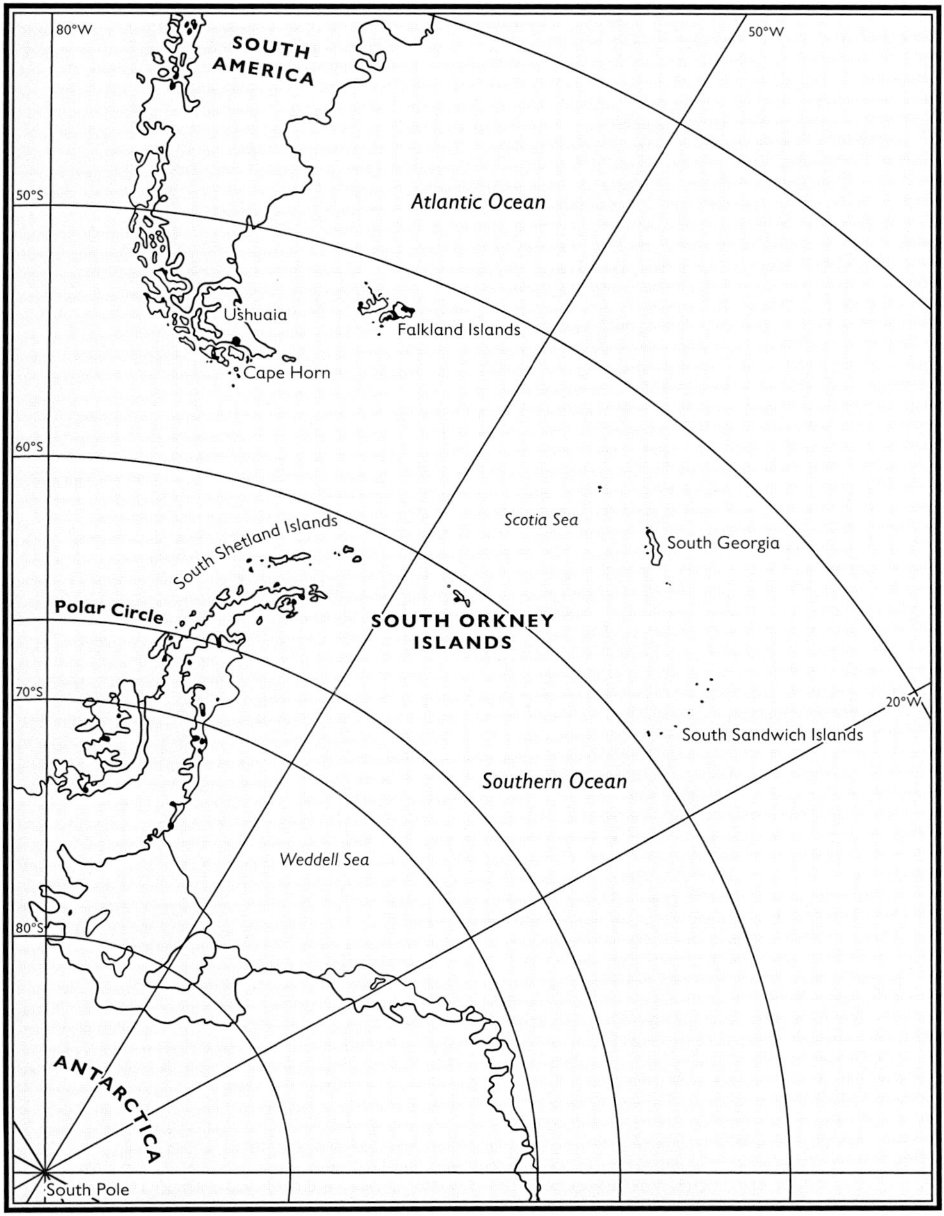

80°W
50°W
SOUTH AMERICA
Atlantic Ocean
50°S
Ushuaia
Falkland Islands
Cape Horn
60°S
Scotia Sea
South Shetland Islands
South Georgia
Polar Circle
SOUTH ORKNEY ISLANDS
70°S
20°W
South Sandwich Islands
Southern Ocean
Weddell Sea
80°S
ANTARCTICA
South Pole

Editorial Foreword

In late 1977 I was stationed on South Georgia by the British Antarctic Survey where I remained for over two years. This provided time and opportunity to become familiar with many of the books, and other literature, on Antarctica and the peri-Antarctic islands around it. The return voyage, in early 1980, visited the South Orkney Islands. Subsequently I was back in the Antarctic regularly, and have been able to visit most of the other islands. Some of these visits have been while lecturing on history and geography to passengers aboard cruise vessels. Thus I came to make several visits to the South Orkney Islands including to the Argentine station on Laurie Island. There I encountered the unique historical book about the observatory: *Cuatro Años en las Orcadas del Sur,* by José Manuel Moneta, an account of early years in the Antarctic but published only in its original language.

The book provided a fascinating and informative description of life on base in the early decades of what became modern Antarctic observatory operations. Records were initiated there in 1903, during the 'heroic age' of polar history, thus it is gratifying to know that meteorological, and other data, continue uninterrupted to the present – the longest such series from the Antarctic. Just over a decade ago the centenary was commemorated in Argentina and Britain when I, then curator, arranged commemorative displays at the Scott Polar Research Institute. The concurrent establishment of a museum in 'Casa Moneta', by Argentine authorities, on the occasion was very appropriate. This is the oldest of the historical buildings of the Antarctic which has been in continuous use and is regularly maintained.

The intention of making a translation, something for which I have been keen after reading the original, languished owing to lack of time. The eventual completion of the work took several years. A colleague, Kathleen Skilton, made the first draft which definitively started the process. An Antarctic friend, Kenn Back, with the advantage of having spent many winters south, commented expertly on the translation. This began the process of 'polishing', including incorporation of maps, illustrations, supplementary notes, bibliography, appendices, and an index.

From the beginning of the project the involvement, interest and help provided by members of the author's family were enormously appreciated, especially for the photographs. Several other friends and colleagues were also involved, as noted in the introduction. I also particularly acknowledge the early enthusiasm of Dr Ricardo Capdevila, Instituto Antártico Argentino historian who, sadly, did not live to see the publication.

As well as the early history of Orcadas the book has details of the whaling industry, shipping, radio communications, and much else in the far south. The complexities of the Anglo-Argentine territorial interests in Antarctic regions were unresolved when the book first appeared. It is gratifying that scientific cooperation continued, and to know that the South Orkney Islands are now under the ægis of the Antarctic Treaty of 1959.

The expertise and enthusiasm of the staff of Bernard Quaritch made the completion of the project a reality.

R. K. Headland.

Contents

Foreword from the Author's Family 5
Editorial Foreword 9
Introduction to the Annotated English Translation . . 19
Prologue to the Tenth Edition: Yesterday and Today . . 25
1: Historical Summary 30
2: Origin of the South Orkney Observatory . . . 33
3: Argentine Expeditions 38
4: Expansion of the Observatory 42
5: Parade of Nationalities 44
6: What took me to the South Orkney Islands . . . 55
7: Equipment and Supplies 59
8: Calling at South Georgia 61
9: Whales 66
10: Harpoon Shot 68
11: In a Cockleshell 72
12: Stuck in a Lift 74
13: *Rosita* and the Crew 76
14: Mountains of Water 79
15: Solid Sea 81
16: Flag Signals 84
17: Our Meeting 86
18: Our Abode 88
19: Impressions 91
20: Panorama 94
21: Farewell 98
22: Our Base 102
23: Water Run 104
24: You will be the Breadwinner 106
25: An Important Man 109
26: Non-rational Inhabitants 113
27: Winter Arrives 117
28: Our Work 121
29: Recording Scientific Data 126

30: Walking on Water 129
31: Ice Fishing 131
32: Argentine National Day, 1923 134
33: By Pick and Shovel 137
34: My Companions 140
35: Lost in the Blizzard 144
36: Tragedy 148
37: Up to the Clouds 152
38: Shags 155
39: Life on Base 158
40: Pharmacy 161
41: An Accident 164
42: Surgical Operation 166
43: Mundane Recollections 169
44: My Birthday 172
45: Infinite Solitude 176
46: Moods 179
47: Spring Arrives 182
48: Waves Under the Frozen Pack 185
49: Blood on the Snow 188
50: Drowsiness 190
51: More about Seals 194
52: Meeting on the Ice 197
53: It is not always Wise to be First 199
54: Reflections 202
55: Moral Turmoil 205
56: The Saga of Four Bars of Soap 207
57: Violence ··· 210
58: Fight! ··· 212
59: Reconciliation 215
60: Break-up of the Sea Ice 218
61: They were Five Men 223
62: All Souls' Day 225
63: Something about Penguins 226
64: The First Egg 229

65: Novel Food 232
66: A Local Meal 234
67: Merciless Looting 237
68: Life Returns 240
69: Possibilities 243
70: End of the Year 246
71: Preparations 248
72: Waiting and Hoping 250
73: A Bet 253
74: Awakening 256
75: The Relief 260
76: Tidings from Afar ··· 263
77: Until We Return 265
78: *Orwell* 268
79: Second Expedition 273
80: In the South Orkneys Again 274
81: Second Winter 277
82: Nothing Without Excitement 280
83: First Entirely Argentine Expedition 283
84: Some Comments 286
85: Voyage via Grytviken 290
86: With the Whalers 296
87: The Ice-Blocked Base 298
88: The Third Time ··· 301
89: Preliminary Work 304
90: Failure? 306
91: General Radio Call 309
92: Emotion 313
93: Contact with the World 316
94: Orcadian Olympics 321
95: Phantom Light 325
96: Aftermath 328
97: Words through the Air 332
98: A Broadcast 334
99: Good Humour 342

100: And Another Year Passed ... 345
101: On the Way Back 349
102: Northbound 352
103: Risky Navigation 356
104: Some Notes of Interest 363
105: Business ... is Business 366
106: Hypothesis 370
107: My Fourth Expedition 372
108: Southern Ocean 377
109: In the Gloom 380
110: In my Home for the Fourth Time ... 383
111: First Things First 386
112: Adoptive Father 389
113: Quack Remedies 392
114: Thus Life Goes On ... 399
115: Farewell! ... South Orkney Islands! 401

Map 1: Geographical location of the South Orkney Islands 6
Map 2: The South Orkney Islands 406
Map 3: Laurie Island 408
Map 4: The vicinity of Orcadas Station 410

Plan 1: Orcadas 412
Plan 2: The basic structure of the dwelling [Casa Moneta] 414

Appendix 1: Toponyms of the South Orkney Islands and South Georgia 415
Appendix 2: Concise Biography of José Manuel Moneta 421

Bibliography 423

Index 427

Illustrations

1: José Manuel Moneta, c.1925 2
2: *Scotia* in the pack ice of Scotia Bay, Laurie Island, 1903 . 34
3: Transfer of Omond House to Argentina, 22 February 1904 39
4: Dwelling house [now Casa Moneta] in 1927, photographed in moonlight 41
5: Dwelling house photographed with accumulated snow 57
6: A group of whale-catchers at anchor in King Edward Cove, South Georgia 61
7: British administrative station with radio pylons, King Edward Point, South Georgia 64
8: Argentine meteorological observatory, King Edward Point, South Georgia 65
9: Grytviken whaling station, South Georgia 68
10: Harpooned whale being injected with compressed air . 70
11: In a whale's mouth at Grytviken 72
12: *Rosita* under way on the Southern Ocean 77
13: Jaramillo measuring the sea temperature in Uruguay Cove 83
14: Panorama of Uruguay Cove, Mossman Hills, Observatory and Mount Ramsay 91
15: Ruin of Omond House, built in 1903 93
16: Panorama; dwelling, radio pylon, magnetic huts, cemetery, and flag pole 95
17: Variometer Hut, housing geomagnetic instruments . 97
18: Frozen Uruguay Cove, Mount Lola and glacier, La Monja, and the Observatory 98
19: Hugo Valentiner, Leader in 1923 101
20: Hauling a heavy sledge loaded with 'solid water' . . 106
21: Life depends on the provisions 111
22: Blue-eyed Shag 112
23: A Gentoo Penguin on its nest 116
24: Snowy Sheathbill 118
25: Brown Skua 119

26: Depot of sacks of coal 122
27: The dwelling in winter 124
28: Devoto and Moneta from the front door of the dwelling 125
29: Stevenson screen with meteorological instruments, dwelling in background 126
30: Absolute Hut housing geomagnetic instruments . . 129
31: Baldoni ice fishing 135
32: At work with pick and shovel clearing snow . . 138
33: Hoarfrost accumulated on the guy-wires of the dwelling chimneys 141
34: Panorama from Mount Ramsay 150
35: Uruguay Cove covered with pack ice, Point Lola and Mount Lola on the left 151
36: Aboard a boat ready to hunt shags 157
37: A colony of Blue-eyed Shags 159
38: Rookery of Adélie Penguins at Point Martin . . 178
39: Giant Petrel 182
40: Pũez Montero with three Weddell Seals 191
41: Monti and Pũez Montero returning from a visit to the seals 193
42: Female Weddell Seal with a pup on the sea ice . . 196
43: Collecting penguin eggs 202
44: The dwelling in the silence of the winter night . . 216
45: Ice stalactites on the eaves of the dwelling from snow melting from the roof 221
46: The grave of Otto Diebel with two Chinstrap Penguins 222
47: The cemetery in summer, 1929 225
48: A rookery of Adélie penguins 231
49: Returning to base with a sledge loaded with fresh meat . 236
50: A sledge-load of penguin eggs 240
51: José Moneta, Ernst Bruhns, and Otto Zieger . . 257
52: The first radio-telegraphy station (1925), later the flammables store 277
53: Georg Piper's islands and the sea frozen to the horizon . 281
54: The dwelling in winter snow 284
55: At sea aboard *Tijuca* under sail 289

56: Shaven heads aboard *Tijuca* 292
57: *Tijuca* in Grytviken whaling station, mid-1920s . . 294
58: Whaling factory *Orwell* anchored off Signy Island . 300
59: Jaramillo working at the desk in the parlour . . . 304
60: Antenna pylon, Casa Moneta in the background . . 307
61: Jaramillo and Fállico with a radio receiver at Point Martin 310
62: Kahl in the radio room 318
63: Communicating with the world 320
64: Becker at work in the kitchen 322
65: A birthday party during the 1927 Expedition . . 324
66: A game of poker in 1927 325
67: Magnetic huts, store house, and post bearing the phantom light 331
68: Variometer Hut during the spring melt 336
69: Base at night; Uruguay Cove and Mount Ramsay . 337
70: Celebrating Argentine national day, 25 May 1927 . . 342
71: Miguel Jaramillo, Assistant Leader in 1927 . . . 348
72: Return voyage aboard *Tijuca* 353
73: The 1927 Expedition aboard *Tijuca* in South Georgia . 354
74: Poster advertising the 1931 film *Entre los Hielos de las Islas Orcadas* 368
75: Winter: Mossman Hills, Point Diebel, frozen Scotia Bay and Uruguay Cove 372
76: Chichilo and Kataplum 376
77: Summer 1929 388
78: Members of the 1929 Expedition 391
79: Devoto and Páez Montero enjoying *maté* 392
80: Monti in his bunk 398
81: Food and other provisions landed for the 1930 Expedition 402
82: Five of the six men of the 1930 Expedition alone on the beach 403
83: Casa Moneta in 2007, now the museum of Orcadas Station 405
84: Orcadas cemetery in 2007 405

Introduction to the Annotated English Translation

During the century following the 'heroic age' of Antarctic exploration most accounts of the early expeditions have been republished, many in English translations. Simultaneously several general books about these expeditions have appeared. Only a few of the books from, or otherwise relating to, this period have had as many editions as *Cuatro Años en las Orcadas del Sur* which has had twelve in the course of twenty-four years (1939 to 1963).

Thus as the 'heroic age' centennial commemorations draw to a close, it is opportune for the first English translation of José Manuel Moneta's work to be published. The decade when he was involved, shortly after the end of the First World War, was when polar exploration, indeed exploration generally, was resuming after peace returned. The early years, and experience on such an isolated station, provide a fascinating glimpse of Antarctic life, especially when it coincided with a major technological development: radio communications which, despite a faltering start, introduced vast changes. The advent of radio may be regarded as a major division between former and modern expeditionary activities, before and after instant communication between the Antarctic and the rest of the world became practicable. The book also includes interesting material about the early whaling industry with personal descriptions of activities. These factors were coincidentally a major facilitator of research and exploration during several decades of Antarctic history.

The translation of the main title of the Argentine editions, from 15 December 1939, was *Four Years in the South Orkneys* [*Cuatro Años en las Orcadas del Sur*]. The first and second editions also had the sub-title: *Narratives and Notable Facts about the Argentine Expeditions* [*Narraciones y hechos salientes de las expediciones argentinas*]. The addition of 'Antarctic' to the English title was made to avoid any confusion for a Northern Hemisphere reader who might be more familiar with the Orkney Islands north of Scotland.

Several of these twelve editions, listed on the verso of the title page, appeared with variant paper and bindings; the first and second editions were in hard and soft bound versions, bindings were variously printed or some embossed, and two later editions appeared close in time to supply school or other educational requirements. The second edition (1940) reproduced twelve reviews from Argentine newspapers. Only the third edition (1944) included a note on the number printed stating the run was 4000, 1000 of which were on a higher quality paper. The type was reset for the third to the eighth editions after which it was maintained in galleys except for the insertion of more names in the *Parade of Nationalities* chapter, addition of the *Prologue to the Tenth Edition: Yesterday and Today* which also appeared in the two subsequent editions, and the footnote in § 44 of the twelfth edition. Consequent pagination revisions became necessary for each of these editions.

Meteorological observations at the South Orkney Islands began on 26 March 1903 when *Scotia*[1] of the Scottish National Antarctic Expedition (1902–04), led by William Speirs Bruce, anchored there and remained during the 1903 austral winter. An observatory was established ashore before the ship left on 27 November 1903, at the beginning of the next summer, for the Falkland Islands and Buenos Aires. Early next year, 1904, by arrangements made through the British Legation in Buenos Aires, the Observatory was to be transferred to the Argentine government. Bruce, aboard *Scotia,* returned to the South Orkney Islands on 14 February 1904 after another call in the Falkland Islands. A formal transfer was made on 22 February when the Argentine Meteorological Service[2] of the Ministry of Agriculture assumed responsibility for continuing the observations and maintaining the station. In 1952 this devolved on the Argentine Navy which currently runs the station. The centenary of its foundation was celebrated on 26 March 2003 in Britain at the Scott Polar Research Institute in Cambridge and the National Museum of

1. *Scotia* was a barque-rigged auxiliary steam vessel built in Norway in 1891, destroyed by fire in 1916; length 42·7 m, capacity 400 register tons.

2. The Argentine Meteorological Service (Servicio Meteorólogico Nacional) was founded on 4 October 1872; it is one of the world's earliest.

Scotland in Edinburgh, and on 22 February 2004 in Argentina at the Instituto Antártico Argentino, by the Armada Argentina, and other organizations in Buenos Aires. In the Antarctic, Orcadas Station was involved in both commemorations. Direct voice contact between the Cambridge Institute and Orcadas Station was made during the former, courtesy of the Argentine Navy which completed the link by radio-telephone from Buenos Aires.

The Observatory has continued to function achieving over eleven decades of continuous records of meteorological and geophysical phenomena. Thus it has the longest continuous scientific data series from, and oldest permanently inhabited place in, any Antarctic region[3]. There is, especially in the time of many Antarctic centenaries, much interest in preserving the surviving historic huts of the Antarctic. In this the Laurie Island station is unique: for it includes the oldest Antarctic building which has remained in continuous use. Now known as 'Casa Moneta', it was constructed in February 1906. It remained the principal building of the base until a replacement was erected forty years later. A modern station is now on the site. In 1992 'Casa Moneta' became the museum of Orcadas Station. As well as continuous maintenance, a programme of refurbishment was undertaken at its centenary, for which this book provided much inspiration[4]. Only three other books describe the circumstances of early wintering on the South Orkney Islands: two refer to the years 1903 and 1904 the first years of the Observatory[5] and one other briefly refers to the visit by *Uruguay* in 1905[6]. Other relevant materials from the early decades of the station include various meteorological, biological, other scientific reports, with several notes in the *Magazine of the Royal Scottish Geographical Society,* most written by Robert Mossman.

Colleagues, Kathleen Skilton in Britain and Kenn Back in Uruguay,

3. The second continuous record started on 17 January 1905 at King Edward Cove, South Georgia. It was initiated by Captain Carl Anton Larsen, using instruments provided by Walter Davis (see § 2), after the foundation of Grytviken. The third continuous record began in 1947, 42 years later.

4. This is described in detail in: Vairo, Capdevila, Aldazábel, and Pereyra, 2007.

5. Brown, Mossman, and Pirie, 1906; Bruce, 1992; and Acuña, 1982 and 2015

6. Mavaroff, 1954

made the translation which Robert Headland reconciled and edited (the latter two have wintered in the Antarctic with the British Antarctic Survey, and the last has also visited Orcadas). Bob Burton, Peter Clarkson, Deirdre Hannah, Jim Heron, and Hilary Shibata have commented on the edited manuscript. José Manuel Moneta's daughters, Dora and Martha in Buenos Aires, his grandsons Gervasio Landívar+ and Julián Moneta, and granddaughter Maria Landívar provided valuable help and information, especially gifts of copies of the latest edition of the book with permission to use the original illustrations. Assistance with *modismos argentinos* was also appreciated. Dr Ricardo Capdevila + (Instituto Antártico Argentino, Buenos Aires), Dr Javier Figueroa (Argentine Embassy, London), Sr Carlos Mey (Museo de la Armada Argentina, Tigre), Brigadier Mayor Rubén Palazzi (Fuerzas Aéreas de Argentina, Buenos Aires), Dr Carlos Vairo (Museo Marítimo, Ushuaia), with several other Argentine friends and colleagues, helped with introductions, historical and other specific information, advice and comments. The facilities of the Scott Polar Research Institute in Cambridge, and the Royal Geographical Society in London, were also of assistance.

Some minor differences between the twelve editions of the book became apparent as the translation proceeded. These have been reconciled to accord with the last, 1963, edition generally and are indicated in the footnotes. End papers, preliminary pages, and map dispositions also varied slightly although, in general, conformity between editions was high. Editorial annotations as footnotes, a total of 311, have been added liberally to elucidate particular points and provide references; although a proportion of these would have needed little explanation for a reader familiar with Argentine history and geography (there were only two substantive footnotes in the twelfth edition, in § 5 and 44). The lists of personal names, which were kept up to date in every edition of the book, are completed to the expedition of 2017 – the editor is confident that the author would have desired this to be done. For the same reason the records of fatalities and of the cemetery are complete to the date of this publication. The translation has endeavoured to retain the author's original style but some sentences have been divided to render them in a form more usual in English than Castilian syntax. The frequent ellipses (···) are maintained as they are another significant aspect of the style to indicate pauses or further thoughts, rather than omissions. Similarly the sentence and paragraph structures correspond closely to those of the original.

The original 115 chapters, in the form of episodes or sections, beginning after the *Prologue to the Tenth Edition: Yesterday and Today*, have been numbered for ease of reference. Maps have been redrawn from modern British and Argentine sources, a general one of the region with a plan of Casa Moneta is also included. A date and a few minor geographical inconsistencies in distance and direction have been silently corrected. An occasional anachronism has been indicated (presumably a consequence of reliance on memory during the 16 years between the author's first arrival and the date of publication). Quotations translated into Spanish from English sources have reverted to the original sources, which appear in the bibliography, to avoid any complications from back translation.

Where practicable all personal names have been given in full on their first occurrence and in their original languages rather than in Hispanicised forms. Where no official British place-names exist toponyms are rendered in the original language, with an occasional parenthetical translation on the first occurrence ('La Monja', 'Isla de los Estados', and 'Río de la Plata' rather than 'The Nun', 'Staten Island', or 'River Plate', for instance). Differences between Argentine and British toponomy are generally minor except for translation of terms for features. All miles specified are nautical miles [1852 m], a knot is one nautical mile an hour [1·852 km h^{-1}], and a barrel of whale oil is 167 litres, about one sixth of a cubic metre; conversions to metric units [in brackets] have been rounded. Linnæan and vernacular names of biota correspond to those of recent taxonomic revisions.

All editions of the book were copiously illustrated with monochrome photographs, including six panoramas, taken by the author during his 1927 and 1929 expeditions. These varied slightly among the editions, the first of which included 74 illustrations and the twelfth had 78 (but not all from the South Orkney Islands). The majority of the 84 illustrations in this edition, where no attribution is indicated, appeared in the first edition. Maria Landívar was kindly able to provide these images from contact prints of the original negatives which greatly enhanced the quality of illustrations. The occasional other plate, mainly of animals, are from either *The Voyage of the Scotia* (reference above) or William Bruce's booklet *Life in the Antarctic*[7] thus they are from Laurie Island in 1903 or 1904. Other images are from the *Scotia* collections in 1903 and 1904, South Georgia in the 1920s, the poster from the Ushuaia Maritime Museum, and the recent ones at the end, provenance is acknowledged in the captions. Six different dust jackets or cover illustrations adorned various editions of the book[8].

The problem of sovereign claims in Antarctic regions is alluded to on several occasions in the book, and José Manuel Moneta was a strong proponent of those of Argentina. Although the South Orkney Islands constitute one of the many peri-Antarctic islands, distinct from the continent of Antarctica, they are south of the 60° south parallel, and thus under the ægis of the Antarctic Treaty of 1959, which came into force in 1961. Argentina and Britain were among the twelve original signatories of the Treaty. Thus, two years before the last Argentine edition of this book, a resolution to any such potential conflicts became practicable, which currently remains in effect.

A gazetteer of Antarctic places mentioned in the text (49 entries), a concise biography of José Manuel Moneta, a bibliography (41 references), and an index have been inserted to conclude this translation.

Kenn Back	Robert Headland	Kathleen Skilton
Translator	Editor	Translator
Carmelo	Cambridge[9]	Cambridge

7. Bruce, 1907

8. These are illustrated on the cover. That of the 1939 edition was also used in 1940, the 1944 version was new, the third version appeared on four editions (1946 to 1951), the fourth on both 1954 editions, fifth on those of 1958, and sixth on the 1963, 12th edition.

9. Correspondence (2017): postal address; Scott Polar Research Institute, University of Cambridge, Lensfield Road, Cambridge, United Kingdom, CB2 1ER; electronic address: <rkh10@cam.ac.uk>.

Prologue to the Tenth Edition: Yesterday and Today[10]

Many years have passed after my first expedition to the South Orkneys in 1923, with subsequent ones of 1925, 1927, and 1929. Although not the first, I was one of the earliest Argentines to spend long drawn-out years surrounded by Antarctic ice thousands of years old. We lived in a small, basic, wooden hut, with a simple diet supplemented by local seals and penguins. We had no company other than four or five companions thrown together by chance, with no news of anything, or of anyone, during our interminable year of voluntary exile. To survive, we trusted solely in God's grace and our robust good health. I recall as well the terrible silence that surrounded us in the early days. That silence was broken on 30 March 1927, when we inaugurated the first radio-telegraphy link between the Antarctic and Argentina[11]. This historic achievement was made possible by the courage and tenacity of the men accompanying me on that South Orkney adventure.

The four years and nine months that I spent in the Antarctic were rewarding and productive; not only for me personally, but also for the idea that began to develop in my mind after completing my first expedition in 1923; namely 'to discover and establish Antarctica for the people of Argentina'. Effectively that is what happened. Four decades ago nobody had a definite concept of what comprised the Antarctic[12]; very few people had seen in action what, at that time, was

10. Written for the 10th edition of 1958, also published in the 11th and 12th (last) editions.

11. This has occasionally been reported as the first radio communication with the Antarctic (Basbous, 1971, Pierrou, 1975). However earlier radio transmissions had been received from, and transmitted to, Antarctic regions: in 1912 from Macquarie Island, in 1913 from Commonwealth Bay on the continent and Deception Island one of the South Shetland Islands, and in 1925 from South Georgia. Many whaling and other vessels also used radio communications before 1927 when they were in the Southern Ocean, as is related in later chapters.

12. The term 'Antarctica' is used to refer to the continent and the broader expression, 'the Antarctic', for the continent and the Southern Ocean with its islands, which include the South Orkney Islands and South Georgia.

a thriving whaling industry[13]. The fingers of one hand would give the number of people who had the foresight and intuition to predict the untold mineral riches that lay in the high frozen peaks which emerge from the vast, ice-covered South Polar continent. There was still no concept that all this was Argentine. Most Argentines knew nothing at all about the Antarctic. That ignorance and lack of awareness of our southern assets was exploited to the full by foreign powers. Covertly they explored, established bases and began exploiting certain resources. Just outside our own front door, the Observatory of our base in the South Orkneys over which our flag flew, foreigners levied taxes on oil from the blubber stripped from whales in our territorial sea, right under our noses[14].

After my last expedition in 1929, I dedicated all my efforts, indeed my whole life, to Antarctic matters, especially to the formal recognition of the Argentine Antarctic[15]. The eminent newspaper *La Prensa*[16] gave me the opportunity to write more than a hundred illustrated articles about diverse aspects of the south polar region. I received awards from the National Commission for Culture that allowed me to deepen my knowledge of the Antarctic in foreign countries. I was subsequently asked by scientific and cultural institutes to speak about the Antarctic. This all contributed, albeit rather late in the day, to our country awakening and taking an interest in the affairs of the far south. Successive Argentine governments, from 1930 to the present, began to emerge from their lethargy. They noted the advantages that other countries were already gaining from incursions into the Antarctic; they fostered various initiatives to try to preserve what was ours and

13. Antarctic whaling did not begin until November 1904 with the establishment of Grytviken on South Georgia which operated until 1965. The subsequent commercial catch of whales in the Southern Ocean was entirely pelagic until the industry ceased to exploit Antarctic regions in 1987.

14. Regulation and taxation of revenue of the Antarctic whaling industry, which was principally a Norwegian enterprise was by the government of the Falkland Islands and Dependencies in Stanley.

15. First specified in 1943; extended and defined as 'Antártida Argentina' in 1947: Palazzi, 2005.

16. *La Prensa,* founded in 1869 and still published, was a major daily newspaper during Moneta's years in the South Orkneys and for many subsequent decades.

which, through our own laziness and naiveté, was being taken from our hands.

In January 1956[17] I had the opportunity to make a new journey to the Antarctic, which of course was very different from those I had made thirty years earlier. On the morning of 31 January 1956 we left Buenos Aires in an aircraft of the Argentine Navy and arrived in Ushuaia, the provincial capital of Tierra del Fuego, that evening. After a four day stay in the Fuegan capital, we sailed south aboard the naval transport ship ARA *Bahía Aguirre*[18] which, during the succeeding days, visited the isolated Argentine bases and detachments on Deception Island, Half Moon Island, Teniente Jubany, Hope Bay, Almirante Brown, and Melchior[19]. On the afternoon of 16 February we docked at the wharf on our return to Ushuaia, and on the following day, at 18:00 hours, our aircraft landed at Ezeiza[20] on the outskirts of Buenos Aires.

In quality and quantity, the facilities and installations of our modern Antarctic stations are vastly superior to anything we had when I lived in the South Orkneys. The amenities currently provided for twelve or more men on each base are extraordinary. The living quarters are ample and comfortable; well-furnished, and with the advantages of central heating, bathrooms with hot water on tap, and an infirmary. Nowadays there are also radio-telephones for base personnel to be able to communicate with their families in any part of Argentina that has a telephone service.

All the bases now have freezers for the preservation of meat supplied

17. After this Antarctic visit the author made another, to the South Orkney Islands in 1973, half a century after his first arrival; see his concise biography in Appendix 2.

18. *Bahía Aguirre* was a transport vessel, in service from 1950 to 1981, which assisted the relief and supply of Argentine Antarctic stations for 30 summers. ARA is an acronym for *Armada de la República Argentina* (Argentine Navy).

19. During the 2017 winter Argentina operated six Antarctic stations: Orcadas, Esperanza (Hope Bay), Belgrano (Luitpold Coast), Marambio (Seymour Island), Carlini (formerly Jubany, King George Island), and San Martín (Marguerite Bay). Decepción (Deception Island), Cámara (Half Moon Island), Brown (Paradise Harbour), Melchior (Melchior Islands), and a few others have been occasionally open during recent summers.

20. Ezeiza is now the international airport of Buenos Aires.

from Buenos Aires for daily consumption. This is a vast contrast to the diet of seal and penguin that we were obliged to consume in times gone by. In the lounges of the modern bases, in the corner reserved for the bar, are bottles of liquors and spirits of well-known brands, both Argentine and imported. These are freely available, without the restrictions or rationing we were obliged to impose on our modest whisky and brandy stocks that had to last the whole year. Our allowance was a single tot of spirits for each man once a week. With cigarettes, it was much the same. As Pedro Casariego, a member of the 1927 winter party, said 'we were ready to smoke the stuffing of the mattresses'. Today on all Argentine Antarctic bases, there is an abundance of well-known brands of cigarette, national and imported.

During the last few years many of our fellow Argentines have lived in or travelled to the Antarctic. Most of them proudly display the distinctive Antarctic decoration with which our authorities honour those undertaking one or more Antarctic campaigns. The only exceptions are the men of the South Orkneys from the heroic years – those who spent a winter there in the most primitive circumstances, who survived there throughout the long year, who depended on seal and penguin meat for their nourishment, who were in total isolation from the rest of the world, but by their unshakeable faith and ardour laid the foundations on which the Argentine Antarctic was built. These men alone have no right to receive any award of recognition or decoration. With the passing of the years many of these pioneers are no longer with us, but the few who remain on this Earth continue their instinctive struggle not to drown in or be asphyxiated by our egotistical civilization.

I have not dedicated this book to any particular person. With respect
and admiration I now offer it to the memory of all those who,
in the long-passed heroic era of the South Orkneys, lived
among the ice-fields of the South in the first permanent
base in the Antarctic and who, by their patriotic
and unselfish efforts, helped to establish,
strengthen, and make known, in our
country and in the world, the
existence of what is now
Antártida Argentina.

José Manuel Moneta[21]

21. Signature copied from the 12th edition of *Cuatro Años en las Orcadas del Sur,* 1963.

1: Historical Summary

The discoveries of the coasts of Antarctica were made at the beginning of the nineteenth century as a logical consequence of the rapidly growing sealing industry[22]. The abundance of seals in northern polar regions was well-known. It was from regions where over-exploitation had so depleted them, that sealing fleets were consequently forced to seek new regions where they could continue their enterprise. This was one of the factors that led to the discovery of new lands on the periphery of the Antarctic. Thus we can begin our historical résumé with the discovery of the South Shetland Islands by Captain William Smith in the year 1819[23].

According to the accounts told by masters of the sealing ships who, accidentally or by pure chance, became explorers, the depletion of populations of Fur Seals[24] and even Elephant Seals[25] had, by then, already become apparent in the south. During the summer of the years 1821–22, Captain George Powell mentioned in his log-book that on Elephant Island[26] there were very few seals, owing to the severe over-exploitation during previous years. To give some idea of the extermination these mammals suffered it will suffice to mention that Captain William Webster, who visited the South Shetland Islands in 1829, said he did not see a single seal[27].

In 1821 a chance meeting took place in the South Shetlands between the Captains Nathaniel Brown Palmer, from Connecticut in the United States, aboard the sloop *James Monroe,* and George Powell, from Britain, born in London, in command of the whaling vessel *Dove*. Because they had been unable to find sufficient seals and

22. Much discovery, including that of the South Orkney Islands, was made by British and United States sealers during the early 1800s. Sealing had been a major Antarctic industry from the late 1700s throughout the entire 1800s and the old form of sealing persisted until 1922. For historical accuracy 'seals' and 'sealing' have been substituted for 'whales' and 'whaling' in this and the next paragraph.

23. Campbell, 2000

24. *Arctocephalus gazella*

25. *Mirounga leonina*

26. Elephant Island is one of the north-eastern South Shetland Islands.

27. He visited during a British scientific expedition aboard HMS *Chanticleer*: Webster, 1834.

whales to make a profitable voyage, they resolved to reconnoitre towards the east in a search for new regions to pursue the industry. That was how, on 6 December 1821, they discovered the Inaccessible Islands, a group situated to the west of the South Orkney Islands. On discerning more land farther east Powell continued his voyage arriving at the south-west of what is now known as Coronation Island, surmounted by Sandefjord Peaks, where he made his first landing. A document sealed in a bottle was buried at the site, which was, in fact, an act of taking possession of these inhospitable lands. Powell wrote about the event[28]: *At this place we landed and took possession in the name of King George the Fourth, leaving a bottle, containing this note, stating the particulars of this discovery; and as I imagined it to be the first land discovered since the coronation of our most gracious sovereign, so I have named it 'Coronation Island'.*

At the same time as Powell was landing and taking possession, Captain Palmer, who had proceeded a short distance farther east, was entering Lewthwaite Strait. He too disembarked and landed on both sides of the strait.

Then, nine weeks after the discoveries by Palmer and Powell, another navigator arrived on these shores, James Weddell[29]. Unaware that the islands had been previously visited, he named them the South Orkney Islands; he landed on Saddle Island and made the first determination of the geographical coordinates.

Sixteen years later the archipelago was visited by the French navigator Jules Sebastian César Dumont d'Urville, in command of the ships *L'Astrolabe* and *La Zelée,* which successfully circumnavigated the Antarctic[30].

After them the South Orkneys remained apparently forgotten for fifty-five years[31], until William Speirs Bruce visited aboard the ship

28. Quoted directly from: Powell, 1822.

29. Weddell, 1827

30. The expedition visited the South Orkney Islands on 26 January 1838: Rosenman, 1987.

31. A German and several United States sealing voyages visited the South Orkney Islands between 1873 and 1882: Headland, 2009.

Antarctic[32]. About the same time, the Norwegian Captain Larsen[33] arrived in the archipelago to obtain reliable information about the size of the seal colonies described by Dumont d'Urville, and to determine the potential for establishing a whaling industry there.

Despite the many voyages made by navigators plying the seas around the Antarctic, by the end of the nineteenth century there still was no firm and reliable notion of what really constituted the peri-Antarctic islands[34] and the South Polar continent. This was especially so during winter; all expeditions to these shores were in the summer, the most favourable season for the whaling industry.

For this reason, well-known scientists of the day agreed that only by dispatching explorers to the Antarctic who were trained to keep precise records for purely scientific purposes, in properly manned and equipped expeditions, could research be accomplished that would extend the sparse knowledge of these regions.

Thus, in 1901, a major international Antarctic programme was organized. It was initiated by the Royal Geographical Society in London with the collaboration of several other countries. In that year Britain[35] dispatched Captain Robert Scott aboard *Discovery*, Germany sent an expedition aboard *Gauss* commanded by Captain Erich von Drygalski, and Sweden with the well-known geologist Dr Otto Nordenskjöld aboard *Antarctic*. The third of these expeditions

32. Actually Bruce was aboard *Balæna* (Captain Alexander Fairweather), one of the four ships of the Dundee whaling reconnaissance, from Scotland, in 1892–93. *Antarctic* made two voyages to Antarctic regions, in 1893–95 to the Ross Sea and 1901–03 with Nordenskjöld, but Bruce was not aboard on either voyage: Murdoch, 1894.

33. Captain Carl Anton Larsen (1860–1924) was leader of Norwegian whaling reconnaissances in 1892–93 and 1893–94 aboard *Jason*. Later he was master of *Antarctic* (1901–03) and founder of Grytviken whaling station, South Georgia, in November 1904.

34. The peri-Antarctic islands consist of 19 groups in, or near, the Southern Ocean although distant from the continent, Antarctica. The South Orkney Islands and South Georgia are two of them.

35. Moneta uses the Argentine conventional terms *Inglaterra* (England) and *inglés* (English) when referring to Britain. The noun Britain and adjective British have been substituted where persons from the several parts of the United Kingdom (England, Ireland, Scotland, and Wales at the relevant time) were involved, with the exception that Scotland (*Escocia*) and Scottish (*escocés*) are used when specifically referring to William Bruce's *Scotia* expedition.

was to explore the south-eastern parts of Graham Land. It was joined by the then 2nd Lieutenant José María Sobral from the Argentine Navy[36]. Subsequently there was a French expedition aboard the ice-strengthened vessel *Le Français,* commanded by Dr Jean Charcot[37].

All these expeditions had to explore the Southern Ocean in the regions assigned to them. They would simultaneously conduct diverse scientific programmes: magnetic observations and other geophysics, meteorology and climatology, geology, topographic survey, studies of the flora and fauna, oceanography, etc.

Argentina cooperated with the international programmes and established the first Observatory, on Isla Año Nuevo [New Year Island], north of Isla de los Estados [Staten Island]. It was placed under the supervision of the Argentine Navy[38].

With the same objects as the expeditions previously mentioned, an expedition was dispatched from Scotland under the command of the well-known and experienced explorer Dr William Bruce[39]. His exploration and research concentrated on the regions south of the South American continent. This expedition left Edinburgh on 2 November 1902 aboard *Scotia.* It was the expedition that established the first meteorological station on the South Orkney Islands, which subsequently became the present Argentine National observatory.

2: Origin of the South Orkney Observatory

During the voyage south Dr Bruce, Leader of the Scottish expedition, called at Port Stanley in the Falkland Islands[40], to refuel and victual. The expedition departed on 25 January 1903 and managed to reach the South Orkney Islands where a landing was made on Saddle Island. A

36. Capdevila, 1993

37. A concise history of all the expeditions of this period, and others to the Antarctic, appears in: Headland, 2009. Specific details of the South Orkney Islands are also in: Marr, 1935.

38. A meteorological observatory was established on Isla Año Nuevo on 31 October 1901.

39. William Speirs Bruce (1867–1921); as well as his Antarctic voyage aboard *Balæna* he had been involved in Arctic expeditions to Franz Josef Land, where he wintered from 1896 to 1897, Svalbard, and Novaya Zemlya: Speak, 2002.

40. Moneta uses the Argentine name 'Islas Malvinas' for the Falkland Islands throughout the book.

2: Scotia *in the pack ice of Scotia Bay, Laurie Island, 1903*
(courtesy Scott Polar Research Institute, University of Cambridge)

few days later, *Scotia* proceeded on her voyage to the south-east and reached 65° south latitude at 17° west longitude, where an extensive field of pack ice blocked progress obliging the ship to turn back[41].

Dr Bruce resolved to spend the winter at the South Orkneys and brought *Scotia* into a wide bay on the south coast of Laurie Island which he named after his vessel. The advanced state of summer, and early arrival of pack ice, resulted in *Scotia* becoming beset and remaining frozen-in all winter. *Scotia* had dropped anchor on 25 March 1903[42].

To facilitate the expedition's scientific observations, a small stone hut was built ashore on solid ground above one of the rocky beaches of Scotia Bay. The interior was lined with wood. In this refuge six of the explorers from *Scotia* resided until such time as the ship should be released from the grip of the ice. This rudimentary little abode was named 'Omond House' in honour of one of the organizers of the expedition[43].

41. Bruce records the attainment of the most southerly position as 74°25´S at 17°12´W, where he was blocked by ice, on 22 February 1903.

42. Bruce, 1992, and: Brown, Mossman, and Pirie, 1906.

43. Robert Trail Omond was the first Superintendent of the Ben Nevis Observatory in Scotland.

This basic refuge was the first construction on the South Orkneys, though sadly only a pile of rubble now remains of it: the harsh weather, avalanches from adjacent hills and the action of the sea during summer months have all contributed to its ruin[44].

The party who disembarked from *Scotia* to live ashore comprised six scientists led by Robert Mossman[45], a well-known meteorologist who had been Superintendent of the principal meteorological station in Edinburgh from 1896 until 1900. He was also, on several occasions, Superintendent of the Observatory on Ben Nevis, Scotland[46].

The members of the Scottish National Antarctic Expedition all devoted their time to their respective scientific work. Shortly afterwards another hut was built[47] to facilitate geomagnetic observations. This was of wood, to house the instruments which would be adversely affected by some of the materials used to construct the first abode[48]. At the same time two screens housing meteorological instruments were deployed near the beach[49].

At the end of the year the ice released *Scotia* from its grip. The Leader of the expedition returned to Buenos Aires aboard *Scotia* to provision for a further year of scientific research and exploration. In anticipation of an early return he left a party of six scientists continuing observations

44. The ruins of Omond House, with the cemetery and several associated relics are protected as Historic Site and Monument number 42, declared by the ninth recommendation of the seventh Antarctic Treaty Consultative Meeting, in Wellington, New Zealand, November 1972.

45. Robert Cockburn Mossman, 1870–1940, subsequently continued a meteorological career in Argentina becoming Director of the Dirección de Meteorologia, Geofísica e Hidrología until his retirement in 1938. The station on Laurie Island was one of his major interests.

46. Several early meteorologists on the South Orkney Islands, and elsewhere in the Argentine Meteorological Service, were from the Ben Nevis Observatory in Scotland which was open from 1883 to 1904; William Bruce had been an observer, but not a superintendent, in 1895 and 1896: Swinney, 2002.

47. Named by Bruce as the Copeland Observatory after Ralph Copeland (1837–1905), the Astronomer Royal of Scotland. It remained in use to 1954 when a replacement was built on its site.

48. That is for the magnetometers which require a non-magnetic (iron free) structure distant from any extraneous magnetic influence.

49. A Stevenson Screen is an instrument shelter, a standard piece of meteorological equipment; a white painted box, with louvered sides to allow air to circulate and protect the instruments within from direct solar radiation. Such equipment is used in meteorological stations in all climates. One may be seen in Illustration 29.

on the South Orkneys, led by Mossman and living in the stone hut.

When Dr Bruce arrived in Buenos Aires he initiated negotiations with the Argentine government to ensure the continuity of the meteorological programme begun the previous year[50]. Dr Bruce offered to provide free transport for the Argentine scientists selected to take charge of the rudimentary South Orkney installations of the Scottish expedition. The offer was accepted by the President of the Nation, General Roca[51]. His decision was based on the opinions expressed by Mr Walter Davis[52], Director of the Argentine National Meteorological Service at that time, who advised him to continue Dr Bruce's Antarctic observations.

For this purpose the National Executive Power issued a historical decree which reads[53]:

Buenos Aires,
2 January 1904.

In view of the Note received from the Director of the Meteorological Service and of the documents and other antecedents relating to the establishment of new meteorological and magnetic stations in the southern sea areas of the Republic; and considering that it is of great scientific and practical importance to extend to the aforesaid regions the observations which are currently being made in the Observatory on Isla Año Nuevo and in the south of the Republic–

The President of the Republic decrees:

Article 1. – The Director of the Argentine Meteorological Service is authorised to take over the installation offered by Mr William S. Bruce in the South Orkney Islands and to establish a new meteorological and magnetic observatory thereon.

Article 2. – The staff will be composed of the employees designated by the Ministry of Agriculture, who may, at a later date, come under the adminis-

50. These negotiations involved William H. B. Haggard, Minister of the British Legation in Buenos Aires.

51. General Julio Argentino Roca (1843–1914) was the Argentine President from 1880 to 1886 and from 1898 to 1904.

52. Walter Gould Davis (1851–1919)

53. Translation quoted from: Bush, 1982.

tration of the Ministry of Marine.

Article 3. - Such employees shall be replaced annually by those appointed to relieve them and they will be conveyed by a vessel of the Fleet.

Article 4. - The allocation of salaries and sustenance allowed for those who are not provided for in the budget, and all essential additional expenditure, shall be determined by the Ministry of Agriculture and charged to the appropriate section of the General Budget.

Article 5. - This Decree is to be promulgated, published and deposited in the National Registry.

Roca

W. Escalante[54]

Concurrently with the enactment of this first actual decree of Argentine sovereignty in the Antarctic, another important event signalled the first stage in the development of an Argentine civil administration in the polar regions. I refer to the establishment of the first Antarctic postal service which was Argentine.

Dr Francisco Moreno[55], the distinguished geographer and naturalist, doctor *honoris causa* in natural sciences, who was the official Argentine expert in the boundary dispute with Chile, was also the protagonist of a tenacious campaign for an Argentine presence in the Antarctic regions.

It is to his spirited initiative and enterprising mind that we owe the success of the negotiations which led to the appointment, in 1904 by Manuel García Fernández, the Director General of the National Posts and Telegraphs, of Señor Hugo Acuña[56], a member of the first Argentine expedition, as Post-Master of the South Orkney Islands. Sr Acuña was provided with an official brief-case containing all the documentation required by the national Post Office, and a cancelling stamp to postmark the Argentine postage stamps used by the explorers[57]. It

54. Wenceslao Escalente (1852–1912), Argentine Minister for Agriculture in 1904

55. Francisco Pascasio Moreno (1852–1919), Director of the Museum of Natural History in La Plata, was a prominent Argentine explorer and academic, his specialities included Patagonia. He was the Argentine delegate to the Sixth Geographical Congress (London, 1895) which adopted a Resolution promoting Antarctic Exploration.

56. Acuña, 1982, republished in 2015 with additional material

57. Interestingly both Argentine and Falkland Islands postage stamps were accepted and can-

should cause no surprise that this postal service functioned well in the South Orkneys. The station personnel were thus able to dispatch various postal items aboard *Scotia,* thanks to the establishment, by Argentina, of this the earliest Antarctic Post Office.

In January 1904 *Scotia* sailed from Buenos Aires, with the first Argentine expedition to the South Orkneys aboard: Luciano Honorio Valette[58], Hugo Alberto Acuña, and Edgar Hans Ulrich Carl Szmula[59] who would be under Mossman's leadership. The Cook was William Smith, who, with Mossman, was one of the members of the Scottish expedition. Thus Mossman and Smith became the first men to spend two consecutive winters in the South Orkney Islands.

3: Argentine Expeditions

Argentine sovereignty over those inhospitable Antarctic territories dates from 22 February 1904; the day on which the Argentine flag was displayed over the hut built by the members of the *Scotia* expedition[60]. From that year, the Argentine government has funded annual expeditions to the South Orkney Islands. The relief voyages are made during the summer months, January to March, which is the most favourable season for them; that is when the solid winter pack ice that covers the ocean has melted and dispersed, making Southern Ocean navigation more practicable.

Until 1922 the expeditions were relieved by the veteran historical corvette *Uruguay*[61], which, in 1903, had rescued the expedition of Dr Nordenskjöld that was stranded on Snow Hill Island after his ship,

celled by the Post Office, which ceased to operate in February 1905. It was the first Antarctic Post Office but was not re-established until 1941: Bagshawe, 1947.

58. Valette (1880–1957) [also recorded as Juan Luciano Honorato Valette] was a zoölogist employed by the Argentine Ministry of Agriculture. His observations on South Orkney Islands biology were published: Valette, 1906.

59. Full name from Bremen passport records; Szmula, born in Prussia, migrated to Argentina in 1896.

60. Flags of Scotland, Britain, and Argentina were flown over the station on this occasion, see Illustration 3.

61. *Uruguay* is a steam-assisted gun ship; built in Britain in 1874 and now a museum ship in Buenos Aires; displacement 550 tons, length 46·4 m. *Uruguay* made 12 voyages to the South Orkney Islands between 1904 and 1922.

3: Transfer of Omond House from the Scottish National Antarctic Expedition to the Argentine Meteorological Service, 22 February 1904. Left to right: William Martin, James Pirie, Alastair Ross, William Cuthbertson, Robert Mossman, William Smith, Luciano Valette, Edgar Szmula, and Hugo Acuña. (courtesy Scott Polar Research Institute, University of Cambridge).

Antarctic, had been crushed by the sheer force of pack ice pressure.

At the end of 1922 *Uruguay* was withdrawn from service and it became necessary to equip another vessel to replace her. In 1923, because of the limited time to organize the relief expedition, it was decided to send an Argentine naval transport ship, *Guardia Nacional*[62]. This was assigned to transport the relief team to South Georgia, from where a whale-catcher, belonging to one of the companies working there, was chartered to continue to the South Orkney Islands. This method was used again in 1924 to relieve the station.

In 1925 the relief expedition went directly from Buenos Aires to the South Orkney Islands aboard another Argentine naval transport ship, 1° de Mayo[63]. In 1926 the expedition voyaged via South Georgia aboard an old sailing ship[64] of the Compañía Argentina de Pesca[65], which was used to transport whale oil. They sailed to Grytviken[66], where they boarded a chartered whale-catcher to continue to the islands. The returning team used the same procedure, but in reverse.

The direct distance between Buenos Aires and the South Orkney Islands is approximately 1560 nautical miles [2890 km][67]; between Ushuaia, capital of the province of Tierra del Fuego, and the South Orkneys is 830 nautical miles [1540 km]. From South Georgia to Laurie Island is slightly shorter [460 n.mi. or 850 km]. Thus for vessels capable of making an average daily run of 240 nautical miles

62. *Guardia Nacional* was a steam powered supply ship, built in Britain in 1900 and in service until at least 1945; displacement 7000 tons, length 1112·5 m. She made voyages to South Georgia to transport South Orkney relief expeditions in 1923 and 1924, and had previously visited Grytviken whaling station in 1905.

63. *1° de Mayo* was a steam powered transport ship, converted from coal to oil firing; built in Germany in 1893 and in service to 1946; displacement 640 tons, length 61·2 m. *1° de Mayo* made three voyages to the South Orkney Islands; see footnote in § 82.

64. *Tijuca;* described in § 84 and footnote.

65. Translated as 'Argentine Fishing Company', but actually a whaling company, the first to establish a station on South Georgia, or elsewhere ashore in the Southern Ocean.

66. Grytviken, in King Edward Cove, is the best harbour in South Georgia, it was the site of sealer's operations during the 1800s, and of the earliest Antarctic whaling station which operated from 1904 to 1965. The island's administrative and scientific centre, King Edward Point, is on its northern shore.

67. These direct distances have been recalculated from modern charts, and minor corrections made.

4: Dwelling house [now Casa Moneta] in 1927, photographed in moonlight

[445 km] a direct voyage from Buenos Aires could be completed in seven or eight days. However allowances must be made for any stops in Patagonian ports, for reducing speed owing to bad weather such as fog or gales, and likewise for meeting any extensive fields of ice or large icebergs. These are all common dangers for vessels venturing into polar seas.

Despite advances in modern shipbuilding design, and safety, with improvements in performance and stability, the best vessels, even those which have good, tried and tested equipment, still have difficulties whenever they navigate in polar seas. Antarctic exploration continues to be a saga of maritime disasters, of shipwreck, destruction, and death. These are hazards confronting anyone attempting to penetrate the unknown and unravel the mysteries that lie beyond the natural ice barriers of the poles.

Perhaps the greatest risk incurred by anyone rash enough to join the adventure of any polar expedition, in our case the one to the South Orkneys, is the voyage itself. I can say, unequivocally, that in all the eight sailings I made to and from the Antarctic, I suffered moments of deep apprehension and danger. Fortunately we were saved, on every occasion, by the skill of the Argentine and foreign mariners to whose reliable hands the safety of the expedition's vessels was entrusted.

4: Expansion of the Observatory

During the expedition to relieve the 1905 winter party a new building was constructed[68] which would continue to provide living accommodation for the staff for the next thirty-four consecutive years. The house, or more accurately hut, was built entirely of wood. It had double walls separated by a ten-centimetre gap filled with saw-dust as an insulator. This refinement, combined with an exterior cladding of two or three layers of Ruberoid[69], stopped the cold from penetrating inside. In fact it is the same system as that used in the manufacture of freezers and refrigerators, except that the insulation process is reversed to prevent the exterior cold from reaching the living quarters.

Over the years this dwelling has been extended several times and is now double the size of the original living quarters. It was originally nine and a half metres long by five and a half metres wide. In that confined space were three small bunk-rooms for the staff, a kitchen, a parlour used as a living, dining, and work room[70], and a pantry. A little store room was built along the north side to supplement the pantry, with a corner partitioned to provide a latrine[71]. The following year two small rooms were constructed against the south wall, to be used for radio communications. In 1928 a further room was added as a store for radio equipment.

With these additions the actual length of the building reached twenty metres. The extensions were all constructed to follow the

68. This, now known as 'Casa Moneta' (Moneta House), remains in use as the station's museum. It was constructed during the visit of *Austral,* from 2 to 23 February 1906: Mossman, 1907. A plan of its original structure appears in Plan 2. It is the oldest continuously occupied building in the Antarctic. Now protected as an Argentine national historic site by Law 26.294 of 14 November 2007 and by the provisions of the Antarctic Treaty (see footnote in § 2).

69. Ruberoid is the trade name of an impermeable membrane of strong fabric impregnated with black vulcanised rubber. It is similar to roofing felt.

70. This room was a combination of a living room and dining room which also had some of the observatory instruments as well as being used for calculating and recording data. It is not easy to name; neither lounge, nor saloon, nor living room are entirely satisfactory. Parlour has been adopted as, although becoming archaic, it conveys the broad function of the principal room of the station dwelling.

71. This addition appears in a Swedish plan of the station from 1913; the original text dated it 1926.

long axis of the original structure. The orientation was north-west to south-east. This detail was considered when the original structure was built so that it, and later the extensions, would offer the least resistance to the severe winds that sweep the area all year round.

Several other smaller structures were deployed in the vicinity of the Observatory for scientific work. They included three little wooden huts[72] to house the instruments for geomagnetic observations; these are situated 150 metres west of the dwelling, and a few metres from the South Orkneys cemetery. In 1925 a fourth hut was erected for the first basic radio-telegraphy station which should have opened that year, but it was never used for its original purpose. It was situated too close to the front of the glacier which resulted in it being covered with ice and snow and becoming completely buried during the winter[73].

The new radio station was inaugurated on 30 March 1927, and continues to function to this day.

Considering the rudimentary structure in which the members of the *Scotia* expedition had lived and survived, and comparing its very basic facilities with those we have today, the improvements and enlargements of the living quarters have made the lives of expedition staff substantially more appealing. However the personal space available to each man is small. It could be said of the station that the only person capable of thriving in these conditions is one who, from his arrival at that bleak spot, feels that he is naturally at home in such surroundings which are not only restrictive but depressing; a white desert with a debilitating climate in which he will have to remain for an interminable polar year.

72. The earliest of these is from 1905, a reconstruction of Bruce's Copeland Observatory.
73. A plan of the station is given in Plan 1.

5: Parade of Nationalities[74]

Expedition records show that men from a range of nationalities, from all over the world but especially from northern Europe, have been employed at the remote Observatory. During the early years very few Argentines were included in the expeditions, and the staff were mainly from Sweden, Britain [England and Scotland], Norway, Germany, and Denmark[75].

The earliest Argentines to live for a year in the South Orkneys were Hugo Alberto Acuña and Luciano Honorio Valette[76] who, under Mossman's command, comprised the 1904 winter party.

The 1906 expedition brought another Argentine: Gastón O. Talamón. Fourteen years later two more compatriots went to the South Orkneys: Augusto Tapia and Alejandro J. B. Boracchia. In 1921 a new era in the composition of winter parties began. From then Argentines were included as follows:

1921: José Reynoso and Elías Reynoso Sosa
1922: Ángel Antona[77]
1923: The author, on his first expedition
1924: Alfredo C. Witt
1925: Alejandro Martín Guerra Boneo, and the author on his second expedition
1926: Alfredo C. Witt
1927: The first winter party consisting entirely of Argentines was led by the author on his third expedition with Miguel Ángel Jaramillo, Pedro Martín Casariego, Luís Fállico, Emilio Alberto Baldoni, and

74. The list of names has been reconciled with Argentine, British, Danish, German, Norwegian, Swedish, Uruguayan, and other sources which have some minor discrepancies. Carlos Mey (2017) has improved and extended the record for many of them. Reference was also made to: Headland, 2009 and Stewart, 2011. Names are given in full, in the original language, where practicable. Only those occurring elsewhere are indexed. The Editor would appreciate further details of full names and other information from readers.

75. There were also men from: Austria, Hungary, Russia, and Uruguay.

76. Valette was born in Montevideo on 20 August 1880; although a long-term Argentine resident his nationality is recorded as Uruguayan in: Acuña, 2015.

77. This name is inserted in the translation; it had been omitted from the lists in all the Argentine editions of the book. He is noted as 'argentino' in editions 1-11 and 'argentino naturalizado' in edition 12.

Konrad Becker[78]

Subsequently a few foreigners joined the expeditions, but they were no longer in the majority.

Chronologically the expeditions were:[79]

1903[80]: William Speirs Bruce, Leader of the Scottish National Antarctic Expedition of 29 men (6 scientific staff and 23 ship's complement) aboard *Scotia,* Capt. Thomas Robertson. Among whom were Robert Neal Rudmose Brown, Robert Cockburn Mossman, William Alexander Cuthbertson, James Hunter Harvey Pirie, Allan George Ramsay +, Alastair Ross, William Smith, and David Walter Wilton (all from Britain)

1904: Robert Cockburn Mossman (Britain), Hugo Alberto Acuña, Luciano Honorio Valette (Uruguay), Edgar Hans Ulrich Carl Szmula (Germany), William Smith (Britain)

1905: Otto Diebel + (Germany), James A. Percy (Britain), Olaf Paulsen (Norway), William A. McKinley (Britain), William Thomas (Britain)

1906: Albin Lind (Sweden), Nils Hessling (Sweden), Gastón O. Talamón, Fredrik Thomander (Sweden)

1907: Angus Fraser Rankin (Britain), Olaf Paulsen (Norway), John Charles Meldrum (Britain), William Thomas (Britain)

1908: William Robert Bruce (Britain), H. Jennings (Britain), Ernest Alexander Miller (Britain), F. Holm (Norway)

1909: Alexander Lindsay (Britain), Karl Falch Wasserfall (Norway), Harald Wiström (Sweden), Guido Salvadori (Austria)

1910: Olaf Paulsen (Norway), Jon Eliesen + (Norway), Sigurd Stranger (Norway), H. Christiansen (Sweden), N. Axel Berg (Denmark)

78. Becker's name was not included in the original list in any edition, he was originally from Germany but naturalised as Argentine (Conrado Becker), see § 83 and § 98.

79. The first name is that of the leader, the last is usually the cook in the earlier years. Throughout the list, where no nationality is indicated, the man is Argentine, as in the lists of later editions of the book. The names are not indexed except for those which appear elsewhere in the book.

80. This entry is not in any edition of the original work but has been editorially inserted to record the first winter, when the station opened in 1903 and meteorological observations began on 26 March.

1911: Harald Wiström (Sweden), Karl Torsten Stāhlhandske (Sweden), Knut Olivecrona (Sweden), Alfred Wiström (Sweden), Ernst W. Andersson (Sweden)
1912: Sigurd Stranger (Norway), Gustav Adolf Söderström (Sweden), Gustav Erasmie (Sweden), Georg Michaelson (Sweden), N. Axel Berg (Denmark)
1913: Harald Wiström+ (Sweden), Karl Torsten Stāhlhandske (Sweden), Anton Stuxberg (Sweden), Johan L. Berg (Sweden), Ernst W. Andersson (Sweden)
1914: Richard Blüthgen (Germany), Wilhelm Kopelmann (Germany), Fritz Fachon (Germany), Carl E. Berg (Sweden), N. Axel Berg (Denmark)
1915: Hartvig Bache-Wiig + (Norway), Anton Stuxberg (Sweden), Thorleif Edgar Hoxmark (Norway), J. Bayon (Sweden), Ernst W. Andersson (Sweden)
1916: Olaf Edgar Lützow-Holm (Norway), C. Staël von Holstein (Sweden), N. Furnheim (Norway), Carl E. Berg (Sweden), S. Strid (Sweden)
1917: Anton Stuxberg (Sweden), Hugo Valentiner (Denmark), Fritz Edvard Curt Stjernblad (Sweden), J. Schultz (Sweden), S. Strauss (Sweden)
1918: Olaf Edgar Lützow-Holm (Norway), N. Hammaren (Sweden), Thorleif Edgar Hoxmark (Norway), Carl E. Berg (Sweden), Otto Hasse (Germany)
1919: Hugo Valentiner (Denmark), Sigurd Schjær (Denmark), Hans Laurits Wange (Denmark), John Rötter (Norway), Robert Plagge (Germany)
1920: Wilhelm Kopelmann (Germany), Bruno Kollasius[81] (Germany), Augusto Tapia, Alejandro J. B. Boracchia, Georg Piper (Germany)
1921: Hugo Valentiner (Denmark), Karl Budde (Denmark), José Reynoso, Elías Reynoso Sosa, Robert Plagge (Germany)
1922: Bruno Kollasius (Germany), Sigurd Schjær (Denmark),

81. Also recorded as Collasius, the Spanish rendering of his surname.

Hans Laurits Wange (Denmark), Ángel Antona, Otto Hess (Germany)

1923: Hugo Valentiner (Denmark), Ernst Bruhns (Germany), José Manuel Moneta, Robert Plagge (Germany), Otto Zeiger (Germany)

1924: Vladimir Dobrovolski (Russia), Carl E. Berg (Sweden), Alfredo C. Witt, Horst E. Irrgang (Germany), Otto Hess (Germany)

1925: Ernst Bruhns (Germany), José Manuel Moneta, Alejandro Martín Guerra Boneo, Albert de Tolnay (Hungary), Georg Piper (Germany)

1926: Carl E. Berg (Sweden), Alfredo C. Witt, Horst E. Irrgang (Germany), Yakob Orlovski (Russia), Edmund Müller (Germany)

1927: José Manuel Moneta, Miguel Ángel Jaramillo, Pedro Martín Casariego, Luís Fállico, Emilio Alberto Baldoni, Konrad Becker

1928: Ernst Bruhns (Germany), Alfonso Chakí[82], Carlos V. Rodríguez, Juan L. Olalla, Fortunato A. Escobar +, Edmund Müller (Germany)

1929: José Manuel Moneta, Félix Celestino y Monti, Jorge Páez Montero, Yakob Orlovski (Russia), Carlos Kahl, Rómulo Devoto

1930: Carl E. Berg (Sweden), Alejandro J. B. Boracchia, Bernardino Matos, Alfonso Chakí, F. Bruno, D. Gaudio

1931: Ernst Bruhns (Germany), José Reynoso, Félix Celestino y Monti, Edmund Müller (Germany), R. Semino, Rómulo Devoto

1932: Ernst Bruhns (Germany), Bernardino Matos, Félix Celestino y Monti, Juan Carlos Lynch, José Bermúdez, Juan Carieiro

1933: Bruno Kollasius (Germany), Juan Carlos Lynch, Armando S. Pico, Lucio Correa Morales, Ian Colin Grant, José María Germán Spika

1934: Carl E. Berg (Sweden), Juan Carlos Lynch, Félix Celestino y Monti, M. Barbieri, Rómulo Devoto

1935: Bruno Kollasius (Germany), Armando Pico, Alfredo de

82. Also recorded as Chaquí.

Martino, José C. Lovisolo, Nicolás Contursi
1936: Aage Jacob Wilhelm Johansen, Félix Celestino y Monti, Carlos A. Fluguerto, Humberto Papa, Rómulo Devoto
1937: Bruno Kollasius (Germany), Delfín Díaz Vieyra, Diego García Goyena, Rodolfo Placke, Nicolás Contursi
1938[83]: Aage Jacob Wilhelm Johansen, Armando S. Pico, Melchior B. Martínez, José Manuel Hualde, Carlos Luís Capart
1939: Félix Celestino y Monti, Rafael Alberto Pastoriza, Guillermo Nieva, Nicolás Contursi
1940: Alfonso Chakí, Pablo Hugo Bisso, José Conchiglia, Ángel Guillermo Nicolás Luís Gusberti
1941: Aage Jacob Wilhelm Johansen, Rafael Alberto Pastoriza, Eulogio González, Carlos A. Pesaro
1942: Alfonso Chakí, Antonio Villalón, Santiago Hidalgo, Ángel Guillermo Nicolás Luís Gusberti
1943: Armando S. Pico, Armando Ferrari, José Gallardón, Carlos Luís Capart, Enrique Poveda, Carmelo Messina
1944: Aage Jacob Wilhelm Johansen, Rafael Alberto Pastoriza, Celso Meneses, Gregorio Villalón
1945: Armando S. Pico, Carlos Luís Capart, José Cestau, Emilio Fermin Ramos, Enrique Poveda, Cesáreo Labusta, Lucio Eduardo Lombardo, Teodoro Nitti
1946: Alfonso Chakí, Jorge Zawels, Fernando Berreta, Luís Risso, Atilio Pincardini, Marcelo Castiglia
1947: Luís Roberto Lombán, Rafael Alberto Pastoriza, Alfredo Torres, Sixto Leyva, José Tedaldi, Lucio Eduardo Lombardo, Gregorio Villalón
1948: Carlos Alberto González, José F. Medina, Reynaldo Soto, Domingo Ángel María Roque Abregú Delgado, Martín M. Marina, Enrique Poveda, Valentín Cicchetti, Antonio Moro
1949: Rafael Alberto Pastoriza, Abel Moisés Horestein, Aníbal Gregorio Llamazares, Homero Alberto Di Russo, Alfredo Torres, Carlos Colson, Luís Oscar Maurice,

83. From 1938 the personnel of Orcadas Station were all Argentine.

Ismael Tomás A. Flores
1950[84]: Enrique José de los Sagrados Corazones Smith Estrada, Lauro Virgil, Reinaldo Soto, Benito Petronilo Rodríguez, Baltazar Silva, Alejandro Dubini, Salvador Américo, Osvaldo Castrogiovanni, Emilio Escobar, Orlando Arturo Angio, Jacobo Leitman
1951[85]: Ifgenio Sanz, Alberto Rubén Torres, José Alvarez, Carlos Antonio Rojas, Armando S. Pico, Eduardo Ramón Leyton, Feliciano Aguaysol, Rogelio Souto, Ricardo González, Enrique Póveda
1952: Guillermo Coelho, Rolando Morgavi, Ramón José Portillo, Aroldo Mansilla, Juan Bossio, Rafael Alberto Pastoriza, Gerónimo Soto, Florentino Loyola
1953: Jorge H. Suárez, Guillermo Angli, Héctor Ullúa, Héctor Sánchez, Italo Rigoni, Abel Moisés Horestein, Eduardo Ramón Leyton, Ventura Correa
1954: Arnoldo E. Cappelletti, Oscar Alberto de Lelles, Miguel Maldonado, Ramón Almirón, Osmar Rodríguez, Rodolfo Andersch, Argentino López, Luís María Fontana, Ángel Maldonado
1955: Luís Messiga, Mario Clemente Cavalli, Raúl R. Longones, David S. Cochagua, Mario Leguizamón, Ismael Barraza, Domingo Villafañe, Sixto Mario Ricarte, Alberto Rosetti
1956: Domingo W. Lucero, Alberto Rosetti, Darío Peralta, Lorenzo Roldán, Emilio Romero, Oscar Málan, Pedro Clemente Mujica, Hiliberto Molinar
1957: Miguel J. Gurusiaga, Antonio Mosca, Roberto Gómez, Claudio Pérez, Pilade Viviani, Juan Molinari, Santos Bravo, Juan Robles, Francisco Rubiolo, José Komar, José Eduardo Jarek, Oscar Moyano
1958: Alejandro José Guisti, Jorge Horacio Villagre, Aldo Julio Irusta, Rodolfo Miksa, Eugenio Vega, Miguel Alfredo Ovejero,

84. From this date the second name is usually that of the medical officer.
85. In 1951 the Argentine Air Force provided personnel and from 1952 the Navy took over the station; the original list included the military ranks of most of the men included.

Luís María Fontana, José Eduardo Jarek, Juan Francisco Morales, Eduardo Cabrera, Ernesto Juan Gaddi

1959: Emilio Filipich, Jaime Laham, Raúl Alberto Macchiaroli, Adolfo Díaz, Gregorio Enelso Soto, Juan Ramón Lukacs, Walter Soto +, Pablo Eber Moreno, Oreste Pellis

1960: Ariel A. Giuntini, Jorge Cirilo Aversano, Luís Alberto Olmedo, Armando Teodoro Guzmán, Teodoro Omar Valenzuela, Sixto Mario Recarte, Pablo Bravo, Luís Alberto Villagra, Eloy José Custodio

1961: Oscar Roberto Padilla, Daud Roberto Oscar Mamet, José Romilo Morales, Oscar Raúl Albarracín, Alejandro Bazar, Oscar Antonio Atienza, Francisco Burzi, César Gustavo Ojeda, Carlos Antonio Astarloa, Juan Miguel Fernández

1962: Carlos Juan Moneta[86], Amílcar César Valle, Demetrio Voznesenski, Roberto Gómez, Angel Miselli, Cristóbal López, Leo Luís Koch, Mamerto Flores, Pedro Clemente Mujica, Luís Alberto Villagra, Fernando de Carmen Tula, Pablo Bravo, José Gerónimo Godoy

1963: Julio César Alvarez, Mario Hernando Soria, Demetrio Voznesenski, José Alcides Sarmiento, Carlos Borrero, Virgilio David Luque, Domingo Agustín Almada, Hugo Raúl Abraham, Ricardo Atilio Aliendro, Nelson Porter, José Eduardo Jarek, Ricardo Giménez, Alberto Teófilo Martín, Rito Idilio López

1964[87]: Antonio Mocellini [14]

1965: Miguel J. Sosa [12]

1966: Rodolfo H. Ludueña [13]

1967: Osvaldo Pedro Astiz [11]

1968: Arturo M. Quintana [17]

1969: Reynaldo Cappa [20]

86. Carlos Juan Moneta is the son of a cousin of José Manuel Moneta. He subsequently specialised in international relations, including Antarctic aspects.

87. From 1964, after publication of the 12th edition of *Cuatro Años*, only the name of the Station Leader is given because the numbers increased substantially. The winter complement of 'Orcadas' is indicated in brackets. The author brought the lists up to date in every edition of the book, thus the Editor has continued it for the translation.

1970: Ernesto J. Tenconi [20]
1971: Vicente Manuel Federici [21]
1972: Raúl E. Comelli [22]
1973: Guillermo Escorihuela [21]
1974: Daniel O. Fitipaldi [20]
1975: Héctor Pascual Ferraro [27]
1976: Jorge Arias [15]
1977: Gustavo D. Richardson [12]
1978: Jorge Luís Laboro [20]
1979: Eduardo E. Nasine [20]
1980: Miguel Ángel Pereyra [20]
1981: Juan Carlos Díez [16]
1982: Mario Raúl Huici [16]
1983: Rodolfo E. Seoane [14]
1984[88]: Jorge Cerqueiro [13]
1985: Guillermo Mariano Palet [17]
1986: Eduardo Miguel Luis Brousson [14]
1987: Guillermo Alejandro Nelson Tarapow [14]
1988: Luís Mario Devicenti [14]
1989: Eduardo Raúl Pisciolari [13]
1990[89]: Jorge H. Garabentos [14]
1991: Carlos Kammerath [13]
1992: Guillermo Flores [14]
1993: Víctor Manuel Pereyra [14]
1994: Josué Alfredo Núñez [12]
1995: Fabián Gustavo Giudice [14]
1996: Humberto Mario Dobler [14]
1997: Alejandro Néstor Olivieri [15]
1998: Alejandro Christian Hormanstorfer + to March, then Marcelo Uberti [13]
1999: Esteban Alberto Russi [14]
2000: Osvaldo Alfredo Toscano [14]

88. From 1984 a few men of the Argentine Air Force also wintered.

89. From 1990 two Park Wardens also wintered, they were involved with the historic sites and visitors.

2001 : Héctor Edgardo Gazzolo [14]
2002 : Antonio Marcelo Serángeli [13]
2003 : Ricardo A. Garde [13]
2004 : Santiago Justo Lescano [13]
2005 : Fernando Christián Paternostro [14]
2006 : Juan N. Di Niro [14]
2007 : Ernesto Eusebio R. Acosta [16]
2008 : Gustavo Adolfo Sánchez [15]
2009 : Gustavo Enzo Lancellotti [15]
2010 : Juan Carlos Zenteno [17]
2011 : Fernando Maglione [15]
2012 : Ariel Gustavo Siles [17]
2013 : Néstor Luján Cassanovas [16]
2014 : Daniel Juárez [14]
2015 : Juan Manuel Garita [14]
2016 : Héctor Daniel Báez [17]
2017 : Marcelo Eduardo Palacios [17]

In Memory of Those Who Died on the South Orkney Islands[90]

1903 : Allan George Ramsay (Engineer of *Scotia*) ; 6 August
1905 : Otto Diebel (Leader) ; 25 September
1910 : Jon Eliesen (Assistant Leader) ; 20 August
1913 : Harald Wiström (Leader) ; 7 May
1915 : Hartvig Bache-Wiig (Leader) ; untraceable, last sighted 30 April
1928 : Fortunato A. Escobar (Radio Operator) ; 27 October

90. Elsewhere on the South Orkney Islands Norwegian, Icelandic, and British men are buried on Signy Island. Five whalers lie in a cemetery near Borge Bay : Henrik Thorstensen (from Norway) died on 6 January 1914, Jacob Jacobsen (Norway) on 19 January 1914, Aksel Olson (Norway) on 21 February 1914, John Johnson (Iceland) on 2 February 1926, and Ragnar Gotfrei Been (Norway) on 14 January 1928 ; British Antarctic Survey meteorologist John Roger Filer died on, or shortly after, 13 February 1961 ; his grave is on Gourlay Peninsula.

1959: Walter Soto (Geomagnetician); 13 October[91]
1998: Alejandro Christian Hormanstorfer (Leader), Daniel Néstor Tavella (Meteorologist) and Ricardo Walter Álvez (Cook); untraceable, last sighted 31 March[92]

The reader might notice that the majority of names listed are native to Scandinavia or elsewhere in northern Europe[93]. This is because living in the South Orkneys was unfamiliar to the majority of Argentines, who are of Latin descent. Recollections of the journeys made by Acuña, Valette and Talamón, and the impressions they made became erased with time. Furthermore, life in the polar regions has always been considered too harsh for anyone not used to polar latitudes. Another consideration is that the scientific programmes required trained technicians. The early Argentine Meteorological Service simply could not recruit sufficient staff with the basic skills who were technically qualified, and who were not foreigners.

Consequently leadership of the expeditions was entrusted to men already accustomed to the rigours of life in a polar climate. In particular cases their experience proved invaluable, as will be seen later. Men who started as ordinary base members went on to become leaders of subsequent expeditions.

Details worth mentioning here are repeated winters made by the same man. The record is held by Carl E. Berg, from Sweden, who until 1934, had taken part in seven expeditions, and was Leader of three of them. There were other men who spent six, five, four, and three winters, while many have served for two. The author has spent a total of four years and nine months in this strange and unusual polar life.

91. Walter Soto drowned off Point Lola. Details of the other deaths are in § 36 (Bache-Wiig), § 61 (first four), and § 110 (Escobar).

92. These three deaths occurred after the last Argentine edition of the book; they are added to keep the list up to date and in the belief that the author would have desired this to be done. The men were lost while at sea aboard an inflatable boat.

93. This sentence is retained from the 1939 edition. At that time, from 1904 to 1939, approximately a third of winterering men had been Scandinavians (most from Sweden), a third from six other northern European countries (mainly Germany), and a third Argentine. The original text reads of Saxon origin.

In more recent years descriptions of various aspects of life on an expedition have been published. However, from the events they describe, it is often clear that the authors responsible for some of these accounts arrived in the South Orkney Islands aboard a relief ship and stayed there just a few hours or perhaps a day or two. Their distorted accounts have been adapted from other books and publications. Their sole aim is to gain credit by saying they 'have been to South Orkneys'.

Many men have made multiple voyages to the South Orkneys, but comparatively few have experienced an entire winter there. According to the statistics[94] only one hundred and twenty-five men have spent one or more winters on the islands. Six of those died there, while many others in the course of time have also passed away or disappeared. To this date the number of men with winter experience of the South Orkneys could be as few as seventy-two, and most of these are now scattered all over the world.

It is worth mentioning that, at the time of my 1923 expedition, two members of the 1904 expedition were still living: Luciano Valette and Hugo Acuña. The former had retired from a senior post in the Ministry of Agriculture and the latter was attached to a leading Argentine bank.[95] Hugo Valentiner, the experienced Leader of my 1923 expedition, is still alive. We have remained friends and today can retell many nostalgic tales of our adventurous time together, of those risky episodes and shared moments on the South Orkneys.

94. The following footnote appears in later editions of the book: 'These statistics are from data compiled from the expeditions dispatched to the South Orkneys until 1939 inclusive. This was the year of publication of the first edition of this book. In subsequent years annual relief expeditions were dispatched to the South Orkneys, as well as to new Argentine Antarctic stations established on the Antarctic Peninsula and islands. Thus many more Argentine nationals have now wintered in the polar regions. Although they have experienced the rigours of Antarctic winters they had better supplies and equipment, and with advantages of technology such as we were only able to dream about during the early primitive period of Argentine Antarctic expeditions.'
95. This and the preceding sentence are omitted after the 10th edition of 1958; their dates were: Valette 1880–1957, Acuña 1885–1953.

6: What took me to the South Orkney Islands

From my earliest childhood I had the good fortune to be closely associated with life in the south of Argentina and in the Antarctic. For me the scarce narratives of polar expeditions were no secret, I knew them all well. I had read several accounts of the unforgettable rescue of Nordenskjöld by the corvette *Uruguay* in 1903. I also had an extensive knowledge of the ill-fated expedition of Captain Scott, and of the achievements of explorers such as Amundsen, Charcot, Powell, Bruce, and others.

I had always nurtured a secret hope that I might, one day, get to know those famed and faraway lands, and this hope bore fruit in my imagination. The earliest opportunity came when I was twelve years old and I went to live in the southernmost part of the Americas, in Tierra del Fuego.

Then, in 1919, I chanced to find myself living in Puerto Militar (now Puerto Belgrano[96]) and had the opportunity to see *Uruguay* on her return from one of the annual voyages to the South Orkneys. The sight of this famous veteran vessel renewed in my mind the dream of going South. Then and there I decided that I would do my utmost to make it a reality.

After my return to Buenos Aires, I was able to join the Argentine Meteorological Service, which, at that time, was directed by George Wiggin[97]. He greeted my application to be included in an expedition to the South Orkneys with much scepticism. It took considerable effort for me to persuade him to accede to my request. My efforts were not in vain as I kept my ambition alive during the four years I spent acquiring sufficient experience to become a meteorological observer.

Eventually, towards the end of 1922, I received notification that I would be a member of the 1923 winter party to be led by Hugo Valentiner on his fourth expedition, and twice as Leader of the station. He would be accompanied by Ernst Bruhns as Assistant Leader, with Robert Plagge as Third Observer and Otto Zeiger as Cook,

96. Puerto Belgrano is now a major Argentine naval station in northern Patagonia.

97. George Otis Wiggin was Director of the Servicio Meteorológico Argentina from 1915 to 1924.

all three German nationals. Valentiner, Bruhns, and Plagge were attached to the La Quiaca Observatory[98]. I did not know any of them but, such was my enthusiasm, I was quite undaunted by the thought of having to spend an entire long year with men whose language I did not know.

I met my colleagues only a few days before the set date of departure. The fact that our Leader had three previous expeditions to his credit and was very experienced, increased my confidence in him.

With all the members of the expedition ready for departure, we were given a very thorough medical examination. We all felt fit to face the rigours of the Antarctic. A sound heart and healthy lungs, without the remotest trace of any incurable infection, contributed to the optimism and excitement we all felt. However, although I was in good physical condition, I could not help thinking of the risks involved in this adventure. Although I was certain about my good health and absence of disease, I could still become a victim of something in those icy wastes, or, worse still, suffer some dreadful accident. In my imagination I was continually recalling the photograph I had seen of the four wooden crosses in the cemetery on the South Orkney Islands.

Nevertheless, I was resolved to go. There was, however, one basic detail that still troubled me. When only twenty years old, even the most adventurous explorer thinks of his mother. This was the only dilemma that cast a momentary shadow on my decision. To be able to leave in peace I had to resolve it.

The expedition was ready; the ship was in port loading provisions and supplies. However my parents were still unaware of the decision I had made.

Eventually I could stand this uncertainty no longer, and I sought advice from the person who had been my best friend for my whole life: my father. I remember his look of surprise when I told him of my decision. He pondered the matter at length and then said to me:

'My duty as a father obliges me to ask that you desist from your voyage, but as a man and as your friend, I advise you to go ....'

98. La Quiaca is high in the Andes, near the Bolivian frontier with Argentina.

5: Dwelling house photographed with accumulated snow

I replied that my preparations were complete, both mentally and physically, and that I was ready to go.

'Well you must do as you wish, but have you given thought to your mother?'

'Yes, I have, and of all my problems, this is the most difficult to solve.'

I recalled a precedent, somewhat similar to the predicament which confronted me. It was the story of a fine young foreigner who had come to Argentina to seek his fortune. He left his mother far away in the village of his birth and maintained her with the help of his savings. He too joined an Antarctic expedition, but just before he left he wrote twelve letters, put them in envelopes, and gave them to a trusted friend, who was asked to post them monthly to his beloved mother so she would think he was working in a safe location and not incurring the risks of such a hazardous Antarctic adventure. In this way his mother would receive a sum of money regularly and would be saved from thinking that her son was somewhere remote and on a dangerous expedition[99]. This man never returned, he remained in the Antarctic buried beneath a white shroud of snow and ice. For several months after his heroic

99. This sentence is omitted from later editions.

death, his mother, at the other end of the Earth, continued to receive monthly letters from a son who was no longer alive[100].

Bursting with optimism as I was, I was unable to believe that my own mark in the South Orkneys could be a cross in the snow. In the back of my mind, however, I still had visions of the accidents and mishaps recounted in the sparse literature about life in those isolated lands.

With the approval of my father, the least I could do was to confess to my dear mother the adventure on which I was embarking. As a palliative to the instinctive sadness she would feel, I reassured her that I was going only to the Observatory on Isla Año Nuevo, just off Tierra del Fuego. From there, as she knew from previous experience, we would be able to correspond monthly.

Departure day arrived. With what reserves of courage I could muster on that occasion, I made it clear that I did not want anyone to come to the ship to see me off and say goodbye. I was quite alone that morning when I joined my companions aboard the naval transport ship *Guardia Nacional*[101], which would take us as far as South Georgia where we would transfer to a whale-catcher[102].

The deck was swarming with people. Then orders were given, lines were cast off, the port of Buenos Aires grew distant, and the bow swung southwards. We were on our way.

I looked for a quiet place on deck where I could linger. I wanted to bid farewell with my eyes and my imagination to everything I was leaving behind, all that had become part of me.

I was alone, completely alone. Minutes passed, then an hour, and little by little I watched the silhouette of my great city become fainter on the horizon, like a mirage.

I was on my way south, to the Antarctic!

100. This incident refers to Georg Odegaard, from Kristiania [now Oslo], who died while aboard *Svend Foyn,* a whaling factory working at the South Shetland Islands, in February 1910. He was aged 16, and is the youngest person buried in the whalers' cemetery on Deception Island. His mother received letters from his colleagues until the ship returned to Norway.

101. Captain Ricardo Vago.

102. Whale-catchers of this time were small fast coal-fired steam vessels designed to work on the Southern Ocean hunting whales.

7: Equipment and Supplies

Each member of the team had been issued with a good outfit of clothing consisting of: four sets of very warm woollen underclothes; twelve pairs of woollen socks of high quality; a pair of felt boots with a sole two centimetres thick; one pair of ski boots; one pair of work boots; two pairs of thick trousers; a type of padded flying jacket; a leather jacket; three alpine balaclavas; a leather hat with ear protectors; a set of windproofs (anorak and trousers); ten pairs of gloves; three pairs of mittens; three woollen jerseys; two pairs of sheets; a woollen mattress; a pillow and pillow cases.

Originally most of these items were purchased from suppliers in Europe who specialised in this type of clothing which was very effective. From 1926 the system changed, and responsibility to supply everything devolved to several firms in Buenos Aires.

Provisions were a vital component of the cargo and occupied much of the hold space. Any nutritional item capable of conversion to a preserved state, was represented: flour, sugar, pasta, coffee, etc. were in hermetically sealed containers to avoid deterioration from the vagaries of climate and weather. The pantry and food store of the base was very well-stocked; even minor ingredients such as condiments, flavourings, and spices for desserts were there. There was even a powder which could be reconstituted whenever eggs were needed. The food store was stocked to the rafters with, among other things, large quantities of dehydrated vegetables. In the pantry were smaller quantities of tin cans of fruit and vegetables which were prone to freezing and subsequent deterioration. For the prevention of scurvy there was an adequate provision of bottled lime juice which had to be prevented from freezing too. Four hundred kilograms of fresh potatoes and a smaller quantity of fresh onions; which, if we were lucky, would just about last the whole year. In the store were also sun-dried fruit, various types of cheese, cocoa, chocolate, coffee, tea, different types of flour, rice, pasta, etc. Strange as it may sound, even the time-honoured and humble *puchero criollo*[103] was taken to

103. *Puchero criollo* is a variety of traditional Argentine beef casserole or stew.

the South Orkneys though, of course, it was in tin cans.

My colleagues commented about the lack of fresh bread, but when I saw the vast amount of flour of many kinds in the inventory, it was easy to deduce that, with a modicum of ingenuity, the Cook would easily be able to provide as many varieties of fresh bread as we desired.

Looking over the lists again I note, with astonishment, that we did not take much alcohol; there was only one case of whisky and one of brandy. My colleagues tell me that this is, apparently, the official ration to be shared among the five of us. This allowed only one small tot of either whisky or brandy per man per week! This restriction was no major impediment to imbibing because each of us was allowed to take one bottle of champagne to celebrate birthdays and other special occasions.

Thirty tonnes of coal were loaded in the hold of *Guardia Nacional* for heating and cooking in the south. In addition we carried a quantity of drums of paraffin for lighting[104], boxes with scientific instruments, a medical kit, new sledges purchased in Norway, several pairs of ski, and a variety of miscellanea.

Added to this were our personal belongings. Generally the official clothing was supplemented by a few personal items; anything that would be of no use in any other part of the world but which might come in handy in the South Orkneys. Those with foresight, or who had been south before, advised the new recruits not to forget minor items such as sewing needles and thread, and even wool for darning, not to mention: soap, shaving kit, a brush and comb, toothpaste and toothbrushes, etc.

Until then I had never smoked, and could have lived quite happily without feeling any overwhelming desire for a whiff of tobacco. Our Leader asked about this before my departure, I informed him that such a trifling matter was of no importance to me; however, he still recommended that I take with me a quantity of tobacco for both a pipe and cigarettes. I am not sure whether I should be grateful to him for his forethought; once in the Antarctic, I did indeed feel the need

104. Lighting at that time was by oil lamps which burnt paraffin, or kerosene, from a wick in a glass lamp chimney; brass lamps were used inside and hurricane lanterns outside.

6: A group of four whale-catchers at anchor in King Edward Cove, South Georgia

to smoke, and subsequently was never able to abandon the habit.

We were departing well-supplied: with sufficient food and good clothing, a spirit of enthusiasm, and a good sense of humour; the general attitude could not have been better. Still all these good qualities were not sufficient to get us through the boundless year in the South Orkneys. Something was still missing. My great friend, and Assistant Leader of my third voyage, Miguel Ángel Jaramillo, once made the following cheerful assertion: 'to live comfortably in the South Orkney Islands, surrounded by ice, there are only two things needed: a good set of clothing, and a sound dose of philosophy ....'

8: Calling at South Georgia[105]

Twenty days at sea. We had already seen our first tabular iceberg, a characteristic feature of the Southern Ocean[106]. Then one morning,

105. Details about South Georgia: Headland, 1984

106. The Southern Ocean is distinguished from the Atlantic and other oceans extending in the Southern Hemisphere. The Antarctic Convergence, or Polar Front, forms the separation. It had

between the fog banks which hid the horizon, we discerned the rocky coast of the crescent-shaped island of South Georgia. We made for Cumberland Bay, wherein lies the small, snug, harbour of Grytviken.

South Georgia is a small island situated to the south of the South Atlantic Ocean, near the northern limit of what has become known as the Southern or Antarctic Ocean. From a distance it has a quite desolate aspect. Snow-clad mountains, sparse low vegetation and coarse grass, were all characteristic of polar regions. Owing to the copious rainfall of the maritime climate these grasses grow thickly in the sheltered gullies and hollows along the coasts. The vegetation is similar in appearance to that of Tierra del Fuego, although that region has sporadic stunted *Fagus* and *Araucaria*[107] trees whereas South Georgia has not one single tree.

As the ship entered Cumberland Bay, a sense of relief appeared on our faces; no longer was the vessel rolling to 30 degrees or more, nor did the fog conceal icebergs which suddenly would loom and tower over the bow. We were no longer obliged to sleep wedged into heaving bunks, with one leg crossed over the other to avoid being thrown on to the deck by a particularly violent wave.

The water in the bay was perfectly calm and reflected, as in a mirror, the silhouette of one of the strangest inhabited outposts in the world.

At the head of the bay is the settlement of Grytviken, and the whaling station belonging to the Compañía Argentina de Pesca. There are four other whaling stations[108] on South Georgia, two belong to the Norwegians and two are British. All are situated in small natural harbours along the north-eastern coast.

Together these whaling stations produce two thirds of the world's whale oil. This simple detail allows one to appreciate the commercial importance of the island that happens to come under the jurisdiction of the British government which, in fact, occupies the Falkland Islands.

---

distinct physical and biological characteristics. South Georgia and the South Orkney Islands lie within it.

107. Beech and Monkey Puzzle

108. In Moneta's years: Husvik and Prince Olav Harbour (Norwegian), Stromness and Leith (British).

Grytviken has an unusual appearance and is situated at the head of the bay from which it gets its name; it means 'Pot Harbour' in Norwegian. Captain Larsen bestowed the name because of the old sealing cauldrons he found near the beach during his first visit[109]. Even today the remains of enormous cauldrons used by early sealers during the mid-1800s may be seen on the island's beaches.

Nowadays the facilities are quite different; substantial structures and installations include large wooden and corrugated iron sheds, there are many smoking chimneys; and the beach is transformed into an enormous sloping plane on to which the whales are winched for processing[110]. There are houses for the workers and management, administrative buildings, stores, etc. Behind all these is a full size football pitch, and next to that a little protestant church[111] which now, in the absence of a clergyman, has become a meeting place for special occasions.

The scene is completed by the magnificent mountains covered in snow, as in a dream or a legend, just like early photographs and paintings of Nordic villages. All this, so unspoilt and idyllic, is offset by the stench of blubber, blood and other detritus of putrefied whale entrails floating in the otherwise tranquil waters of the cove. This stench pervades the pure clean air which, from a distance, one would have expected to find there.

Along the foreshore and moored to the piers are huge ships ready to take the precious cargo of whale oil and sacks of guano[112] destined to improve the now infertile soils of the old world. Anchored close inshore are several small vessels known as whale-catchers, with their

109. This was in May 1902 after Johan Gunnar Andersson discovered the trypots after which Larsen, Captain of *Antarctic* during Otto Nordensjköld's Swedish Antarctic Expedition (1901–03), named the bay. Several of these cast iron sealers' try pots remain at Grytviken, they have a capacity of approximately 400 litres

110. Termed the 'flensing plan'.

111. A Norwegian Lutheran church, built in 1913, which appears in illustration 57. It still stands, but resident clergy were few and sporadic.

112. Guano is a whalers' term for both meat meal and bone meal, used as fertilizer and stock food (it is not related to the more familiar bird guano, except that both may be employed as fertilizers).

7: British administrative station with radio pylons, King Edward Point, South Georgia (photographed by Alfred Saunders, Discovery Expeditions, *1925, courtesy Scott Polar Research Institute, University of Cambridge)*

characteristic big cannon mounted on the bow, from which harpoons are fired. They are waiting their turn to go to sea and catch the great whales.

On a peninsula to the right, lies another small community formed by the government authorities: the British Magistrate, whose powers are equivalent to those of a provincial governor in Argentina, and who is directly responsible to the Governor of the Falkland Islands. The *Discovery Expeditions,* also operated by the British, has an oceanographic laboratory there[113]. There is also a police station, with a single policeman, to deal with the two thousand men who come to work during the summer months. Rising behind the buildings are the very high radio antennae pylons of the radio-telegraphy station of the British government[114].

113. This building, known as Discovery House, was completed early in 1925 at the same time as the radio station opened; it still stands on King Edward Point.

114. This station began operating on 1 April 1925. Neither Discovery House, nor the radio pylons, was present in 1923. This anachronism is probably due to confusion resulting from five

8: Argentine meteorological observatory, King Edward Point, South Georgia

At the tip of this peninsula, near the beach, is a small white weatherboard house with a red roof, on which there are a wind vane and the cups of an anemometer. In front of this building is a pole where the sky blue and white flag flies[115]. This is the meteorological station maintained by the Argentine Meteorological Service. Thorleif Hoxmark, who now works there as observer after his previous experience in the South Orkneys[116].

On the other side of the cove is a small cemetery surrounded by a white fence; it has a few graves, among them is that of the famous explorer Sir Ernest Shackleton, who, before his tragic death, had requested that his remains be laid to rest there[117].

---

different visits made by Moneta to King Edward Cove.

115. This is the Argentine flag; the observatory building was constructed, by the Compañía Argentina de Pesca in 1907 which operated it until 1949, several observers were from the Argentine Meteorological service. In 1950 the Falkland Islands Dependencies Meteorological Service took over and expanded its operations to include forecasting. A new building later housed the observatory, and the 1907 one was demolished in 1974. See notes in: Blythe, 1954.

116. Hoxmark served at the South Orkneys observatory during the 1916 and 1918 expeditions.

117. Sir Ernest Henry Shackleton (1874-1922) died aboard *Quest* in King Edward Cove, on 5 January; he was buried in Grytviken cemetery on 5 March. His widow had requested that he be

From the ship anchored in the centre of the cove I contemplated this view which was rendered even more beautiful by the tranquil summer weather, and I thought it might easily be mistaken for any one of a thousand images of Scandinavia.

This was no fantasy; in real life, we were transported to a piece of Norway, a land of sealers, legendary fishermen, and of Vikings. As we came ashore we were greeted by the manager[118] and staff of the whaling station and the only thing we heard was snatches of conversation in the soft lilting cadences of Scandinavian languages.

9: Whales

These huge mammals abound in the Southern Ocean. They were formerly found in massive numbers in the Arctic regions, until steady and ruthless hunting almost exterminated them. As a result the whaling industry had to move to other regions where whales were still abundant.

Whaling became a huge industry in South Georgia. The five companies operating there catch thousands of whales every season. The oil extracted from their carcases fills an amazing amount of barrels. The demand for whale oil has always been insatiable, although it appears to have diminished slightly in recent years. The primary use of the product is in the manufacture of margarine or artificial butter, for which there is a vast demand in many parts of the world.

The British government monitors and taxes all whaling activities particularly the production of whale oil and by-products. Not only does it tax the shore based factories but also the floating factories that each have a flotilla of three or more whale-catchers. In summer these work on regions of the Southern Ocean delimited by Britain wherein her far-reaching hegemony is exercised.

As one can imagine, the British authorities levy a tax or duty on every whale and every barrel [$^1/_6$ m^3] of oil produced. These funds are put to a good use by the British government in maintaining the *Discovery*

---

buried in the Antarctic, the site of his greatest endeavours.

118. Captain Wictor Esbensen, Manager of Grytviken whaling station from 1921 to 1930

Expeditions, a scientific organization dedicated to oceanographic research and exploration in Antarctic and surrounding regions, and to specialist studies of whales[119].

It is quite common to find whales twenty to twenty-five metres long in these waters. They are of several species, with much variation in size and colour. The blue whale[120], for example, is prized for the quantity and quality of the oil obtained from its blubber.

With this information you can imagine that my curiosity was aroused and I wanted first-hand experience of a whale hunt in all its gory detail. So one day I embarked aboard a whale-catcher to get a personal impression and proper understanding of these unusual activities.

Whale-catchers are small vessels, some thirty metres long, with a steel hull and are driven by steam engines powering the two propellers[121]. Their manoeuvrability is surprising and their seaworthiness has been tested in ferocious storms and the generally severe weather that strikes these latitudes. They are manned by a crew of twelve to fifteen men, who have known no other occupation throughout their working lives.

We were under way before dawn to reach the open sea at first light. On the highest point of the mast is the characteristic crow's nest, a kind of barrel in which a crewman is stationed to scan the horizon.

There is nothing ahead to see except water on every side. Astern South Georgia was slowly fading on the horizon.

I am on the bridge of the whale-catcher when, suddenly, the lookout in the crow's nest shouts something that I do not understand[122] and points in a specific direction. I look where he is indicating but my unpractised eyes fail to distinguish anything at all in that endless expanse of ocean.

At once the whale-catcher changes course and we set off in a new

119. The *Discovery Expeditions* began in 1925 and investigated the entire Southern Ocean; the last of a series of voyages it organized was in 1951.

120. *Balaenoptera musculus,* the largest animal of any species

121. Only the whale-catcher *Don Ernesto* had two propellers, one is normal for a whale-catcher, see § 85 footnote.

122. Probably *hvalblast!* in Norwegian, equivalent to 'there she blows' in an English-speaking whaler's vocabulary.

9: Grytviken whaling station, South Georgia

direction that I do not yet understand. Minutes go by; fifteen minutes and still I see nothing different. Then suddenly, in the far distance, what looks like a cloud of white smoke or a jet of steam appears from the sea. I fix my attention on the spot and the same thing happens again, and then a third time.

We steer at full speed in the direction of these blasts. As we approach I see a rhythmic pattern with a definite interval between every whoosh of vapour. Suddenly, at a range of fifty metres and directly ahead, a big dark grey body appeared at the surface. It was the back of a whale with its dorsal fin which, after a few seconds, disappeared again, but not before expelling a blast of high pressure breath which, on contact with the freezing air, remained suspended and looked like a jet of steam.

10: Harpoon Shot

An ingeniously shaped harpoon cannon is mounted on the foc'sle[123] of the catcher, above the bow. Protruding from its muzzle is something, resembling a solid 'square arrow', containing a grenade with a time-fuse. After it is launched this terrible projectile penetrates the whale's

123. A contraction of forecastle, the part of the vessel at the bow.

10: Harpooned whale being injected with compressed air

body and the grenade explodes inside it. Attached to the harpoon is a strong Manila line[124] which is coiled beneath the cannon. When launched, at the proper instant, the line uncoils freely with the momentum of the harpoon. Attached to its end, at about fifty metres, is a strong steel cable which runs to a winch drum equipped with special brakes which allow it to be released, as much or as little at a time, to 'play' a wounded whale according to its remaining strength.

I pay close attention to all that is going on. The whale, still swimming at a speed of five or six knots [about 10 km h^{-1}], suddenly dives almost under the catcher's bow. I am astonished when the catcher then turns in the opposite direction, because this manoeuvre was contrary to what I had expected, especially when my intuition told me the whale would continue in the same direction as when it submerged and would reappear a bit farther away.

On the bow, the harpoon gunner looks vigilant with his finger on the trigger. Suddenly, in the direction that his cannon is pointing, the back of the whale appears. A loud bang is heard and the Manila line rapidly spirals out. The fatally wounded whale, with the harpoon now attached, dives and swims off at great speed. The steel cable unwinds from the winch, the catcher stops engines, and the winch brakes start to take the strain of the tightly stretched cable. The whale's blood has stained the sea red but, though mortally wounded, it still has sufficient strength to tow the catcher as it attempts to swim away.

Gradually the steel cable begins to slacken and is wound back by the steam-powered winch. The Manila line reappears and finally the whale, now dead, is hauled alongside with the harpoon embedded in its body.

The whale is hauled alongside the catcher where two or three men wearing high boots with sharp spikes on their soles, jump nimbly onto its back. An instrument resembling a hypodermic needle two metres long with a rubber hose attached to the upper end is passed down from the deck. This is stabbed into the body of the whale then compressed air is pumped into the carcase so that it will float, instead of sinking to the bottom of the sea. This air accumulates in the stomach and

124. The *foreloper,* Manila fibre was used because of its flexibility and strength.

intestines, inflating the whale's belly. The dead whale gradually turns upside down becoming afloat, mouth upwards.

Another man then jumps on the whale to claim this catch and stabs a bamboo pole bearing the distinctive company flag into its belly. Thus marked, the corpse is left and the catcher is free to leave in pursuit of the next whale within range.

Once the catcher has taken and flagged three or four whales it is easy to find and collect them. Then it returns to the whaling station with this valuable catch in tow.

After a whale has been flagged in this way it is utterly unknown for any other catcher to misappropriate it. There is mutual respect among whalers in this difficult and dangerous work, and though the men have rival interests they will never hesitate to cooperate and help each other in a crisis. This is clearly documented in all the stories of shipwreck and rescue in which they have been involved.

All these operations and manoeuvres to catch even one whale, are conducted far from land, sometimes hundreds of miles[125] from the coast in squally weather and rough seas. Whale-catchers put to sea in all weathers to hunt and catch. The only thing that stops them is fog. If there is no visibility they cannot work.

The returning whale-catcher enters harbour and secures its catch to a whale buoy. Shortly afterwards a launch tows each whale to the shore where cables from the factory's winches and capstans haul it from the sea onto the sloping wooden boards of an inclined plane. There stand several men equipped with strange-looking tools shaped like hockey sticks or sickles with the blade curved outwards[126]. They begin butchering the animal by cutting off the blubber in long strips.

Every part of the whale is used: the meat and the bones are cut by mechanical saws and the heavy pieces dropped into boilers to make guano. Only the offal and entrails are thrown back into the sea. To reduce the stench of putrefaction a launch tows these far away from the cove.

125. In the 1920s 150 km was exceptional (300 km return), as whales became scarcer in later years catchers went much farther.
126. Termed 'flensing knives'.

11: In a whale's mouth at Grytviken; left to right: Baldoni, Becker, Casariego, and Jaramillo

11: In a Cockleshell

With the help of the crew of *Guardia Nacional* we transferred all our baggage, provisions, building materials, and coal into one of the company's small whale-catchers named *Rosita*[127]. She has been specially chartered by the Argentine government to convey us to the South Orkneys.

Navigation on the Southern Ocean has always been considered dangerous for any kind of ship, not only because of the storms that lash these latitudes, but also the icebergs that calve[128] from the ice shelves around the Antarctic ice sheet and then drift at the mercy of the elements. In dense fog it is impossible to make progress; the risk of suddenly encountering one of these huge white monsters is too great, some of them may reach the size of an island many kilometres wide.

That is why it was deemed inappropriate to dispatch *Guardia Nacional* all the way to the South Orkneys. The ship was simply too long to manoeuvre comfortably in ice, and her hull too ancient to survive a sudden collision with it.

127. *Rosita* (*Little Rose*) was built in Oslo in 1905 and sold in Panama in 1957; 166 gross register tons and 31 m long.

128. The verb 'calve' is used for the action of ice breaking off from ice fronts. It is derived from 'cave' as in 'cave in' rather than any reference to cattle.

One afternoon our departure from South Georgia for the South Orkney Islands was announced. It was to be at first light on the following morning.

We gathered on the pier at the appointed time accompanied by authorities from the company, some Argentine employees, and a few curious onlookers. They bid us an enthusiastic farewell, wishing us a good voyage, then we cast off and set course directly for the South Orkneys.

As we began the voyage aboard the whale-catcher our first concern was to organize our personal effects and occupy the sleeping quarters assigned to us in the stern cabin.

This little vessel was similar to the other whale-catchers but was now retired from whaling and deployed as one of the auxiliary fleet plying between the several whaling stations on South Georgia. The facilities aboard were not merely meagre; they were practically non-existent. The crew consisted of the Master and Navigator combined, two engineers, and four or five sailors who took turns as helmsmen. The seamen had to share the few available bunks in the foc'sle cabin. This was actually a store room which had to double as a mess and common room, as well as providing sleeping accommodation for this skeleton crew.

In the stern was another little cabin or saloon with six small bunks, three on each side, a central table and a coal-fired stove; with a companion ladder leading up to the deck.

As passengers we were treated as 'first class'; occupying the luxury of the aft end of the vessel. The Master and the First Engineer occupied the sixth remaining bunk, taking it in turns to snatch an hour or two of sleep. Our cabin was tightly sealed; it was impossible to open any of the portholes as they were almost always below the water line. The only hatch, which led to the deck, also remained tightly shut, to keep out the intense cold, not to mention the seas which constantly swept the deck.

Directly below the open bridge there was a tiny room four metres square which was euphemistically called the 'dining saloon'. In it were a table and two long benches all so cramped you could sit only by taking it in turns to get in one at a time. A hatch communicated with the galley, which was no larger than the dining saloon. Through this the duty cook, one of the seamen, proffered whatever he had prepared in the way of customary sailors' fare.

While I tried to find my way around the ship, I was naive enough to ask the Skipper where the latrine was. He laughed and replied:

'We use the most convenient place aboard - over the side!'

The temperature I felt was glacial, and the only way to relieve the numbness in my arms and legs was by rubbing them vigorously or slapping myself around the chest.

We couldn't lie in our berths all day and someone discovered that the best place to lurk was the boiler room. It was there, and in the engine room, where I spent most of the voyage. In fact I found these were the only places where I could stay warm and not feel the icy blasts on my skin which were beginning to welcome me to the Antarctic.

12: Stuck in a Lift

We continued the voyage, not quietly and calmly as aboard a trans-Atlantic liner but with turbulent jolts and shuddering in a very choppy sea. The farther we went from land, the worse the pitching and rolling became.

Aboard *Guardia Nacional* we had rolled quite a lot, but that was nothing compared to what we were experiencing now. On the former vessel it was possible, with a little dexterity, to stand upright and even walk on the wide, slippery deck. Now, in the whale-catcher, we were condemned to total immobility. It was nearly impossible to go on deck as the waves swept it incessantly, while, for anyone unaccustomed to this way of travelling, the steep angle of the heaving deck made it difficult, even dangerous, to try to keep one's balance.

Movement, incessant movement, with no respite, was debilitating, shattering. Slowly we were overcome by a strange malaise which brought an odd taste in the mouth and a total loss of appetite, ···

We stared at each other in silence, inwardly laughing to ourselves at the sight of the others.

'How are you feeling?' Valentiner asked me.

'Me? Never felt better ··· and how about you?'

'Perfect, but I wish we were not cooped up in such a confined space, it is like being stuck in a lift.'

'... and the stench of whale blubber is awful', I added.

I heard the clang of a bell and was told that lunch was ready. I made my way to the dining saloon, but I think it was more for the sake of it and to avoid losing face than from any desire to eat. On the little table was a stack of enamel bowls, some stained cutlery and a metal water jug for everyone. I sat down with my companions and, shortly afterwards, the cook passed through a soup tureen containing a steaming brew. Valentiner helped himself, but all we could do was stare at his plate in disbelief. I had never seen anything like it; a compote of stewed dried fruit in which swam whole cinnamon sticks, and from which arose the smell of cloves. This was explained as genuine Nordic sailor food; indeed it was very common in Scandinavian households.

I wanted to try a spoonful, but the nausea I felt was so strong I was unable to swallow it. I opted for the next course which wasn't slow in arriving.

'Lobscouse', announced the chef.

It consisted of a purée of tinned meat with potatoes and onions; I was informed that this too was a classic seaman's meal. It looked very runny and quite unappetising. I decided I preferred simply to swallow the saliva which was accumulating in my mouth.

Even though my companions had all been to sea before, there were some who felt just as bad as I did. One by one they all stood up and left the table: there is nothing more nauseous than the smell of food to anyone on the point of being seasick.

One of the company then made an untimely remark; 'lobscouse', he said 'was greatly relished by sailors on their long night watches. It was easy to digest and did not sit heavily on the stomach'. Taking a good look at that runny fluid (which I was later to agree, had a quite pleasant taste) I found the sight so disgusting that I had to make a hasty exit to deliver to the fish the small quantity of food that remained in my stomach.

As luck would have it, after a few hours I began to improve. Making a huge effort I was able to chew some dry biscuit and swallow a few gulps of water. This settled my stomach and the following day I was able to face the sweet soup and runny stew which did not taste quite so bad after all, though it still required much will power to hold them down.

As we steamed onward, the catcher's pitching and rolling seemed to become more violent. Even remaining seated was becoming an exhausting exercise, and if we lay on the bunks we had to strap ourselves in so as not to be thrown out by the violent lurching.

Our cabin was in the worst part of the ship. We were not only flung from side to side but also lifted vertically; only, in the same instant, to be dropped into what seemed a bottomless abyss. All this happened at great speed without a moment's respite.

I managed to get used to it, eventually. On that first voyage of initiation I suffered severely from the confined space and the fundamental upheaval in my whole way of life.

13: *Rosita* and the Crew

Everyone, at least once, has seen paintings of typical Scandinavian seamen: suntanned faces, weather-beaten by wind and sea of their home land; heads protected from the elements by a typical sailors' sou'wester[129]; their bodies clad in oilskin jackets, trousers, and thigh-length boots; countenances furrowed by deep wrinkles; some with patriarchal beards and light blue eyes, the same shade as the firmament in which their serene gaze is lost.

Strong, intrepid men, indefatigable in the labours of tying and loosening lines and cables worn and stiffened by sea water, by the freezing weather, by the ice and snow. Always wet, often soaked from head to toe, but ready and willing to climb nimbly to the topmost shrouds of the extensive rigging. Often they could be seen swinging dangerously aloft at the highest sections of the masts, furling a sail or securing a recalcitrant line through a block. Occasionally, by some stroke of fate, they fall, and are inevitably lost in a raging sea, for no human hand can possibly save them.

129. An oiled cambric hat with a wide brim, worn in wet windy weather. Curiously in Argentina and Uruguay this style of hat is known as a sou'easter owing to the strong onshore winds from the south-east.

12: Rosita *under way on the Southern Ocean (note the angle of the horizon)*

With a small band of men of this calibre, ready for anything, we were on course for the South Orkneys. Our Skipper, named Larsen[130], looked as if he had come directly from one of these Nordic pictures depicting men whom we would consider exceptional. Tall, with a somewhat narrow face, he looked about forty years old. His features were covered by a trim blonde beard. I could not fully believe his Herculean strength until I saw him lift, with one hand, a crate containing two drums of paraffin. I remember brushing against him once and feeling his muscles through his shirt: they felt as hard as tempered steel.

The rest of the crew consisted of several strapping young blonde lads, helpful and attentive, good-natured and always ready to take a turn at the helm, or to be roasted alive while shovelling coal through the small circular stoke hole while fuelling the fires under the boiler.

On the Southern Ocean the sailors perform all the essential work aboard; their attitude and mild nature are such that they are content with the simplest things of life. Even the pay is not enough to tempt anyone, yet this is the life they have led from the day when they were born.

130. Captain Karl A. Larsen (not Captain Carl Anton Larsen, founder of Grytviken in 1904, who had left South Georgia in July 1914).

Their work is very hard with few moments of relaxation, especially during the whaling season which lasts for seven or eight months of the year.

While all these men are frugal and realistic in their expectation of the standard of comfort while at work, and even of the food they receive, the vast majority of whalers whom I have known would be capable of trading a large proportion of their wages for a single shot of grog.

In all South Georgia whaling stations the sale of alcoholic beverages is strictly forbidden, and hefty fines are imposed on anyone found in possession of a bottle of brandy, even if the culprit has no intention of emptying it in one bout during his off-duty hours.

From this prohibition stems the craving they have to wet their parched lips with a glass of whatever fire-water they can lay their hands on.

In my baggage I had a bottle of eau-de-cologne. When I was seasick I often had to sniff it to settle my stomach.

Larsen saw the bottle and asked me if he could have a drop. I passed it over and he spent a moment looking at the colourful label. Then he uncorked it, took a sniff which he seemed to enjoy, and then, to my astonishment, he put it to his mouth and took two enormous swigs[131].

I later learned these men had such a craving for alcohol, they had no qualms about drinking anything that contained it. However, these outwardly rough and ready men have noble hearts, and are generous and sensitive.

A common sight at the head of their bunks, are photographs and drawings of beautiful women with long golden hair and blue eyes. One would guess these were the companions of the valiant whalers whom they must imagine are with them during their voyages.

In all the navigation charts that record the first landfalls, female names are given to the bays, peninsulas, rocks and mountains of all the Antarctic islands, in memory of the owners of their hearts.

131. Eau-de-Cologne is composed of about 80% alcohol (ethanol), and was unexpectedly popular with the whalers.

We ourselves were aboard *Rosita,* the name with which this cockleshell had been baptised to perpetuate the memory of a woman.

14: Mountains of Water

Third day at sea; we felt completely exhausted in this bucking 'lift' that leapt around like a mountain goat and did not relent for a moment. It was early morning and I was just awakening from a fitful sleep due to the catcher's constant rolling. It seemed as though there was a different rhythm, a more measured movement, not quite as violent as earlier. In the cabin the only detectable sound came from the high-pitched vibration of the propeller blades and the occasional severe juddering as it raced in mid air through the troughs between the waves.

Valentiner came down from the bridge looking for his binoculars. 'Hey' he called out 'if you want to see huge icebergs, get dressed and come on deck'.

I quickly put on my oilskins and came on deck carefully, half-opening the hatch to avoid being drenched by the seas washing aboard, and there before me was one of the most magnificent sights I had ever seen.

Vast walls of water, huge high waves, passed the beneath the catcher, from one side to the other. One moment we were rising interminably to the crest of a mountain of water, and next came the descent on the other side of the wave.

I had seen such seascapes in pictures depicting high-masted vessels weathering a gale in high seas with their masts half-hidden in the troughs between the waves, much higher than them. Every time I looked at the illustrations I was too sceptical and ignorant to appreciate that the artist was not exaggerating, but the scene I now beheld convinced me that I was wrong to reject them as implausible.

I had to cede before the evidence, for there in front of my eyes was *Rosita* rolling easily over these vast mountains of water.

When we reached the highest point of the wave, we were struck by the fiercest icy wind whipping plumes of water from the crest, but as the vessel descended into the troughs the strength of the wind abated.

I was too scared to negotiate the narrow gangway from the deck to the bridge. Whenever the vessel slewed to let a big wave pass under the keel, the lee rail and scuppers disappeared under water.

Eventually I saw what I thought was the right moment, but another wave broke green over the deck with such force that it soaked me through and through then propelled me away from the railings to which I was desperately clinging, and I found myself sliding across the deck towards the poop. This was highly dangerous, as another wave could have easily cast me overboard without anyone seeing me go. Regardless I made another attempt to reach the companion ladder that led to the bridge. This time I made it and clambered up in four bounds to reach the wheelhouse.

There I found Valentiner and Larsen, accompanied by the helmsman.

Through the ports, directly ahead about two miles away [3 ½ km], I could discern the gigantic mass of an iceberg – so white that it looked like an enormous oblong block of sugar afloat on the sea. It appeared to measure five kilometres long and a height which I judged as two hundred metres[132]. To have an idea of the volume of these masses of ice, it is sufficient to know that there is about seven times more bulk submerged beneath the sea than what is visible above the water-line.

The majority of icebergs is calved from the vast barriers of floating ice shelves which encircle Antarctica, their characteristic tabular form reveals their origin. Their dimensions may be extremely variable; later I will tell you of my encounter with one of these monsters nearly one hundred kilometres long.

As we advanced other icebergs, equally grand and majestic, appeared on the horizon. When the catcher passed close to one, you could see the spray and furious backwash as the waves smashed into its vertical flanks. This constant battering by the waves erodes the base of the iceberg occasionally causing large sections of them to detach and fall into the sea with a loud crack. As this happens the centre of gravity of an iceberg moves, which may make it overturn and an unusual swell will spread from it.

132. This is an over-estimate frequently made in poor visibility; tabular icebergs rarely exceed 40 m high.

15: Solid Sea

The gale which had been lashing us during the last two days was abating. It was no longer blowing a hurricane, although the wind still cut our faces with its icy blast. The huge waves, known to sailors as 'swell' still rocked our small vessel, but not quite so violently as we had suffered earlier.

As we advanced southwards, the icebergs were increasing in size and number; they could be seen in all directions. We would approach one, pass close to it, and leave it astern as the catcher progressed, only to find that others emerged on the horizon which was now clear of fog. The sky continued to be overcast furrowed by lines of thick dark cloud scudding past.

From the bridge I was staring towards the distant horizon, there dead ahead amidst the bergs, I discerned a thin white line which appeared to float on the sea[133].

'Pack ice,' Larsen commented, and added 'that is very bad news.'

'Why?' – I inquired.

'If its extent is widespread, we will have to make a large detour to reach the South Orkneys.'

'Are we still far from land?' – I asked.

'About a hundred nautical miles [180 km] off.'

'So in that case how long will it take to arrive?'

'It depends on how we get through the pack ice, or the time lost in making the detour', replied Larsen.

Then, scrutinizing the sky, he added:

'Now we have good weather and need to take advantage of it. In these seas one can never predict how long it might last.'

'Good weather!' I exclaimed, 'here we are in a rough sea and severe swell, with an overcast sky, and glacial cold and you call it "good weather"!'

'Much worse' Larsen replied 'would be to be caught in another thick fog amidst this city of icebergs; in that case we would find ourselves in serious danger.'

133. This is the phenomenon of 'ice-blink' which indicates the presence of pack ice because it reflects sunlight on to cloud.

He was right; it was much better to navigate with good visibility, to be able to look ahead and weave through those 'floating reefs' which interposed themselves in our way.

For these sailors 'good weather' means 'good visibility'. Neither storms nor tempest trouble them, they are accustomed to navigating with the fury of both. They only call it 'bad weather' when lack of visibility forces them to reduce the vessel's speed.

These thoughts were in my mind while we were advancing towards the edge of the pack ice. I could discern it now, although not very clearly.

For those who have ventured to enter it this ice has caused more than one tragedy. It is formed by the sea freezing and solidifying near the coast, or even farther out in the open sea when it is very calm, and when the air temperature drops below minus fifteen degrees or more. In these conditions the surface of the sea turns from a liquid to a solid state and is covered by a soft layer of ice which becomes thicker as snow falls on it as long as the temperature continues to drop. Its total thickness may reach ten metres[134].

In spring and summer, when open water reappears, the sea begins to receive the Sun's heat, the swell works beneath the ice first cracking it then splitting it into large floes. These crash into each other reducing them to ever-diminishing fragments each time.

From the crow's nest very large floes of pack ice were visible. Nearer the bow were smaller ones, like an advance party. The swell penetrating within this field of ice produced a strange undulating motion of its solid surface. Farther in, this movement diminished owing to the distance travelled and the decreasing force of the waves moving beneath the floating pack.

The catcher's speed was reduced. A lookout wedged in the crow's nest monitored the movements and signalled a course through the least dense areas of pack.

134. This thickness is exceptional and might indicate a severe pressure ridge; such a phenomenon is very uncommon. In the Southern Ocean normal pack ice is rarely more than one metre thick.

13: Jaramillo measuring the sea temperature in Uruguay Cove

Thus we continued, constantly advancing towards the South Orkneys.

A huge floe of ice came into contact with the bow. With the engines on full-power, the catcher could not move it out of the way.

Suddenly an alarm sounded and that same instant the engines were stopped.

A thick piece of ice, at least ten metres across, had wedged itself directly over the propeller. Had it continued to turn the ice could have snapped off one of its blades. In this case we would have remained imprisoned in the ice without any possibility of rescue, as the catcher had no radio equipment to alert anyone to our plight.

Still it was necessary to continue forward, and with great skill, using the engine, Larsen was able to manoeuvre this massive block of ice out of the way. No sooner was this done than others arrived to fill the gap. There was so much pack around the sea surface appeared white as far as the horizon.

It was at moments of danger such as this, forcing a passage through solid sea, that the expertise of the crew to whom we entrusted our lives, was evident. These uncomplicated seasoned men were inured to battling the elements whatever the conditions which presented themselves.

16: Flag Signals

At six a.m., on our fifth day at sea I was awakened with a shout; land, our destination, was in sight. I came on deck and before my eyes lay the bleakest and most desolate panorama I had ever witnessed.

From the horizon of the sea rose tall pyramids of blackish rocks, splattered with large white patches of ice and snow.

Very low clouds furrowed the sky, at no more than three hundred metres above sea level, and hid the summits of those sheer peaks with their almost vertical walls. They seemed like giant shards emerging from the depths of the sea.

Vast conglomerations of ice and rock were evident. In places, where the mountains were less precipitous, the snow that had accumulated over thousands of years had compacted into layers of solid ice forming glaciers in the ravines and ice caps around the summits; the immaculate heights of the latter stand out majestically.

From some of the glaciers which had descended and reached the sea, enormous blocks of ice had broken off and floated away, progressively forming growlers and bergy bits[135].

With my binoculars I carefully examined that coast in its rugged detail; it looked completely inaccessible and revealed no sign of any living thing.

The rock which could be seen exposed through the ice was completely smooth; just a few small areas had some sparse greenish stains indicating the diminutive South Orkneys vegetation, a kind of excrescence which grew, encouraged by the high humidity. Much later I discovered some lichens, with a maximum height of only ten centimetres[136].

Very slowly we neared the coast, keeping a sharp lookout from the bridge for any treacherous reefs concealed below the surface.

135. Icebergs calve, split, and break progressively forming growlers (dangerous as they are awash and difficult to see in rough weather), bergy bits, and eventually brash ice.

136. Terrestrial vegetation in the South Orkney Islands consists of two species of vascular plants, approximately 25 of mosses, 2 liverworts, 42 lichens, and 2 algae. Some of these have a very local distribution. Occasional cultivated plants were on base, one appears in the photograph with Miguel Jaramillo in 1927 (Illustration 71).

I could now distinguish the innumerable bays, headlands, and peninsulas which formed the topographical irregularities of this archipelago.

Aloft, a few albatross and petrels flew past to see what we were up to; like aerial messengers they circled the vessel a couple of times before flying onwards, to be replaced by others.

A seal popped its head out between the ice fragments but rapidly dived out of sight when it saw it was not alone.

Slightly farther ahead, on another ice floe, a group of twenty or so penguins clustered together and gravely watched us passing. Instead of fleeing like other animals, they came closer as if genuinely inquisitive about our indiscreet disturbance.

My attention was suddenly diverted by the laboured flight of a shag. Valentiner pointed it out and commented: 'there goes our Christmas turkey'.

With all this happening my impatience to know the location of our base, and to actually see that icy place in which I would exile myself for an entire year, filled me with obvious apprehension.

'We are close to Saddle Island' said Larsen, pointing to its characteristic mountainous outline, and added:

'In a bit more than an hour, we will reach our destination.' Shortly after that, my veteran leader took my arm and pointed in the direction of the coast saying:

'There it is now. Can you see that black dot visible on the coast? The rock in the middle of that broad white band on the coast.'

'Which one?' 'That one below the glacier?'

'Yes that is it! That point towards which we are now directly steering is the *Manchón Austral*[137] the name given by the first officers who reached this place aboard the corvette *Uruguay*. It appears on all the maritime charts as the navigation mark for the entrance to the natural harbour we are now approaching.'

'When will we be able to glimpse the Observatory and, more importantly, meet the lads?' I asked.

137. *Manchón Austral* may be translated as the 'Southern Mark'; it is a prominent conical rocky outcrop, 40 m high, at the north-western entrance to Uruguay Cove from Jessie Bay, shown on Map 4.

'As soon as we pass the promontory, Point Lola, which we will leave to port. We will first see the cemetery, and immediately after that the magnetic huts will come into view, and then last of all, the living quarters will appear.'

'What about the men? Will they know, or have any idea, that we are arriving?'

'It's possible they will sight us before we anchor, and in that case ….'

'In that case what?' – I interrupted.

'In that case,' answered Valentiner, 'we will know at once whether they are all alive or if anyone is missing.'

'How will we know?' – I questioned again.

'If they have seen the vessel coming in, they will fly a flag on the pole, there near the beach. If it is flies from the top, it means everybody is alive. If, on the contrary, it is at half mast, it indicates we can expect to find one man fewer.'

17: Our Meeting

We were just clearing Point Lola. The cemetery, and then the magnetic hut, came into view; then two other huts simultaneously appeared. Ultimately came the long awaited flag pole … but it was empty, no flag of any kind on it.

The catcher's speed slackened. On the bow two or three men struggled to free the anchor. We were in Uruguay Cove; just to port were walls of ice that appeared to have been split with an axe; to starboard rock and ice. We had left the *Manchón Austral* astern.

Nobody was looking at the surroundings, we were all waiting to see the base[138]. Then this came into view: bright green walls, a gabled roof black and shiny. Smoke, coming from two chimneys was blowing horizontally on the wind.

Valentiner looks at me and smiles:

'It looks as though the men are all asleep!'

Then he looks at Larsen and says a few words in Norwegian. The

138. The term 'base' is used for all the installations of the observatory, the principal building includes the living quarters and some stores which also have laboratory and radio functions, it is termed the 'dwelling' when a distinction is required.

Skipper tugs a line that sounds the catcher's steam whistle; there is a shrill blast, loud enough to deafen me.

It took barely a couple of seconds for three men to appear, struggling to be first through the front door. At the same instant a fourth man climbs head first out of a window. All four run to the beach to meet us.

A fifth man appears in the doorway as if about to follow, but he must have forgotten something and goes back inside then reappears with a package under his arm. He runs quickly to catch up with his companions. They gather at the flag pole and it doesn't take long before our national flag begins to rise, bit by bit until it reaches the top.

'Everyone alive!' ... we all breathe a sigh of relief; a great weight is taken off our minds.

We see the men on the beach simultaneously raising their arms and we think we hear the sound of their 'hurrahs'.

We answer in the same manner, this time for them.

Larsen, with a voice of command, orders:

'Let go!'

The anchor drops rapidly, rattling its heavy chain to the bottom of the bay.

A little boat is lowered. We embark, Bruhns and a sailor row to the beach.

They all await us smiling and literally jumping with joy. Two of them wade in the water to make our landing easier.

They are all big men, robust with weather beaten faces. There are smiles all round and they embrace us warmly. We need to double our numbers to answer all the questions they fire at us.

'How is Buenos Aires ?'

'Has another war started anywhere in the world?'

'How was the voyage?'

'Have you brought fresh potatoes ?'

Valentiner hands Kollasius, the Leader, a thick packet containing letters from the families who for a whole year have had no news of these solitary, courageous men.

As we walked together towards the dwelling, our companions gradually began opening their letters; they did more than just read them, they devoured the words before them, in that way they caught up with news sent by mothers, wives, fiancées, girlfriends, friends and relatives.

No person who has not actually lived there, in such total isolation from the inhabited world; buried alive amid the solitary confines of polar ice, with no means of communication with the rest of humanity, and in utter ignorance of all the things that could possibly happen in the course of a year at home, can imagine what it meant to those men to receive news from that distant civilized world.

Vast satisfaction and contentment was expressed on everyone's face. As we walk towards the base I take a closer look at them. I noticed how they treat each other like close friends, using familiar terms to each other[139] and cracking flippant jokes which are greeted with laughter, so infectious and overwhelming is the irrepressible happiness to have letters.

In a few words they tell me what it has been like for them during the last year. They ask if I have any experience of snow and ice, and they soon realize that my sentiments are very well-disposed to them, that I have arrived with sufficient integrity and strength of character to confront whatever will come.

They treat me like an old comrade and are amused by the nickname 'Kleine'[140] bestowed on me by my companions.

We enter the building which is to be my abode, my home, or my prison for as long as life or destiny chooses to incarcerate me therein.

18: Our Abode

The living quarters of a polar expedition …. I was part of one at last, in one of those basic huts which is covered in snow and ice up to the eaves, sometimes even higher; then it becomes completely submerged in a flat white plain of engulfing snow. Its presence is indicated only by the smoke pouring from the blackened chimneys which emerge above a snow surface blasted by icy gales.

139. In Spanish the formal second person singular pronoun is *usted* (a contraction of the antique *vuestra merced*, an honorific equivalent to 'your mercy'); in Argentina the familiar term *vos* is used between close friends and some relatives, *tu* is an even less formal personal pronoun.
140. *Kleine*, an affectionate nickname, means little one, or 'titch' in German.

However, on this occasion, the base was not covered in snow as I had imagined, but they lost no time in telling me that soon it would be.

We entered a narrow corridor where I saw several shovels and picks which, later, I would find myself wielding constantly to clear the snow and chip away the ice which imprisoned us. In one corner there were several sledges and, on the left wall, a large rack with numerous pairs of ski.

Directly in front was another door which opened into the interior. We went through and I found myself in a confined room two and a half metres across, the kitchen. In it was a basic stove with a couple of pot rings on the hotplate which were glowing red. Different pans were steaming, and from them wafted a pleasant aroma of herbs and spices.

Alongside the stove I saw a large galvanised iron tank in which ice and snow are melted to provide fresh water.

The walls are all covered with shelves which hold pots of many sizes and cooking utensils of every kind.

At the far end of the kitchen, to the left, is a door which leads into a bunk-room for the cook and one other man to sleep. Farther along is another door through which one passed and entered a small parlour, two and a half metres by four metres. In it is a large stove which is lit and, like that in the kitchen, glows red hot. There is also a wooden table and several chairs. The walls are covered with shelves replete with books, instruments of all kinds, a gramophone[141] well-stocked with records, precision chronometers and a barometer in its glass case.

Everything is very clean and tidy. I am told that the building is painted every year by the men who are departing. They are responsible for leaving it in good condition for the next winter party.

To the right of the parlour, which will be our combined living, dining and work room, are two windows to the outside. These windows, like all the others in the building, are double glazed, one section opens outwards and the other inwards, which helps prevent an ingress of frigid air.

To the left of the parlour are two doors: one that leads to the

141. A spring powered mechanical device for reproducing music from shellac discs (records).

leader's bunk-room and the other is the bunk-room which Bruhns and I will share. Opposite these is another door which leads to a tiny dark-room which only became a space when they blocked a door to the outside[142].

I go to the bunk-room assigned to me and have to laugh! This is not a bedroom, it is the size of sleeping compartment on a train, or a cabin of a small ship. In it are two wooden bunks, one above the other. Opposite is a long wooden table which folds. You can tell by looking at it where it came from; it is similar to a mess-room table from one of the whale-catchers which brought us here, and was probably exchanged by an earlier skipper for a bottle of whisky. Above the table, running the full length of the wall up to the ceiling is wooden shelving; this too is full of books of all kinds, predominantly by English, German, French and Scandinavian authors. They are mostly novels; I can see several by Spanish authors; and there are just a few Argentine ones.

In the right hand corner, in the space beyond the bunks is a rural wash stand, with a basin and jug of white enamel.

At the front of the bunk-room is a window, double glazed like the others, which seems to be semi-locked, and as I try to open it, they warn me it is better not to do so as it is practically impossible to close when the frames become clogged with ice.

This will be my room; a two and a half metres-long space in which bunks, table, chairs, and wash stand all have to fit. Not so much a living room or bedroom, more like an actual prison cell.

The thought passes through my mind that the Leader, who will live alone, will have more space to himself, as the Cook will be sharing quarters with Plagge. However I am wrong; Valentiner's bunk-room is crammed with suitcases, instruments, books and other equipment.

Available space is extremely limited but, somehow, everything just fits into it. There is an eminently practical expression here: 'A place for everything, and everything in its place.' If this were not so there would not be space enough to turn around.

142. The dark-room, attached to a former outer door from the parlour, appears in a 1913 Swedish sketch plan of the dwelling.

14: Panorama of frozen Uruguay Cove, background Mossman Hills, the Observatory on the horizon, and Mount Ramsay glacier to the right (note Moneta's long shadow on the left)

We are gathered in the parlour and every man has brought a bottle of whisky or brandy which he has been saving for a year. We are celebrating the grand annual event; the meeting of the two groups. With the warmth of the stove, always glowing hot, we chat about a thousand different things, while my eyes wander slowly and take in all the nooks and crannies of the little dwelling. I find it quite congenial, thanks to the heating, but also very confined and small. I will just have to become accustomed to it until I feel as comfortable here as in the most spacious mansion.

19: Impressions

Two hours after the Sun passed over the meridian of the South Orkneys we begin to feel hungry. Otto[143], our stalwart cook has put a white cloth on the table around which we are gathered in a warm comradeship. The table is quickly laid: cutlery, glasses, plates, a jug of cold water, and, to look like red wine, a bottle of currant beverage concocted from dried red-currants specially brewed in the South Orkneys and called 'saft'.

Accompanying these drinks is a bottle of lime juice, which is an obligatory beverage for all men of the expeditions[144].

We take our seats and the main, and only, course soon arrives: penguin steaks, garnished with chips, made from the potatoes we have

143. Otto Zeiger, the Cook, is the only colleague of the author who is referred to by his first name in the book.

144. As a source of ascorbic acid, vitamin C, for the prevention of scurvy.

just brought ashore, and for each man, two unusual looking eggs fried in butter.

The veterans all tuck into this amazing meal at great speed, Bruhns and I look at each other discreetly in silence.

'Please eat, do start', we are told.

Very much against our better judgement we are obliged to do likewise ··· and give it a go; my first mouthful of penguin. I chew it while thinking all the time that I am about to experience a thousand different nuances of flavour. So many people, who supposedly knew what they were talking about, told me what to expect; a slightly fishy taste or rather like molluscs. For me this is not really the case; it tastes somewhat more of a hare that has been well-hung. Dark flesh, almost black, very fibrous, not tough, no really definable taste. I find it edible, very passable.

'This penguin is not the best,' I am told 'it is a Chinstrap[145]. Gentoos[146] are far more palatable, but this morning none passed this way'.

'Well, fancy that ··· fresh meat delivering itself daily to our door.'

'Ah yes, but only in summer' I am told 'in winter you will have to walk far to find it'.

The penguin eggs attracted my attention. The yoke is much darker than that of hen eggs and somewhat larger, but there is a greater proportion of albumen. The albumen, even when well-cooked, does not set in the same way as that of hens; it hardens, of course, but then turns gelatinous and translucent, with its edges slightly browned by the heat of the stove.

The appearance of this delicacy is grim. The others see my prevarication and taunt me:

'Go on, eat with confidence, don't worry about it.'

I try these weird fried eggs; they are not dissimilar to *Tero*[147] eggs. I am pleasantly surprised.

'They must sit a bit heavy on the stomach', I comment.

145. *Pygoscelis antarctica,* Chinstrap penguin

146. *Pygoscelis papua,* Gentoo penguin

147. The *Tero* is a lapwing, *Vanellis chilensis,* a ubiquitous South American bird whose eggs are occasionally eaten.

15: Ruin of Omond House, built in 1903

'Possibly, but we will soon "walk them off" after lunch'.

The expression amuses me; I can't help thinking, as I savour this novel food, that it is all very palatable. Even if the idea of serving such a meal in the midst of civilization might seem repugnant, down here in the middle of nowhere it is the perfect meal, it comes from fresh produce, with a high proportion of vitamins essential for men who have to undertake hard physical work.

Dessert consisted of an appetising cake, also made with penguin eggs. That first South Orkney meal could not have been a more typical 'local' one, with which to welcome the new team.

After lunch we went outside to walk around the base area. First we made our way to the remains of Omond House or Scotia House, the famous historic refuge which had served as the base for the first explorers. All that is left now is a shapeless mound of rocks. The ruin was not usable for anything but later, in 1927, I took parts of it to make the foundations of the first radio-telegraph station which I inaugurated.

From Omond House we went on to a cluster of huts called 'Absolute Huts', so named because they housed the instruments which measure the absolute variations of the Earth's magnetic field[148].

148. The plan of the station shows three huts for geomagnetic observations: two are for abso-

From there we were only a few paces from the cemetery. A small dry stone platform indicates where the body of each of the four explorers lies who lost their lives in the South Orkneys. Four black wooden crosses, symmetrical, with the names and dates carved by hand, stand at the heads of the graves[149].

'The memorial plaque for Hartvig Bache-Wiig is still missing', they inform me.

'Why,' I enquired, 'was he someone else who died here?'

'Yes, but Bache-Wiig was lost and his body has never been found.'

I did not feel like asking any further questions. During the long winter Valentiner would tell me the full story.

A cold shiver ran down my spine. I thought about what lay beneath each of those four stone mounds, of the accidents and tragedies responsible, and of the men whose lives were lost in this remote icy land.

20: Panorama

The next day the departing winter party required me to take my first lesson in Orcadian geography. On the table a map showed the archipelago where I saw the three main islands composing it[150].

To the west, the large Coronation Island; in the middle Powell Island, its name bestowed to commemorate the British navigator of the same name, to the east lay Laurie Island, on which the Observatory is situated.

Two straits separate the three islands: Lewthwaite Strait, between Coronation Island and Powell Island, and Washington Strait between the latter and Laurie Island.

The three main islands of rock and ice are surrounded by numerous islets, capes and peninsulas. The whole archipelago is mountainous with many irregularities in its formation.

---

lute measurements (Absolute Huts) and the third (Variometer Hut) for diurnal variations in the terrestrial magnetic field. The huts are remote from the dwelling and other structures to isolate them from extraneous magnetic fields.

149. This observation dates from the 1923 winter. § 5 with its footnotes include all subsequent graves and memorials in the South Orkney Islands.

150. See the maps at the end of the book.

16: Panorama left to right; dwelling, radio pylon, magnetic huts, cemetery, and flag pole

The predominant colour of the landscape is white; ice and snow everywhere, on land and at sea. On the sea are large ice-floes swept along at the mercy of the wind, or icebergs propelled by the invisible forces of strong ocean currents.

The Observatory is situated on a typical low-lying isthmus, some three hundred metres wide, between two beaches. Crossing The Isthmus is a distinct area of relatively level scree and gravel, where buildings could be constructed. North and south of this isthmus are Uruguay Cove and Scotia Bay, respectively. The former is a wide bay commemorating our veteran naval corvette. It opens towards the horizon and, in the middle distance, breaking the monotony of the landscape, lies Saddle Island.

The eastern limit of this bay is marked by a headland, Point Lola, rising vertically from the sea. The main feature to the west of this bay is Mount Ramsay, five hundred metres[151] high, with its inaccessible flanks and ice-cap. Skirting it is the slope of the glacier above the *Manchón Austral.*

To the west of The Isthmus lies the perpendicular Mount Mossman, which marks the natural limit of our private property. To the right is the cradle of La Monja[152] Glacier, which farther up joins the glacier on Mount Lola, where the two coalesce in a huge plateau of ice[153].

151. Mount Ramsay, 537 metre elevation, is the highest point on Laurie Island; see § 114. (Mount Nivea on Coronation Island, 1265 m, is the highest peak on the South Orkney Islands.)
152. Literally *The Nun*
153. These two confluent glaciers have retreated, as have the other glacial phenomena on the archipelago. In 2017 the glacier edge lay about 150 m back from its 1920s position. A steep mo-

We turn around to face south where Scotia Bay appears with its wide horizon. To the left, the east, a wall of ice of about ten kilometres long terminates in Point Rae.

To the right, almost five hundred metres away is Point Diebel. What lies beyond this point is anybody's guess. Only by ascending La Monja Glacier would one be able to get a better view of that part of the bay. On that same coast, at a distance about six or seven kilometres, they point out a beach which is less steep than the other ones. Its strange chestnut colour makes me ask:

'What is that over there?'

'A rookery'[154], they reply.

'Rookery?' – I ask, 'What is that?'

'A penguin rookery, a place where they gather when breeding.'

On the outer limits of the Bay, in its middle, at about fifteen kilometres from The Isthmus is a pyramid of rock known as Ailsa Craig[155]. To the right of this the flat headlands near Cape Murdoch can be seen.

All the shores are vertical, like vast walls of rock or ice; desolate, without any signs of life; whipped by fierce winds which chill us to the bone, despite our thick winter coats.

'Overcast, always overcast,' they tell me, and add:

'Seldom will you see blue sky here, it's very rare. Generally it is just cloud, cloud, cloud; always dark, dirty grey, and very low.'

Looking southwards, over the sea I can see much ice stretching across the bay, including icebergs of every shape and size.

We have completed our tour of the territory in which our lives and work will evolve. Let us look more closely at our immediate surroundings.

Twenty paces from La Monja Glacier, running north-west to

---

rainic ridge now lies north-east of the base beyond which is a 120 m wide shallow lagoon across most of the isthmus, before glaciers are reached.

154. The term 'rookery' is generally used for penguins while groups other species may be described as a colony.

155. Named by William Bruce after a rocky island of similar appearance in the Firth of Clyde, Scotland.

17: Variometer Hut, housing geomagnetic instruments

18: Panorama: left frozen Uruguay Cove, background Mount Lola and glacier with La Monja behind, the Observatory is in the centre of the photograph

south-east, is our dwelling. It stands alone and isolated from the other buildings which are situated near the base of Mount Mossman some two hundred metres away. From the base it is a shorter distance to the beach of Uruguay Cove than to that of Scotia Bay.

My eyes do not tire from gazing at all this, everything is so new for me. I make a promise to myself that, even if it is only for my interest, I will try to discover something new beyond those hills and ice fields which, in fact, are outliers of that vast Antarctic continent.

Amidst all my thoughts and ruminations, especially concerning that frigid white grandeur, that boundless abandoned solitude, on occasions some irresistible internal force takes my eyes back to those isolated crosses which mark the place where other men like me, lost their lives, men who, like me, possessed exactly the same healthy optimism and boundless faith in the success of their future.

21: Farewell

Two days after our arrival, we had managed to unload all our cargo and equipment. It was a job which required an inordinate amount of labour. I remember having to fill many sacks with coal, lifting heavy crates and stacking them on the beach, only to have to lift them all again and carry them farther up to save them from the incoming tide and away from the voracity of the waves. Apart from that we had to perform just about every other task that one could possibly imagine which needed to be done in such circumstances.

We took possession of our home. The departing team had embarked and installed themselves aboard the whale-catcher to give us some space. We were engaged in writing our last-minute news to families and friends, all those whom we would not see again for the duration of our voluntary exile.

The hours seemed to flash by, and each one of us instinctively felt a desire to delay that ultimate moment of departure.

The happiness of those leaving contrasted with the obvious pensiveness of those of us staying. This was apparent in the facial expressions and gestures of each one of us.

Kollasius, the departing Leader, and his team had donated all the spare tobacco and chocolate they had managed to save from their ration. In truth it was a grand gesture, which we learnt to duly appreciate during our year of isolation.

Silently we contemplated everything that lay in piles around us: particularly the provisions, the coal, the clothes, and the books. We possessed everything needed to face a polar winter and make it a success, and to live our solitary life in the lost world of the Antarctic.

Scientific work would absorb most of the time we spent there. That was what would prevent us becoming bored, and even ill, from the tedium that afflicts sailors during long sailing voyages plying the oceans of the world.

I knew, once the returning team had departed, we would be left alone to our fate, with no possibility of help from any source; no obvious way of contacting the outside world; unable to look at any other faces except the familiar ones of our companions; unable to hear any voices apart from those of my colleagues, most of whom spoke a language I did not know. Just the company of four very different souls left to the utter mercy of our fate and God's good grace ….

I wrote a long letter to my parents and others to my friends. At dawn, as I was just finishing my correspondence, Larsen called to say farewell.

Shortly afterwards the departing team joined us. Then we all walked very slowly from the base to the beach of Uruguay Cove where the whale-catcher had just re-anchored.

We reached the pole where the flag was flying and began our farewells, hugging each other effusively while exclaiming:

'Bon voyage.'

'We wish you a very good winter.'

'Greetings to Buenos Aires ··· to everybody!'

'See you when we return, we will be sure to come and look for you.'

'Do not forget to see my parents and give them news of me.'

We went to the water's edge where the boat was waiting. I am sure our thoughts were selfish, we were trying to detain them for a few more minutes, even a few seconds longer.

They were heading north, to a world that was so far away it seemed like Utopia, like a dream ···. One day, perhaps, we would rediscover that world; yes perhaps we would.

Nimbly they clambered aboard the boat, which was given a push at the right moment to speed it through the breaking waves.

The oars were set in the rowlocks and the boat pulled away from the shore. We watched in complete silence, as if petrified by the magnitude of the sacrifice we were making.

We were renouncing the world, everything, and beginning to start our submergence into oblivion, a whole year of life there alone.

The departing team reached the whale-catcher. The capstan began taking in the chain with its heavy anchor.

Valentiner touched my shoulder and, pointing to a particular spot, said:

'Let's wait there, beneath the flag.'

Looking back, towards the catcher, every moment, we went up the slope from the beach and slowly walked towards the flag pole.

Rosita gradually got under way, and, at the same moment, a plume of steam issued from the whistle followed by a strident sound which shattered the silence and reverberated around that impressive whiteness that engulfed us.

Then came another whistle, even more piercing this time.

Valentiner, with the halyards in his hands, slowly lowered the flag half way, then paused, then up again to the top of the mast.

Three whistles from the catcher: three salutes from land with the sky-blue and white flag.

19: Hugo Valentiner, Leader in 1923

From aboard they shouted three cheers for us and we answered the same way, for a safe journey, and a joyful return home.

They were going back to life, back to that distant world, from which we would be excluded for an entire year.

The catcher, northward bound, grew smaller and smaller while we remained motionless beneath the national flag which flew horizontally on the polar wind above our heads as if it were protecting us from the multiple dangers that lurked behind that immaculate white Antarctic ice.

Now we were really on our own, confronting our destiny.

22: Our Base

A few minutes passed before the whale-catcher was lost to view behind Point Lola. Our first job was to move all the provisions and coal that had been left high on the beach of Scotia Bay to the base. Obviously it was dangerous to leave such things there for any length of time; a high tide could wash away the crates of provisions and sacks of coal, making it impossible to recover any of them.

We moved all the cargo to higher ground; a cheerless and exhausting task, as it began to snow. The intense cold chapped our faces, and our hands started blistering from carrying the rough heavy loads.

When we considered that everything was safe from the tides, Plagge and I were assigned the duty of transporting it onward to the dwelling.

No snow had accumulated on the shingle and rough ground that covered The Isthmus. The only vehicle capable of carrying provisions, sacks, crates and other heavy loads was a sledge. Its runners had been greased with seal blubber[156] to make it easier to slide over the rocky ground. Even so we had to use all our strength to drag this heavy weight. Eventually, item by item, we managed to move everything to a depot in front of the dwelling[157].

156. Weddell seals, a source of fresh meat, have oily blubber, a few centimetres thick, under their skins. This would have been abundant on base, and an expedient lubricant for sledge runners.

157. After 1930 a Decauville railway was introduced to make this task easier. It appears in station plans and is described in Plan 1.

This laborious task took several days. Afterwards we began opening the crates and a vast assortment of tin cans and jars of food appeared. Otto, the Cook, took charge of storing them separately on the shelves in the rudimentary pantry.

The problem arose as to where to store the large quantity of fresh potatoes and onions. The food store was too cold, there they would freeze and then decompose, making them inedible.

'Experience from previous years,' Valentiner explained, 'has proven that a better place for storing these provisions is an area that is not too cold. For this reason, we will have to resign ourselves to storing the crates of potatoes and onions in our bunk-rooms, under the bunks or anywhere else which we could find that was suitable'.

This is what we did. The same procedure was used for the bottles of lime juice and other liquids, such as vinegar. These liquids expand when frozen, then, on becoming solid, crack their bottles and release the contents.

The chocolate ration was distributed equally. This is a very useful food to carry on field trips for its large energy content in proportion to the quantity, and it is easy to carry.

Most of the alcoholic drinks were also distributed individually; each man could then decide when to have some. The remaining bottles were kept for communal use on special occasions.

The coal was stacked outside, a huge pile of sacks one on top of another as tall as practicable to make it easily accessible, especially during the winter when snow began to cover the isthmus. Inside the base, all our personal belongings were sorted: clothes, books, indispensable tools, instruments, some new furniture, even records for the gramophone which would enliven long dark winter evenings. A place for everything was found.

The arrangement of the bunk-room which I shared with Bruhns, took much of our spare time. I had the top bunk on which I placed two mattresses. These bunks were not well-sprung, and I had to get used to their hardness.

I had some good quilts and, like my companions, I slept between woollen blankets.

With every day that went by, these cramped quarters became more habitable and more comfortable despite the small amount of space available.

Every space was utilized, even the most inaccessible corners. Actually, when inside, we did not have any notion of the conditions outside; the freezing cold, the wind and the snow. The stoves in the kitchen and the living area were constantly burning, maintaining a pleasant working temperature of about eighteen to twenty degrees above zero.

Every man did his share of the housework. There was a strict roster for tidying the parlour and communal area, for filling the coal scuttle beside the stove, for cleaning and adjusting the paraffin lamps on which we depended for light. Steadily we became accustomed to this methodical and orderly lifestyle in which we all collaborated to make our coexistence more comfortable.

23: Water Run

Very good fresh water was available throughout the year; it was sometimes so cold that we had to add hot water to make it easier to swallow.

I have previously described the kitchen area. A large galvanized iron tank, about two metres high and sixty centimetres wide, with a tap and an open top, stood next to the oven of the kitchen stove. From this we derived our water supply for general purposes; that is for cooking, drinking, washing, etc.

One morning, Valentiner remarked:

'Lads, we will have to make a water run, the tank is nearly empty.'

We dressed with our thick outside clothes and rubber boots. As we would be working with ice we wore two pairs of gloves; woollen finger gloves protected by outer leather mittens.

Outside each man carried an ice-pick in one hand. We took the long sledge with three empty crates tied to its frame. Using the ropes we hauled this strange contraption to the beach at Uruguay Cove.

The strong swell had deposited a large block of the purest glacial ice on the beach. It was faintly pale blue in colour, furrowed by many striations and cracked in several places. The cracking was a consequence of the very low air temperature, about ten degrees below zero, and subsequent contraction of the ice.

This particular block of ice had been washed up on the middle of the beach. The sea had lapped its base, but it could not float until the tide had risen by at least another metre.

Bruhns and I clambered on to the upper part of the block and began, using picks, to hack off chunks that shattered into haphazard pieces. Plagge picked these up when they landed on the beach; Valentiner and Otto formed a human chain to pass them into the crates on the sledge.

The crates slowly filled with this 'solid water' as we continued to wield our picks.

At each blow that I struck there came a sound like breaking glass. I found it easier to work from the centre of the block, striking the sharp pick into what was the remnant of a piece that had calved from a glacier.

I had just given it an almighty blow, when I heard a hollow sound beneath my feet. As quick as lightning a crack opened from one extreme to the other, warning that the block was about to split.

I repeated the blow, but this time, without warning, the crack widened under my feet and the ice block split asunder. Bruhns and I simultaneously fell into the water.

The depth here was negligible, only about half a metre, but it was enough to drench us completely.

I felt an intense cold, like piercing needles through my body. As soon as I recovered from the shock I managed to dash back to base followed by laughter and comments from my companions.

I changed into dry clothes. When I returned to continue the task, I found the job completed and the crates were piled full of solid water.

We began hauling the heavily-laden sledge to the base, but instead of taking it to the front door we went to the opposite side of the building and stopped by the kitchen window. Otto opened this from the inside and we formed another human chain to pass him the blocks of ice which were put in the tank until it was full. The heat from the stove would do the rest.

20: Hauling a heavy sledge loaded with 'solid water'

Our task of obtaining water being finished, we went back inside and clustered around the stove. There we philosophised about people living farther north who, in the height of summer, went to much trouble to chill the tepid water that came from their taps. Down here, as we approached the middle of winter, we were learning, with more primitive plumbing, how to convert Antarctic ice into water, so essential to sustain human life.

24: You will be the Breadwinner

While we were working Otto was performing marvels in his tiny kitchen. From early in the morning he could be seen wearing his broad white apron, sleeves rolled above his elbows, stretching and kneading the dough, making fresh bread, while covered from head to foot in flour.

There was an incessant clanging of saucepans, frying pans, and dishes. Delicate whiffs of all kinds of tasty morsels wafted around.

At eight thirty in the morning we sat around the table for breakfast- The table was laid as if for a banquet[158]; not just the ordinary breakfast

158. Breakfast in Argentina is usually 'continental' in style and not a major meal (perhaps only coffee and croissants or 'media lunas').

of coffee with hot milk, but plates and cutlery were laid as if for a full meal. There was fresh white bread, wholemeal bread, and black bread. Several kinds of jams, honey, and *dulce de leche*[159] were decanted into dessert dishes; also cheese, hot sauces, a bottle of crystal clear water and the usual bottle of lime juice cordial.

A steaming dish of *milanesas*[160] of penguin breasts and fried penguin eggs soon appeared. Shortly afterwards, Otto passed a piping hot coffee pot for us and a jug in which he had warmed some diluted condensed milk.

All this accompanied by thick slices of bread, with plenty of butter which, like the milk, was from a tin can, with a choice of several varieties of jam, honey, etc. Quite honestly it was a more than adequate repast with which to rouse our minds for the scientific tasks that lay ahead.

'*Mein Herr!*' Otto exclaimed to Valentiner 'there is no more fresh meat'.

'Well,' replied the Leader, 'we will have to organize a visit to the beach to get some'.

Then he addressed Plagge:

'Take Kleine with you, find some penguins and deal with them in the customary way.'

I put on my usual outside clothing. I had never hunted penguins in my life and, thinking I was doing the right thing in the eyes of the old hands, I took a shotgun, slung it over my shoulder, and filled my pockets with ammunition.

'What on Earth are you doing?' – Plagge enquired.

'Well, aren't we going hunting?' – I replied.

'Indeed, ··· but not with that shotgun.'

'Well, how do you kill penguins then?'

'Just wait and see; put the shotgun back in its place and take this instead.'

So saying he handed me a heavy club made from the handle of one of the picks we used for splitting ice.

159. *Dulce de Leche* is a caramelised sweetened condensed milk, a popular accompaniment to any Argentine meal.

160. A *milanesa* is any fried meat or fish coated with breadcrumbs.

Going after penguins with clubs was something I never could have imagined. Plagge was in charge so there was nothing for it but to resign myself to his experience. I followed him brandishing this primitive-looking weapon.

We headed for the beach of Scotia Bay and there spotted a group of about twenty penguins. Plagge went straight to them in a direct line, without even bothering to take the usual precautions for not scaring the prey away. About three or four steps away from them he stopped and signalled me to come forward to where he stood.

The penguins watched us silently, hardly moving, completely oblivious of what was about to happen. Contrary to expectation they did not flee in the first instance. From their behaviour, I imagined that perhaps a glimmer of distrust of those two outsized penguin-like creatures (Plagge and I) who were visiting them was developing.

'These ones are useless,' Plagge told me, 'they are Chinstraps, and their flesh has a decidedly strong flavour ··· those over there are better.'

'Which?' – I asked, 'those with a very black head and neck?'

'No, those are Adélies[161], their flesh isn't very good either, though it is more palatable than Chinstrap. It is from the Adélies that we get our annual supply of eggs.'

'Well, where are the good ones then?'

'Those over there, the Gentoos, the ones with the reddish beak and a white blaze on the head.'

We approached this group, and I noticed they were slightly larger than the Chinstraps or Adélies, but when we were within ten metres of reaching them, they seemed to guess our intentions and began to walk ahead rapidly in the same direction as us, with the obvious intention of gaining the sea.

'Careful!' – Plagge shouted 'don't let them anywhere near the water or we will lose them.'

I ran to the beach to cut them off from the water, while Plagge raced after one penguin that had become separated from the rest and on which he then unleashed his devastating weapon.

161. *Pygoscelis adeliae,* Adélie penguin

The penguin reeled to the ground as if struck by lightning and I saw my companion pick it up by the neck and swing it, in the same way as one would a hen.

It was pitiful to pursue those poor defenceless creatures. For a moment my sentiments were repulsed by such a wanton killing of innocents; but, almost immediately, I thought again that it was an important contribution to our common survival.

I steeled myself to make a sacrifice of the innocuous victims, and began applying blows left, right, and centre with the malevolent club.

At least we now had fresh penguin meat ….

25: An Important Man

We made our way back, each of us carrying four penguins. We walked a hundred metres and had to stop to rest; these creatures are surprisingly heavy. I looked towards our dwelling, which was still three hundred metres away, and there was the unmistakable figure of Otto standing by the front door waiting for us.

We arrived and handed over our catch which he immediately began to prepare, taking off the leathery skin with its thick layer of fat and exposing the blackish bleeding flesh.

I watched this operation in silence as the knife, deftly deployed by Otto, separated the breasts, each of which weighed at least a kilogram. Similarly the legs were equivalent in size to those of a well-fed turkey.

That is as much as can be used from these web-footed creatures; each one provides sufficient food for two meals for the five of us.

Otto then washed the meat and placed it in large enamel bowls, covering it with vinegar, salt, pepper, Worcester sauce, and several spices. Penguin meat cannot be used at once; it has to remain marinating for at least two or three days to rid it of its strong, gamey flavour.

A few minutes after midday, Otto called us in for lunch. After we were seated he served an entrée of preserved meats garnished with thin slices of fresh onion. The next course appeared; it was a penguin casserole, served with potatoes and rice.

This was a typical mid-day menu and actually there was no necessity for more, the servings were ample and each course was accompanied with thick slices of bread and butter, sometimes with thick cheese slices as well.

Coffee followed, then finally a well-packed pipe of blonde tobacco.

It is a simple matter to list the different dishes that comprised a meal served at the table in the South Orkney Islands.

At night, at 20:00 hours, there was some minor variation of the mid-day meal. Otto prepared a substantial soup, either a commercial brand from a tin can or sometimes one made from fresh meat; followed by a course which would leave the most demanding person satisfied; and finally a dessert; a pudding or a cake, a sweet such as crème caramel, cheese, tinned fruit salad or dried fruit, or anything else to sweeten the palate. Lastly a small black coffee was served, accompanied by a pipe or a cigarette.

We were accustomed to the routine of gathering round the table four times daily to indulge ourselves in the meals that Otto, a master of his trade, would prepare for us. Up to now he demonstrated a vast repertoire in the different menus he prepared every day. Whenever practicable he dismissed tinned or dehydrated ingredients, preferring to prepare good fresh food which we devoured with huge appetites, sharpened by the cold weather and the tough physical work we did outside.

Otto quite literally spent the entire day in the kitchen; he was always working, opening and closing pots and pans in which steamed the most exquisite food.

For afternoon tea he never failed to offer us some buns or pastry, coated in *dulce de leche*, or delicious *pan dulce*[162]. There was always something ready for us.

One day, smiling, he said to me:

'*Mein Kleine*, you will have to eat more in order to grow a bit more, but perhaps it is my cooking that is not acceptable to you?'

Otto had a good-natured face; a wide forehead and grey hair. He had a twirled moustache, Prussian style, which gave him a martial look typical of his race. He was about fifty years old, tall with a pronounced belly, which he invariably hid behind a broad apron.

162. *Pan dulce* is sweetened bread, usually incorporating dried fruit.

21 : Life depends on the provisions

I made a private agreement with him; he would do my laundry twice a week, and in return I promised to present him with cash when we returned to Buenos Aires.

'This man is good at all these things,' commented Valentiner and added: 'one has to say that, down here, the best cooks are those who have had catering experience in international shipping, especially aboard sailing vessels, where they gain invaluable practice and skill in disguising the sameness of flavour of all tinned food'.

Possibly the only mental and physical suffering endured by Otto in the South Orkneys was the inability to realize his dreams. These involved a bar with inexhaustible quantities of beer from a keg, and a large table in the middle with countless glasses full of froth-topped ale.

'Everything would be fine,' he told to me, 'if we had enough beer'.

He enjoyed drinking but although he had finished the ration of two bottles of whisky we all had been issued, long before the rest of us; he nevertheless had the strength of will to sip them gradually, and not quaff a whole bottle in one go which was his natural predilection.

I learned to gain his goodwill by offering him a glass of grog from my supplies, on the days that we celebrated some special occasion or event.

22: Blue-eyed Shag

Good food makes a happy heart, they always say, and there is truth in this.

We could put up with the elements of that cruel climate, go through moments of profound melancholy in our voluntary isolation, suffer anything at all, but all was forgotten once we were seated around the table. Invariably we were confronted by delicious meals, affording us the one pleasure capable of making Antarctic life more bearable, which translates as 'eat well'.

26: Non-rational Inhabitants[163]

The non-rational inhabitants of the South Orkney Islands are, in fact, very few.

They are limited to the three species of penguins which I had already mentioned; with seals, petrels, shags, gulls and a few other birds which were spotted flying over our domain.

I still had not been able to visit the large penguin rookeries. The easiest way to reach the nearest ones, in Scotia Bay, was by walking across the sea ice after it formed, but this had not yet happened. It had not been practicable to use the boat because of the prevailing bad weather and the rough sea which made a voyage in such a frail vessel too hazardous.

Talking this over with the Leader he told me:

'Undoubtedly the safest and most practical way to reach the rookery is across the sea ice. If you try to go overland across the glaciers it is dangerous because the slopes are too steep and there are innumerable crevasses.'

'When the sea freezes, will the penguins still remain in their rookeries?' I asked.

'Possibly not; when winter begins, they migrate northward, far to the north, and base themselves in waters which do not freeze.'

'This migration, is it due to the low temperatures which prevail here during the winter?'

'That could be one reason, but the most probable is that because

163. That means animals other than man (which is presumed to be a rational species).

they live off things they catch in the sea, mainly crustaceans plus some marine algae, they are obliged to go to where the sea is clear of ice so they can dive into the water to get it.'

'In winter we will not have the genial company of these friendly creatures?'

'We will not see a single one', Valentiner replied.

'Where will we get our stews and steaks?'

'All that will be taken care of. In a few days we will fill the food store with the last stragglers as they cross the isthmus on their way north.'

I had to be content with postponing my intended visit to a future date.

Instead I dedicated myself to observing an intriguing family of Giant Petrels[164] flying swiftly overhead, with their huge wings extended. They were of a range of shades, anything between black and white, but the majority were brown.

I never saw any of them resting in the vicinity of the base, although I carried my shotgun at all times with the intention of shooting one down to examine it closely. Sometimes they flew very low, less than ten metres from the surface, invariably gliding and following the air currents.

On repeated occasions I fired at them with a series of rounds, but was astonished to see a few feathers fall but yet they continued their majestic flight as if nothing had happened. One morning I shot one and brought it down with a broken wing. It looked like a miniature glider and measured more than two metres between its wing tips. Once I had skinned it I found the shot pellets embedded in the layer of blubber that surrounded its body, none had penetrated the vital organs of that enormous bird. Its layer of fat served to make it substantially bullet-proof.

There are other petrels in the South Orkneys as well, but smaller than the Giant Petrel.

The Pintado or Cape Petrel[165] is abundant in South Georgia, but scarce in the South Orkneys. The Snow Petrel[166] is completely white

164. *Macronectes giganteus*
165. *Daption capense*
166. *Pagodroma nivea*

and of the same shape and size as the Cape Petrel, with a wingspan of sixty to seventy centimetres. Finally there is Wilson's Storm Petrel[167] which has a wingspan no bigger than forty centimetres.

Before winter sets in, you can also see the Dominican or Kelp Gull[168], Antarctic Tern[169], and Brown Skua[170]. These last ones are invariably found in places with remains of decomposing seals, penguins or blubber.

Shags[171] are rarely seen. They always fly laboriously with their necks extended forward and in perfect V-formation, from one bay to another.

These few birds, and a few seals[172] which hauled out on the beaches to sleep, were the only companions we had at that latitude.

After one has been on base a while, the appearance of any of these creatures caught your eye and broke the monotony of the severe landscape which we were condemned to see every time we went through our front door.

There were very few local animals, but those present provided a most interesting subject to study. When I mention these thoughts to Valentiner, he said:

'It is too late to do anything now; like the penguins, the other birds and the seals will also migrate north as the winter begins. In spring and summer there will be ample time to observe them at leisure.'

'Does everything migrate?'

'No, not all species; we will soon see the Sheathbill[173], as they are commonly called.'

'What kind of a bird is that?'

'In English they are called Snowy Sheathbill, in fact it looks a bit like a dove.'

167. *Oceanites oceanicus*
168. *Larus dominicanus*
169. *Sterna vittata*
170. *Stercorarius antarcticus*
171. *Phalacrocorax atriceps,* Blue-eyed Shag (recently classified as Bransfield Shag, *P. bransfieldensis)*
172. These are Weddell Seals, *Leptonychotes weddellii*
173. *Chionis alba,* which has entirely white feathers and is about the size of a large dove.

23: A Gentoo Penguin on its nest

'What, doves in the South Orkneys?'

'Yes, you will soon see for yourself. When there are no more penguins in the rookery, they will come to find the company of the only living beings in this area. You will see how, in the midst of our winter solitude, the Sheathbill will be our only companion, they never abandon the area where they hatched.'

27: Winter Arrives

May went by. One morning our Leader announced:

'With the winter beginning a bit earlier this year penguins have already become very scarce. We must be ready to intercept any group we see crossing the isthmus.'

'Does that mean we must gather our meat supply for the rest of the year?' - I enquired.

'Exactly, that is the only way we will fill the food store.'

We did not have to wait long; by midday we had about thirty penguins. Otto, diligent as always, took the breasts and legs.

The meat was hung on hooks from the rafters in the food store. I should add that this store had no heating so everything in it was at the same temperature as outside. Meat froze immediately and was preserved in this natural way until we were ready to consume it.

I don't remember exactly how many penguins we caught for the store, but it must have been well over a hundred. When we finished the job we were confident that we would not run out of fresh meat for regular meals before the end of September or October.

As well as our jobs as butchers, we also had to apply ourselves to the dirtier, but equally important, work of coal-men.

Directly outside the dwelling we made a huge pile of coal for burning in the kitchen stove and parlour fire. We also stocked about three tonnes of coal in a corner of the food store. We would start using the inside stock after the outside one had become completely buried by snow and ice and difficult to reach.

Another priority task was to install new galvanized iron pipes on the chimneys of the stove and fire, then secure them with guy wires and cables which could withstand the strong winds.

24: Snowy Sheathbill

We made a depot of drums of paraffin[174], for the lamps, outside, but conveniently close to the base. This was required for general illumination, as well as for the heating lamps to maintain the temperature above freezing in the huts with the magnetic instruments.

Each of us put the things that would be in regular use during the winter in easily accessible places. The virtues of foresight and tidiness are incalculable. The severity of the climate and the harshness of polar living conditions become easier to bear once all the proper precautions have been taken, to make individual physical effort during those winter months no more difficult than necessary.

We were getting the idea of what the critical period of winter was going to be like. Already it was snowing incessantly and the sky was

174. Paraffin, or kerosene, is a hydrocarbon (decane) used in wick lamps and pressure stoves for lighting and heating. It is a potential fire hazard and is thus stored outside; see the footnote in § 7.

25: Brown Skua

overcast with a mantle of thick dark clouds. Gradually the snow was accumulating on the rough ground of the isthmus and concealing all the gullies and ravines on the slopes of the mountains.

The temperature had dropped to more than twenty degrees below freezing and although the thermometers and thermographs had graduations which register below this, they had been replaced by instruments that could even register temperatures below minus fifty degrees.

On 10 May 1923, I happened to be working on my meteorological data. I was sitting next to the stove, when Valentiner came in and said:

'Get your outside clothes on and come with me: I want to show you something very interesting.'

I did not have to be asked twice. Although I would have preferred to remain warm inside, I donned my leather coat and woollen hat, and followed him outside. A strong wind was blowing from the south-east and it was freezing cold. It was not snowing at that moment, but the sky was overcast as always.

'Can you see,' the Leader said to me pointing south, 'out there on the horizon a long white band reflected on the underside of grey cloud?'

'Yes, I can see it clearly, quite different from the rest of the sky; there on the horizon that strip of very white cloud.'

'It contrasts with the darker ones that form the stratus layer above us.'

'Indeed,' I replied, 'but I am at a loss to explain that white line in the sky.'

'Well that white line denotes ice, floating sea ice; a huge field of pack ice[175].'

'Now I understand,' I replied; 'the white of the snow on the surface of the ice floes, is reflected to the underside of the clouds, causing this change of colouration.'

'Exactly' – affirmed Valentiner. 'so it is easy to track extensive fields of pack ice, especially for anyone who has an idea about cloud classification. In instances such as this it is quite simple, you have seen how the sky is completely overcast with a dark grey pall, unbroken to the horizon, and on this the distant ice is reflected, as clear as day. It is still far distant, we are unlikely to actually see it on the horizon for at least a few days more, despite the fact that it is moving towards us at quite a speed, driven by the south-easterly wind and the north-westward flowing current.'

'That ice,' I asked, 'will it completely surround the South Orkneys archipelago?'

'Yes, completely! We will remain blockaded by the pack ice until the temperature rises in spring and the thaw begins. The winds will then sweep it north-eastwards, that is seawards, and gradually it will melt and become water again.'

I saw before me, its clarity reflected in the permanently overcast sky, the advance guard sent in from the circumpolar ocean.

The freezing cold wind, twenty degrees below zero, was searing my

175. This is ice-blink, see the footnote in § 15.

face, while my thoughts were on the immense onslaught of ice, which ineluctably would imprison us among those mute and lifeless rocks, and would maintain our isolation from the outside world, through which no other human hand could penetrate, to assist or help us, if, on any occasion, we should need it.

28: Our Work

Our scientific observations were monotonous, punctual, and accurate; there was absolutely no variation. I say scientific to distinguish it from the manual chores in which everyone participated and which were an essential part of our existence. Science was, however, the primary reason for the expedition.

We were recording twenty-four meteorological observations every day, hour by hour, which we divided into shifts. The Leader, from 13:00 hours to 15:00 hours; then came my turn from 16:00 until 22:00 hours, when I handed over to Plagge who would remain on duty until four or five in the morning. Bruhns then took over until midday, followed by Valentiner again, and so it went on successively in the same order every day.

These shifts involved recording data purely of a meteorological nature. To give a typical example:

I am on duty, seated at the work table in the parlour, it is five minutes before 18:00 hours. The spring which sets the chime of the clock, has made its characteristic sound which prepares the hammers to strike the bell in five minutes.

Mechanically I get up from my chair, habituated to the routine; take a pencil and the daily observatory register from the desk, and turn to record the instruments in the room.

I first look at the anemometer, an electrical apparatus which measures the external wind velocity.

I note seventy-eight kilometres an hour,Force 9 on the Beaufort Scale[176].

176. A nautical scale of strength of winds; 'Force 9' is a 'strong gale'; the maximum on the original scale was 'Force 12'.

26: Depot of sacks of coal outside the dwelling

From there I go to the other side of the room, to the automatic recorder of wind direction, which moves with the weather vane on the roof of the building.

I observe its recording pen is indicating south-west, and record this also.

It is now one minute to the exact hour so I go over to the standard barometer which is in a glass case, on the wall in the middle of the living area in between the doors to Valentiner's and my bunk-rooms.

I read the temperature on the thermometer next to the barometer tube. We have two barometers, this is the one used as standard, the other is in reserve in case of breakage or failure. I record that temperature: eighteen degrees and two tenths (above zero).

This temperature helps me regulate the room temperature where we are living. Anything above eighteen degrees is comfortable, and, instinctively, I turn around and make sure the stove has enough coal in the grate.

I proceed to the barometric observation; setting the index and adjusting the Vernier scale and record the atmospheric pressure without reduction for temperature: 745 millimetres of mercury [993·3 millibar]. I then look at the barograph and note the pressure that it is registering; the figure is similar to that of the barometer and I record both.

I look at the trace on the barograph chart and see it is sloping downwards quite sharply. 'This is a sign of bad weather coming' I say to myself, while thinking that of all the observations, I must now make the ones that are uncomfortable, and go outside.

Inside the dwelling I am wearing my woollen inner clothing, over which I have a thick pair of trousers and a heavy knitted jersey, two pairs of socks and clogs with wooden soles to keep my feet warm.

I need to dress to face the blizzard that is raging outside.

I take off my clogs, put on my thick windproof and waterproof trousers, and heavy boots. On top, a second jersey, my leather jacket, a broad woollen scarf, a balaclava, and last my gloves.

As I go outside I hang a hurricane lantern on my arm, and am ready to face the rigours of the weather. I pass through the kitchen wearing this polar outfit which feels more like a diving suit. I go through the corridor and open the door and then I am outside. I close the door immediately to prevent snow blowing in.

A terrible wind is howling. Fine ice crystals, similar to needles, strike my face. Total darkness. I grope my way forward through the blizzard. Thirty paces ahead I find myself outside the meteorological screen, which houses the psychrometer[177], maximum and minimum

177. A combined pair of thermometers; a dry bulb and a wet bulb (in summer), or an ice bulb (in winter) set for determining relative humidity by rate of evaporation of water or sublimation of ice.

27: The dwelling in winter

thermometers, hygrograph and thermograph. I read the thermometers and note, as best I can, the measurements of the other instruments. Just as I am about to turn back an extra strong gust of wind blows out my hurricane lantern, but at this stage I no longer need it.

I want to look at the sky to observe the clouds types, but the ice crystals sting my eyes and almost blind me. Nevertheless, I have gathered the data I need and can now return to the warmth.

I hasten back to the dwelling and in a few seconds I find myself once more beside the fire. I doff my outdoor clothing, stoke the fire and begin to tabulate and check my observations. This is always done in the standard order and procedure. I completed everything in only four minutes including inside and outside observations.

This is our straightforward routine, day and night, week after week, month by month, throughout the entire year.

Every chime of the clock announces the hour; we already know what has to be done. On no occasion is an observation missed. We never fail: we have to be observant and alert, noting the delicate movements of the recording indices of the instruments.

It is simple and straightforward enough, but very demanding, especially at night; even more so for the unfortunate man who has to

28: Devoto and Moneta from the front door of the dwelling

get out of bed at four in the morning, only to be chilled to the bone by the icy wind.

The man on night shift is also the night watchman, responsible for maintenance and safe operation of the fires in stoves. It is a justifiable reason for gratitude towards the fellow who does his job conscientiously, enabling the rest of us to get a good night's sleep.

29: Stevenson screen with meteorological instruments, dwelling in background

29: Recording Scientific Data[178]

It is not only the hourly meteorological observations which comprise our technical work. Every month we have to compile climatological data for each of the parameters observed: atmospheric pressure, air temperature, relative humidity, vapour pressure, cloud cover and type, wind speed and direction, visibility (that is the horizontal distance at which known objects may be seen) from which is determined the transparency of the atmosphere, hours of daily sunshine, etc.

All these tabulations take considerable time to prepare; they must be totalled and averaged by the number of hours or days to calculate mean values, ten-day means, and monthly means for the different phenomena observed.

It is generally during the quiet night shift, that Plagge and I work on the statistics for which we are responsible. Subsequently Valentiner and Bruhns check our figures.

As well as all this, every four or five days, the Leader and Bruhns have to make direct observations of the Earth's magnetic field. On

178. The data from Moneta's years, with a brief account of the history of the station, are summarized in: Argentina; Dirección General del Servicio Meteorological Nacional, 1951.

these occasions I see them going out in the morning, after our eight a.m. breakfast, in the direction of the three magnetic huts. Valentiner walks to the little hut which houses the Kew magnetometer, and there he records the declination and horizontal intensity of the magnetic field. Bruhns goes to the other hut which has the inclinometer, or dip circle, to record the angle of incidence of the magnetic field.

They return about noon, have lunch, and then repeat the series of observations from two p.m. till four or five in the afternoon. After this they perform the complicated mathematical calculations to coordinate all these terrestrial magnetic observations.

Every four hours (at 04:00, 08:00, 12:00, 16:00, 20:00 and 24:00 hours) the observer on meteorological duty also has to go to the third hut which houses the photographic register and the autographic recorders, that is to the magnetometers. This last of the little huts, distinguished as the Variometer Hut, is isolated and situated about two hundred metres west of the base, close to the cemetery[179].

Daily, between nine and ten in the morning, Bruhns must go to the beach on each bay to lower a thermometer in the sea to measure the surface water temperature. I have been told that when both bays freeze, or become covered with pack ice, it is necessary to bore a hole through the ice to be able to make this measurement.

Every day I have to measure the depth of the snow at ground level, for which I have made a ruler with a scale measuring in centimetres. At that time, when winter had just begun, the scale was already buried to half its height.

This is the total of our technical and scientific work, supplemented by a few additional observations of the sea ice and any other phenomena of interest.

This all takes time. It is easy enough to list, but the fact is that one lives continually with this work. Numbers and tabulations the whole time; countless figures recorded in every notebook and on every sheet of paper. Generally we do not write words, but occasionally when

179. The plan, number 2, shows three geomagnetic huts are all in the vicinity of the cemetery, isolated by about 150 m from the other station buildings.

someone makes a supplementary note in the 'remarks' column, it always comes as a surprise to see what the manuscript of a companion looks like; it had become so unfamiliar.

One lives to a rigorous schedule, completely dependent on the orders from the chimes of the clock. At the appointed time we move with mathematical precision, no longer does anyone ask what is expected of him, by now we are all habituated to being automatons, certainly as far as the labour of recording observations is concerned, which forms our principal employment.

We hand over our shifts leaving the coal scuttles full, the stoves glowing hot, the grates clean and empty of ash.

This routine still leaves a little free time. Even on duty we can still get on with whatever minor job needs to be done. The meteorological observations take only four or five minutes of every hour. The most complicated ones take place in the Variometer Hut, one of these takes twenty minutes to complete from when the clock chimes.

Sometimes, though rarely, the sky clears briefly. When this happens Valentiner, assisted by one of us, will go out and make astronomical observations to determine the exact time for the chronometer.

As we are situated somewhat to the east of the sixty degree meridian[180], which is the one that governs the official time in Argentina, we keep a local time which follows astronomical time. This is one hour and fifty-two seconds ahead of that kept on the Argentine mainland.

Bruhns is in charge of the three chronometers, the accuracy of which he checks every day.

As you can see, we live to a clockwork routine in every sense of the term; at any given minute of the hour there is a different task to be done.

Thanks to this we are always busy and, after a few months, we are so accustomed to this routine, that, without a second thought, we find ourselves performing functions which, in any other circumstances, would become tedious, monotonous, or difficult to perform: namely to live in a world of numbers and calculations.

180. The longitude of the station is 44°44′ west; Buenos Aires is 58°23′ west.

30: Absolute Hut housing geomagnetic instruments

30: Walking on Water

Ever since we first noticed 'ice-blink', I had been scrutinizing the horizon continually in the hope of glimpsing the arrival of the pack ice. The brilliant whiteness, reflected on the clouds, became more apparent as the hours passed and showed that the ice was advancing rapidly towards us.

On 23 May a distinct white line of ice on the horizon was clearly visible. Many icebergs, some monumental in size, loomed in the distance and, with the pack, were drifting closer towards us.

At that time penguins became very scarce, only one or two could be seen running rapidly across the isthmus, heading north as if fleeing from that huge white solid obstruction which would close their path to the sea, where they sought their food.

Petrels and other birds of the region had also migrated north, looking for open water, and after that the seals instinctively followed them.

Increasingly we were becoming the only creatures left, totally isolated and trapped by the ice progressively surrounding the islands.

All signs of life were slowly departing from our world. Almost without considering it, without our being fully conscious of the utter abandonment into which we were sinking, we came to be the only living creatures for thousands of kilometers around.

Now we had the ice, that great white solitary vagrant from Antarctica, only a few kilometers from our patch. It had reached Ailsa Craig, closing Scotia Bay from Cape Murdoch to Mackenzie Peninsula. A few hours later it would reach the beach on Scotia Bay itself, almost at our front door.

That afternoon the wind ceased after blowing for many days from the south, and a complete calm halted the northern movement of the ice at the line mentioned above. The outdoor thermometers read minus twenty-four degrees, and the sea, which was calm, started to solidify at such a low temperature.

The sea first turned an opaque, greyish-white or milky colour throughout all the large areas of open water that I could see. This formed the first layers of what is known as 'young ice'; these then welded together forming a compact mass. Then this entire mottled white surface gradually hardened and turned into solid ice.

Six hours later, when it was completely dark, I walked to the beach with my electric torch to test the solidity of that frozen sea.

Valentiner had told me that it was not a good idea to walk over the bay ice for at least twenty-four hours after the freezing had begun, and my intention was to not disobey him, I just wanted to see with my own eyes if, what I had imagined so many times, actually happened to sea water after the freezing started.

I arrived at the beach, and with help from the dim starlight, all I could see was an extensive white surface, completely smooth and uniform. It appeared like an enormous powdery plain extending as far south as I could see, until it became lost in the darkness.

I tested the resistance of the ice with my foot, but it was still not very strong, only a light pressure broke the thin layer and water appeared.

I had partially satisfied my curiosity, and resolved to wait until the next morning. I returned home to rest after the day's work.

The next day, 24 May, as soon as I got up I went to the beach with Plagge and Bruhns.

Plagge, who was already familiar with sea ice from his experiences in previous years, headed straight onto the ice and before I realized what was happening, he was already walking on the water.

Bruhns and I did likewise, but not without slight trepidation. At a sign from Plagge we shouldered a pick and shovel to make a hole so we could measure the temperature of the water beneath the ice.

As we walked across it I tested the strength of the ice with my own body weight and gave it a good kick with my heel, it was reassuring to feel its solid resistance.

We began crossing the bay and halted about two hundred metres from the shore. Bruhns brandished his pick three or four times and lifted out a chunk of the hard frozen surface, in its place there appeared crystal clear seawater.

We opened a hole about a metre square through which we could clearly see the seabed with large rounded boulders. Between these were a few algae, and some little starfish.

We would use this hole for our daily temperature readings; Bruhns would be in charge of what would prove an increasingly difficult task, as the ice became proportionally thicker as winter advanced.

At the moment we had no more than five centimetres of floating ice to stand on, but I felt as safe as I would have done if I had been stepping on solid rock.

There we were, strolling calmly over the sea. Though all that separated us from a watery abyss was a thin wafer of ice, we felt no fear.

31: Ice Fishing

With the solidification of the sea surface we were presented with an unsuspected change to our communal diet. Until then every meal had been based on well-seasoned penguin meat. This marinade, with strong spices and condiments, ends up fatiguing any stomach that is unaccustomed to such a diet; even the strongest digestion suffers.

Consequently, any variation in this diet is very welcome, and that is why, in every expedition, the Cook has an essential role to play in its success. It is his onerous responsibility to keep the entire team in good health.

These important men, whose names often come at the bottom of the many lists of expedition personnel, cannot fulfil their work adequately unless supplied with the proper ingredients with which they can vary the menu.

In our case, after the sea froze, a new ingredient was added to our otherwise abundant South Orkneys cuisine; fresh fish of excellent quality.

I have had opportunities, in several parts of the world, to see many different ways of catching fish, but I had never seen anything like the actual way it was practised in the South Orkneys, and therefore I devote this special chapter to ice fishing.

Before Scotia Bay was closed by ice, I noticed Plagge preparing a line with hooks of a distinctive design.

It consisted of a lump of lead approximately the thickness of a finger and about ten centimetres long. This weight was in the shape of a little fish with a hole through the tail for threading the line.

At the free end, the head, there were four or more hooks all pointing upward.

The length of the line is variable, but ten metres is adequate for any of the local bays.

One night after dinner we talked in general terms about the possibility of catching fish to eat. It was decided that one of the veterans should be appointed instructor, to teach us novices the art of fishing in the Antarctic, so we agreed it should be Plagge who was to give us our first lesson.

That afternoon we took a long sledge, a pick and a shovel to clear the ice from the hole we had previously made in Scotia Bay. Our only equipment was a fishing line with the hooks, or mini-harpoon described earlier, and the lead weight that had been carefully polished until it shone.

We were just about to reach our destination, when I suddenly thought 'what about some bait for the hooks?' Plagge's reply:

'Bait is not needed here; the fish will hook themselves without it.'

I thought I was about to be the victim of one of his usual practical jokes but, as I was fascinated to see the result of his preparations, I kept my mouth shut. I was wondering how he intended to catch any fish without bait.

Eventually we reached the ice hole. I had to remove the thin layer of ice that had formed after Bruhns and I had visited that morning. After this was done Plagge placed the sledge to straddle the open water and then he clambered on top of it. Tall as he was he manoeuvred his full length and lay along the sledge. The surrounding ice, now more than twenty centimetres thick, supported his weight with no problem.

Plagge unwound the line and with no further ado, dropped the weighted lure in the water.

At the same time I went down on my knees to get a better view of the sea bed.

We found ourselves suspended about eight metres over the sea bed and could clearly see every detail. The lure reached to within a metre off the bottom of the sea. Such was the clarity of the water that the line looked as though it was rigid and suspended vertically like a plumb line.

Fewer than twenty seconds had elapsed before I saw a black outline pass quickly brushing the lure. Another immediately followed it, about thirty centimetres long, and then several others came from all directions. These were the denizens of the Southern Ocean. The largest ones came in a straight line and collided with the lure, pushing it in several directions.

Now I realized: the shine of the polished lead weight attracted them and they rushed over to see what it was.

The most audacious ones would bite it, and even the smaller ones, gaining courage from their bigger brethren, would nudge it.

Suddenly Plagge gave the line a sharp pull upwards, and I saw that two fish were hooked, one below the head and another half way down its body.

He pulled them to the surface and, after lifting them from the water, placed them on the ice beside the hole and removed the hooks. There he left them. After one or two convulsions, they lay rigid with cold until they were frozen solid. All this happened within a minute of leaving their natural element.

They were about thirty centimetres long and covered in scales. They had large heads with very wide mouths, their backs were grey and bellies a dull shade of yellow[181]. Their outward appearance was not very attractive, but later I discovered their texture and taste compared favourably with that of a fresh *Pejerrey*[182].

In a short while we filled two buckets with our successful catch, and walked back to base discussing the gullibility of the fish being thus tricked. Fancy mistaking a shiny lump of lead for their natural food of delicate crustaceans!

32: Argentine National Day, 1923[183]

Today is the National Day of Argentina in the South Orkneys. Yesterday evening Otto received instructions to prepare a special menu for lunch and dinner for 25 May. At Orcadas national days, traditional religious festivals, and our birthdays, have one thing in common; a special meal.

I always maintain that, in South Orkney Islands, sitting at a well-provisioned table and having a feast is the only pleasure readily available. That is why, a special mid-day or evening feast and sharing a bottle of wine with everyone, is an unforgettable experience, especially on a date no Argentine will ever overlook.

I had been living at Orcadas now for three months and a few days in the company of four men who had been born abroad, although some of them had adopted Argentine nationality.

We got on well, but I was particularly interested to see what would be their reaction to celebrating the national day of the country they had adopted. As the sole Argentine present I preferred not to take any initiative, and let them follow their own inclinations.

On the evening of the 24th, after dinner, they proposed we stay together and wait for the arrival of midnight. I gladly accepted and we sat around the fire, sipping our glasses of whisky very slowly. We

181. *Notothenia coriiceps,* the Bullhead Notothenia

182. The *Pejerrey* or Silverside (*Odontesthes bonariensis*) is a common fish around southern South America.

183. Argentine national day, 25 May, commemorates the overthrow of the Spanish vice-regal government of Río de la Plata, in Buenos Aires, in 1810. This was the beginning of the separation of the Argentine Republic, and several neighbouring countries, from the Spanish empire.

31: Baldoni ice fishing

told, and listened to, anecdotes on many subjects.

A few minutes before midnight (Bruhns had set his pocket chronometer to a quarter to midnight) they asked me to get dressed and go outside. They then handed me a Mauser[184] rifle to fire the salvos with which we would celebrate the arrival of 25 May.

Everyone had a rifle or a shotgun, and we walked towards Uruguay Cove, to congregate where the flag pole stood.

The wind was relentless, freezing cold and blowing from the southeast. The snow, falling from the clouds, was being carried horizontally so forcefully that it stung. The night was completely overcast and dark.

We reached our destination and prepared ourselves. Bruhns was carefully looking at the movement of the second hand on his watch and suddenly shouted:

'Ready!'

Simultaneously we opened and closed the bolts of the rifles.

'Aim!'

We pointed our rifles north.

Then came the command:

'Fire!'

184. Mauser was a German manufacturer of bolt action rifles and semi-automatic pistols as military and civilian firearms.

Five simultaneous gunshots rent the air, five reddish lightnings flashed from the barrels ….

It was exactly zero hours on 25 May 1923.

That is how we ushered it in, in the far south.

We ran back home at the double as the cold was becoming even more intense. We reached the parlour, my emotions knew no bounds when I heard the first chords of the Argentine National Anthem[185] that Plagge had started on the gramophone.

I could not contain the tears and although I had not really felt abandoned in that isolated speck in the world, at that moment I felt the full responsibility of being the only Argentine present in the Antarctic representing our great country.

The next morning my companions awakened me at nine and as soon as I sat up in my bunk I heard the martial beat of the March of San Lorenzo[186] on the gramophone.

I opened the shutters of my bunk-room window and instinctively looked north, as I always did and saw a big sky-blue and white flag with its gilded Sun billowing straight as a board from the pole on the beach.

As soon as I had dressed and entered the parlour I found there was another surprise. A fine portrait of the immortal General San Martín[187], draped with another Argentine flag, had been placed on one of the walls.

We sat down at the table at the normal time. On that particular day our menu consisted of tinned partridge preserved in *escabeche*[188], instead of the usual tinned stuffed turkey.

For dessert, Otto presented another surprise. He brought to the table a monumental cake which he had decorated around the top with little

185. *¡Oíd Mortales!* (Hear, Oh Mortals!)

186. An Argentine military anthem composed in 1901, by Caytano Silva of Uruguay, to commemorate the defeat of Spanish forces at the Battle of San Lorenzo in 1813.

187. General José de San Martín (1778–1850) a South American patriot who had a major role in securing the independence of Argentina, Chile, and Perú from the Spanish empire. See Illustration 70.

188. *Escabeche* is a marinade of brine, lemon juice, onion, and herbs.

flags of all the different countries of the world with that of Argentina as centrepiece.

Our lunch was truly delicious; and there was no shortage of wine, with Otto and Valentiner donating bottles from their scarce personal supplies.

The whole day was spent amicably. Although we had a break from the additional tasks, important though these were, we continued our routine meteorological shifts, the work that was common to the three observers, Bruhns, Plagge, and I. These observations justify the existence of the expedition; there must be no break in the record. Not even the most important commemoration of the year is enough to permit a day off.

I walked silently to the flag pole at dusk.

There was no wind and the sky was, as always, overcast with a uniform layer of stratus cloud.

The ground was covered with snow, a metre deep, in which I sank to my knees.

I stopped by the pole where my country's flag was flying.

I looked toward Uruguay Cove, which was still clear of pack ice; and farther to Saddle Island, then beyond to the infinite horizon of the ocean.

Farther, very much farther away, and still to the north lay life, civilization, humanity ….

The darkness of the night of 25 May descended over that immense white mantle, which was merely a speck on that endless polar ice sheet, but on which, for the benefit of all mankind and for science, fluttered a sky-blue and white flag, gallantly impervious to the relentless onslaughts of Antarctic wind.

33: By Pick and Shovel

As days went by the leaves of the calendar were inexorably torn off every morning until the first of June appeared. Winter was upon us. We had no respite from the wearying work of shovelling away the snow that was continually accumulating against the walls and doors of the base.

32: At work with pick and shovel clearing snow

We measured the snow depth, and every afternoon a few centimetres less of the scale were visible. The snow accumulating across The Isthmus became deeper, covering everything; there was absolutely no trace of tools or other gear that had been accidentally left and forgotten outside.

Day and night the snow came down continuously, forming an increasingly deep and compacted blanket over the ground. Artists invariably portray it as falling vertically in large flakes in their paintings, but none of them has ever experienced an Antarctic blizzard.

On the contrary, snow did not fall here in the usually accepted sense of the word. Rather it was hurled and swept by gale-force, eighty kilometres an hour, gusts of wind. The flakes disintegrated into small particles of ice that fell obliquely or were sometimes swept along almost horizontally.

It was impossible to face squarely into such a blizzard, the ice crystals stung one's eyes and face, and sometimes, through any gaps in the clothing, penetrated to the bare skin.

Only by walking side-on to the wind was it possible to make any progress through such a ferocious combination of squall and snow.

Many times I had to duck quickly to avoid being struck by a box or plank flying through the air as a projectile that had been torn from its moorings by the force of the hurricane.

On other occasions we had to stop in our tracks, remain still and use all our agility to keep balanced, like a tightrope walker, so as not to be blown over by the violence of the gusts.

Whenever a lull in the storm permitted us to go outside, we would grab a shovel and immediately begin to clear the snow, as far as practicable, from the front door of the base. As the winter progressed this became more and more difficult. We had to clear not only the door, but also the windows, through which the sparse half-light of a winter day tenuously infiltrated, for an hour or two anyway.

The most troublesome result of all, caused by the fury of the natural elements against which we struggled, was when we were sitting beside the stove after an evening meal. We were enjoying the warmth and comfort, listening to the gramophone as if we were at home. Then it happened; quite without warning, the dreaded sound of twisting metal would be heard and then the 'Flight of the Chimney Pipes'.

Suddenly above our heads we would hear a sharp deafening noise, of heavy objects banging against the roof as a chimney blew off. All at once the room would be filled with smoke which could no longer escape.

When this happened there was no alternative but to don our outside clothing at once, fill our pockets with tools, bravely climb to the roof to install a new chimney. This was much easier said than done. We had to remain for ages, perched precariously atop the roof, while we secured the cables and frosted wires. It was miserable work for which we had to remove our gloves and work with bare hands.

Hands became stiff from cold; one's entire body felt chilled to the bone. In a mælström of darkness, wind and driving snow, we tried to unravel a tangle of wire cables and iron pipes all blackened with soot.

It required hard labour, as was apparent, just to maintain a comfortable home in which we could take refuge from the elements and recuperate from the never ending daily work.

Eventually we became inured to anything and everything. Whenever the wind blew a hurricane, our entire attention was focused on what might happen to the roof, especially at night. Although we took every precaution to secure the chimney pipes properly, the slightest failure, however insignificant, might result in one or more chimneys being torn away, thus depriving us of the vital artificial warmth from the stoves.

All this was one part of the constant battle against the natural elements which have no respect for anything and subjugate everything in their path.

34: My Companions

June has arrived. Winter is here in full force surrounding us in an endless white mantle lost to infinity in any direction. Even Uruguay Cove, which was the last to take on a solid form, has frozen; the ice that covers it has thickened and is blanketed by a deep layer of snow.

These are the surroundings in which we live. Until now there hasn't been anything exceptional to record in my journal, but in the long night watches spent sitting round the table after the evening meal, I became familiar with many South Orkney anecdotes related by Valentiner, some of which I will retell at the right moment.

I have also had the opportunity to observe, at leisure, the four men with whom I lived. From their slightest gestures, looks and behaviour, I came to understand their most intimate thoughts.

During the previous five months I had seen none other but these four faces, all very different. Engraved on my mind is an image of each one. It is not just physically that I have come to know them, but being obliged to listen to them fairly constantly, even when they spoke in their native tongues, I reached a point when I could gauge intuitively their innermost thoughts.

For now I will limit myself to describing them just as they appeared.

Hugo Valentiner, the Leader, was forty-two years old. He had a few grey hairs and a weather-beaten face. He was born in Denmark but had lived in Argentina for many years. He was of medium height, and though the weight of years had not taken any major toll of his physical appearance, you could detect traces on his face of the strain of three years he had already spent in the South Orkney Islands.

33: Hoarfrost accumulated on the guy-wires of the dwelling chimneys

He was a man of wide education and culture. Besides mastering Spanish he spoke good English and German. His manner was very refined even in times of stress when he had to use all his energy organizing and accomplishing the programme of work imposed on the expedition.

For me he was a good leader; in subsequent years the recollection of his experience and example were always present when chance took me back to fulfil the same role.

Ernst Bruhns, a German, thirty-five years old, was the Assistant Leader. He would be the substitute for Valentiner in the event of the latter falling ill or being otherwise incapacitated.

This man was my cabin mate. Tall and sturdy, with large robust arms, he was a former officer of the German merchant navy. His whole bearing and manner showed him to be a well-seasoned man of the sea. German to the hilt, even his physical features denoted pure Prussian origin.

I now know this man very well. On the outside he is all flesh and muscle in a slightly pinkish skin; but this gruff exterior conceals a truly innocent and guileless soul.

In the style of most foreign sailors, he has tattoos all over his body. I happened to notice these when he undressed for his weekly bath.

On his chest is a sailing ship with seven masts, on one arm there is an anchor and above it the inevitable silhouette of a woman.

He has mentioned his family all live in Germany and he has no close relatives in Argentina. He has spent many years in the Observatory at La Quiaca from where he went directly to the South Orkney Islands. He was determined to learn Spanish and asked me several times to tell him if he expressed himself incorrectly.

On my part I had volunteered to undertake the task of organizing our room and keeping it tidy. From the beginning, it was a struggle demanding substantial effort. My cabin-mate, so self-disciplined in his work, had no clue about domesticity. Jacket, trousers, boots, underwear, tools, and other essentials lay scattered all over the place in that small living area. This caused huge confusion and was a constant source of annoyance when one of us had to find something in a hurry.

Eventually I managed to get him accustomed to putting the soap for the wash-basin back in its correct place; to leave the nail brush on the small bracket I had made specifically for that purpose, of hanging his clothes on their respective hooks, and to put his books, with the rest of his tools, etc., back in their proper places.

All the time we were together we got on well, although there were occasional disagreements, inevitably over trivialities.

On the other hand, it could not have worked in any other way. From the beginning of the expedition, I had assessed the situation and adopted a disposition somewhat akin to that of the trainer of an athlete. His outward appearance concealed a gentle and simple soul; his heart was as meek as that of a lamb.

Robert Plagge was also German but had lived in my country for many years. He was my work associate and spoke Spanish and English very well. He was the only one of my companions with whom I used the familiar terms, which says much for our friendship.

He was an excellent lad, loyal and straightforward in everything, and a man who had experienced more than a fair share of the vicissitudes of life.

Otto, the Cook, shared a room with him. Plagge was a couple of centimetres taller than me, and, except for his blonde hair and blue eyes indicating Teutonic physiognomy, he could have passed for a *criollo*[189].

These are the men, with Otto previously described, with whom I lived in the South Orkneys.

One feature of this confined life, to which we were condemned, was that I gradually became familiar with certain aspects of even the more private lives of my companions.

The frightening solitude which hemmed us in, little by little, broke the barriers behind which they preferred to keep the innermost secrets of their lives. Unfailingly there came moments when every topic of conversation had been exhausted, and they drifted off into subjects which, in a more normal life, we would never have considered broaching.

I learned by experience the best way to survive in the South Orkneys was to become self-centred as far as my most profound thoughts and feelings were concerned, and keep them exclusively to myself.

It is rare for someone to go to the South Orkneys without, at one time or other, divulging something that he meant to keep to himself.

Inevitably there comes the moment for mutual confidence. Such absolute isolation, the civilization with which we are naturally imbued, brings the spirits and minds of men together. It is as if they are searching for a feeling of mutual solidarity to help recall distant memories of events, of things long passed, in places so remote from the South Orkney Islands that they seem part of another world.

The one who is guarded with his sentiments, and has the resolve to keep his intimate thoughts and recollections to himself, is the one who, in the end, will triumph over all the others. Gradually he can get to know everyone well, without the others really getting to know him.

It is a real triumph to be able to be like that. A man will be able to excuse the failings of everyone else, adapting his own life and making it agreeable, without causing the slightest trouble to the other inhabitants of that strange and forsaken world bordering on the South Pole.

189. That is a person of Spanish origin born in Latin America.

35: Lost in the Blizzard

For the duration of winter, life pursued its monotonous course: scientific observations, calculations, tables full of numbers. At regular times we gathered round the table to eat the meals Otto had cooked and served.

From time to time we would haul the sledge to the beach and return with a full load of ice.

Always the same routine with no abrupt variation; even the landscape remained exactly the same; white, white everywhere ….

During my new shift I had to make two visits to the Variometer Hut, which was close to the cemetery, at midnight and four in the morning. It was my duty to regulate the temperature maintained by the heat from paraffin lamps[190] to keep the instruments above freezing point.

In the depths of winter, Valentiner decided a third visit to the hut was necessary. This was to be at two in the morning, in the middle of my shift. It was also important to maintain the temperature inside the hut as constant as practicable.

The return trip was only one hundred and fifty metres each way, but even that short distance, from the dwelling to the Variometer Hut, was extremely arduous owing to the great depth of snow and the gustiness of the winds which threatened to overbalance anyone rash enough to attempt to walk across the isthmus in these conditions.

In the darkness of night, when you could not see a couple of paces ahead, or at times when it was snowing more heavily than usual, it was an arduous task to reach the huts with any degree of comfort.

At that time we lacked sufficient batteries to use an electric torch whenever we needed. The hurricane lantern, used for the outside meteorological observations, was not suitable for taking to the magnetic huts as it could not always be kept alight in the storms. Even if the flame was still burning at its destination, we were forbidden to bring such a lamp anywhere near the magnetic observatory, because its metal body

190. Presumably brass paraffin lamps because they have no effect on magnetic instruments.

would disturb the high precision needles of the instruments[191].

To some extent, however, I was able to solve the problem of lack of light, by leaving a lamp burning inside the window nearest to the front door of the dwelling. From outside it looked a bit like a navigation light – a beacon to guide me back home.

I went out as usual one night. All my companions were sound asleep, the clock had struck four in the morning.

Before setting out for the Variometer Hut, I put the lamp in its usual place at the window, as I was accustomed to do.

As soon as I went outside, a blast of icy cold air, charged with snow, bombarded my face.

I launched into that pitch black impenetrable darkness and began to walk in what I thought was a straight line. Instinctively I passed close to the meteorological screen; I could not see two metres in front of my face, but it served to keep me on course.

After a few minutes, which felt eternal, I reached my destination, but guided only by personal intuition: the blizzard was so severe the snow and wind prevented my seeing anything at all.

I quickly adjusted the two heating lamps, and was preparing to leave the Variometer Hut to return home. I had to be back in time to wake Plagge and hand over my shift before the five a.m. meteorological observations.

Before leaving the hut I took one last look at the chronometer, which read four hours eighteen minutes. I had enough time to get back and, as the storm seemed to be at its maximum strength, I decided to wait a few minutes hoping that it would abate.

At four thirty I was prepared to go outside, I felt tired and sleepy; moreover the thought of a nice warm bunk made the risk of falling, being struck by something, or any other hazard, seem worth taking.

I left the hut, and as soon as I did, I realized the visibility had not improved and the elements were still lashing everything with unabated violence.

191. Metal parts of hurricane lanterns, being iron, would influence the magnetic field near sensitive instruments.

Even though it lay only a hundred and fifty metres ahead, it was impossible for me to distinguish the direction of the illuminated window of the dwelling. Guided only by instinct I started to walk in what I thought was the right direction, while the wind buffeted me from one side to the other in my course.

The storm was blowing from the south-east, which meant all my efforts must be to keep the wind striking my right side, this was the one and only way I could get back to the vicinity of the dwelling. Once I had located the building I could consider myself safe.

I continued walking on what I thought was the same course, but it suddenly dawned on me that I had already gone too far without finding the slightest sign of anything that would indicate that I was anywhere near my destination. I went some fifty metres farther, and much to my surprise found I was beginning to walk uphill.

'Surely I must have walked right past one side or the other of the dwelling' I told myself.

I started to back-track and, with some guesswork, could just discern my deep boot prints in the snow. I turned slightly to the left, but with no success. I then did the same in the other direction, but again nothing.

It seemed quite impossible that I could have become lost in that small area delimited by glaciers and the frozen sea, but I had to believe the evidence. I concentrated on carefully scrutinizing the snow to find my boot prints and retrace my steps back to the Variometer Hut. Luckily I found them and managed to get back to where I had started.

I looked at my watch and saw I had been walking for fifteen minutes, sufficient time to have gone back and forth to the dwelling at least five times.

There was no time to lose; I needed to get back as soon as possible. I was beginning to feel the effects of the cold seeping through my thick clothing, and I did not have the slightest inclination to spend the night out there.

I re-emerged from the Variometer Hut and started walking again. This time I did not try to follow my previous footsteps which were rapidly being covered by the drifting snow.

I walked for a significant time and gradually found I was going uphill again. I came to the following conclusion: 'Undoubtedly I am on La Monja Glacier, once again I have passed the dwelling by some distance; if not, I must have gone around in a circle and now be on the slopes of Mount Mossman'.

I continued uphill then, briefly through the storm, I saw the unmistakable black rocks of Mossman Peninsula.

I could never have believed myself capable of going around the same circle twice and ending in the wrong direction, the complete opposite of where I intended to go.

I was now scared to retrace my steps as the snow was covering my boot prints in a matter of seconds. I decided to continue skirting the hills towards the ruins of Omond House. This took me away from the dwelling, but I had a plan ....

After a few minutes I reached the rubble and ruins of the first expedition hut and a few paces away I could distinguish the cracks in the ice marking the junction of land and frozen sea[192].

I groped my way towards these cracks, walking in an easterly direction and when I reckoned I had gone about two hundred metres, there to the left about ten metres away, I could distinguish the cairn that marked the astronomical position of the Observatory.

I sheltered in the lee of the cairn and took a few moments to regain my strength, sapped by my sustained struggle against the elements. Then I took courage and launched myself in the direction that seemed to me to be the most accurate course back to the dwelling.

The wind pounded my back and swept me along, my feet barely touching the ground. Suddenly to my right I detected a large dark obstacle, it was my home, and I breathed a sigh of relief.

I entered and looked at my watch: two minutes to five in the morning.

Everyone was sound asleep and oblivious of my adventure.

As I was still dressed in my outside clothing I decided to do the five

192. The cracks in the ice are tide cracks near the shore, barely perceptible when the tide is high but wide open when it is low.

a.m. meteorological observations before waking Plagge.

Silently I recorded the observations. While doing so I couldn't help thinking that even the most experienced man could have become disorientated in the field when it is impossible to find any reference points to check the route. I had become lost in such a ridiculously small area that, had anybody else recounted the story to me, I would have had great difficulty believing him.

36: Tragedy

The following day, at lunch, I recounted to my companions what I called 'my adventure' of the previous night. They listened and showed no surprise for, in the circumstances in which we lived, it could have easily happened to anyone.

The conversation turned to the accidents which resulted in those four crosses in the cemetery. Valentiner raised the subject of the irreparable loss of Bache-Wiig.

He began his story thus:

'It was during the expedition of 1915. It was a Scandinavian group, led by a fearless and hardy Norwegian, Hartvig Bache-Wiig. The Assistant Leader was Anton Stuxberg, a Swede and the other assistants were Thorleif Hoxmark from Norway, and J. Bayon from Sweden. The Cook was Ernst Andersson, also Swedish.'

'With the exception of Stuxberg and the Cook, the others were all first timers in the South Orkneys. Bearing in mind their nationalities, they were all experienced polar men.'

'Bache-Wiig had been a ski champion in his country. His companions were used to watching his dangerous jumps and precipitous downhill skiing on the steepest slopes of the glacier hills.'

'Apparently snow and ice held no secrets for him and he did not think anything of startling his colleagues by performing the most astonishing and riskiest exercises.'

'His confidence on ski was unlimited, and perhaps it was this that was his undoing.'

'You must assume,' Valentiner continued, 'that down here it is impossible to forecast the duration of the rare spells of good weather.

A calm period may last one hour, perhaps a day, but it is very rare for it to last for forty-eight hours. It is not a good idea for that reason to go far from base when making an excursion, especially to little known places or to areas which no one else has visited.'

'Bache-Wiig never listened to advice! Even that from the most experienced men, and would head off in any direction in search of adventure.'

'One morning he put on his ski and began to ascend La Monja Glacier. He was seen later heading in the direction of Mount Lola, and farther up in the distance, he was just a black dot on the white slopes of the glacier.'

'At midday the inevitable happened, another storm blew in from the south-east, and the usual low clouds completely hid from view all the peaks and slopes of the mountains.'

'It started to snow, but none of this alarmed his companions as they were accustomed to Bache-Wiig returning at dusk every time he went on an excursion.'

'About four in the afternoon, when the first shadows of dusk appeared and the visibility was only a few hundred metres, alarm began to spread with concern for his safety.'

'By six in the afternoon, it was already completely dark. Stuxberg and Hoxmark decided to set out and look for him, and following the direction he had taken they began to ascend the glacier.'

'As they needed extra light, they took a hurricane lantern with them but they had to frequently relight it as it kept blowing out in the wind.'

'About a kilometre from the base, they halted: visibility was reduced to zero because the snow began to fall thickly. With the risk of themselves becoming lost, they decided to return and begin an intensive search at first light.'

'That night no one slept, they spent the entire time conjecturing about his fate or the luck their absent companion might be having. Very early the next morning they all set out to continue searching.'

'At the top of the glacier, they divided into two groups; one went directly to Mount Lola, where Bache-Wiig had last been seen, while the other veered to the right, in the direction of Jessie Bay, some-

34: Panorama from Mount Ramsay: from the left: foothills of Mount Lola, frozen Uruguay Cove, Casa Moneta and the antenna pylon, near the centre with the three magnetic huts and store, frozen Scotia Bay on the right (see the station plan in Plan 1)

one thought he had seen some ski tracks in the snow leading in that direction.'

'The entire search proved fruitless; no one found or saw anything because within a few paces any ski tracks were obliterated by the thick layer of snow that had fallen during the previous night.'

'Faced with the evidence of the terrible accident suffered by their comrade, they decided to return to base. No subsequent trace of him was ever found, not one shred of evidence of his fate despite search parties sent out to look for him during the following days.'

'Bache-Wiig never returned.'

'There is much speculation about his fate. Even though it is easy to become disorientated in a blizzard it is unlikely that he should have become lost during the return journey. One possibility is that, having wandered so far, he may have lost his bearings, at the time when the blizzard struck suddenly at nightfall. Being disorientated by a night in the open in such harsh conditions, he could easily have perished from exposure'.

35: Uruguay Cove covered with pack ice, Point Lola and Mount Lola on the left

'The most plausible theory is that he lost control on the almost vertical face of the glacier and fell from a great height into the sea. This is feasible for there is a precipice more than two hundred metres high behind Mount Lola; there were some vestiges of ski tracks leading in this direction, but it was impossible for his companions to follow them in safety.'

'No single clue could be found; no ski, no ski-stick, no glove, absolutely nothing to indicate his sad fate.'

'Down here it doesn't do to be too self-confident', Valentiner finished. 'The example of what happened to Bache-Wiig is conclusive proof, that even if he was very experienced in snow and ice through having been born amongst them, it is still difficult to fight the elements face to face when confronted by the full ferocity of this hellish polar environment.'

37: Up to the Clouds

I was keen to make an excursion up Mount Lola. From the summit I wanted to admire the white panorama. From that elevation[193], the view had to be spectacular. For several days the weather had settled, it was not snowing much and the wind strength was decreasing. The sky, nevertheless, remained covered by the impenetrable stratus layer and the temperature remained between twenty-five and thirty degrees below zero.

I had learnt to ski. This cost me some hard knocks at first, but my perseverance was rewarded, and after a month of practising every day, I was able to keep my balance while hurtling down the glacial slopes near our dwelling.

Until then I had not been able to venture far for the simple reason that, on most occasions, I had to go completely alone. None of my companions could come with me because they all had work to do.

Another reason was the spectre of Bache-Wiig and the crosses in the cemetery which curbed my natural urge, indeed my longing, to uncover the unknown or to aim for uncharted places where I would have no definite idea of what I would find.

Eventually a day dawned with a promise of a bit of sunshine and although the sky had not completely cleared, I envisaged an ideal opportunity for my planned excursion to the summits partly veiled by cloud and drifting snow.

After lunch I put on my ski and started to ascend the glacier. Nobody made any comment as they all thought that, as usual, I was off to practise a downhill run.

When I found myself at the top I took a direct line to La Monja, and skied with long strides, constantly ascending the wide inclined plain of the glacier. As I went I took the greatest care not to deviate from my heading and run into the dangerous crevasses I saw to the left and right.

Every hundred metres I turned to contemplate the view behind me, endeavouring to memorise it for when I came to return.

I arrived at the base of the slope of La Monja and took a left turn.

193. The elevation of Mount Lola is 170 m.

Continuing to ascend I found myself to be some three hundred metres above The Isthmus. Slowly I neared my destination and on one of the last occasions on which I turned to look back, I could discern the magnetic huts of the Observatory. From this elevation and distance they seemed like little black dots on that enormous relief map, entirely white.

I was close to the ice-cap beyond Mount Lola and from there I would be able to look over a vast panorama.

The slope became steeper by the minute; I took off my ski and stuck them vertically into the snow next to one of the sticks. The other stick was to serve me as a support for the climb.

I was quite breathless by the time I reached the last rocks covered in ice where, after finding somewhere safe, I sat down to rest.

I was four hundred metres above sea level[194] and keen to admire all the details of this strange and marvellous view. First I looked at the sky to see the state of the weather and reassured myself that I could expect it to stay fine for several more hours.

Far below was the curving outline of The Isthmus, where the little black dots of the station buildings were distinctly seen.

The outlines of the two bays, white as far as my eyes could see, were clearly discernible. The level ice in the bays contrasted with the irregular contours of the rough terrain along the whole coast.

To the north, some ten nautical miles [18 km] distant, I could see Saddle Island emerging from the frozen sea. To the north-west end of Laurie Island lay Cape Robertson and the entrance to Washington Strait.

Eastwards, below where I stood, was Jessie Bay and farther round Pirie Peninsula which terminated at Cape Mabel.

Everything was white; it resembled a dreamlike panorama composed of mountains of icing sugar or meringue snow caps.

It was only the dry, icy wind, cracking the skin of my face, which, momentarily, brought me back to reality.

No living thing was anywhere to be seen; not a single bird; neither

194. This altitude indicates he had reached the Melville Highlands, beyond Mount Lola, which exceed 500 m in places.

petrel, nor even a shag, beat its wings in the air of that marvellous panorama.

Inevitably my eyes were drawn northwards. That north so very distant, but so long coveted ....

On the seaward horizon, very far off, I thought I could discern a darkish line; possibly it was sea water which still had not yet frozen, but it was very distant, at more than seventy kilometres away.

I suddenly recalled something: at that precise moment I was probably roughly in the area where Bache-Wiig was presumed to have been lost. I couldn't avoid casting a glance downwards to the base of the precipice over which he had probably hurtled.

If that was indeed where he met his demise, he must have met it instantaneously; the height was so great that, after going over the edge and falling, the fatal blow as he struck the ground must have been extremely violent.

I retraced my steps, descending to the place where I had left my ski. I tried to memorize all the details of that grand white landscape which very few men have had the privilege of seeing with their own eyes.

Once I put on my ski, the descent was precipitous, continuously downwards. There were moments when my speed was so great that I seemed to be suspended in the air with the wind pressure braking my descent.

The frozen surface sped beneath my feet. I reached the final slope and I was close to the base. At full speed on that last descent I prepared to excel myself.

The surface disappeared under my feet and tears were forced from my eyes as I tried to keep them wide open.

I began the final descent; my speed increased and I was struggling to remain upright.

Suddenly I suffered a powerful blow, it seemed that my body was folded in two like a hinge and the two parts had separated in different directions. When I stood up I saw my companions laughing and gesticulating, as they ran to greet me.

I was satisfied. I was beginning to gain experience: already I could recount what I had seen from the ice field beyond Mount Lola.

38: Shags

The weeks went by, provisions of penguin meat hanging in the food store diminished with our daily consumption. We were at the end of July and there were still four months remaining before we might see those genial creatures again.

It was desirable to vary the menu which was beginning to become monotonous, despite the outstanding efforts of Otto who took great care in preparing distinctive meals.

On several occasions I saw a few shags flying rapidly over the isthmus, passing from south to north, but I didn't believe their flesh was edible until Valentiner informed me to the contrary.

'Their meat is very tasty' he affirmed, 'I have had the chance to taste it on previous expeditions and it seemed that I was eating common duck.'

'What about the fishy taste or flavour of crustacea?' I asked.

'I have not noticed it, the ones prepared for the table were marinated in the usual strong spices.'

'Do you have any idea where these birds live?'

'I have never visited a shag colony, but in Washington Strait, not far offshore, I believe there might be one. That is where my companions from previous expeditions went to hunt them.' 'These birds,' he continued, 'live all year round in the South Orkneys, although in winter they are very scarce. Because they fly quickly it is relatively easy for them to travel far to places to feed where the sea has not frozen'.

After this conversation, on almost every day when I left the base to go for a walk, or an excursion farther afield, I carried a shotgun under my arm. Although I took an occasional shot at the shags flying overhead, I was never able to bring one down, as their flight is so quick, and the wind so strong, it disturbed my aim.

With much effort I persuaded Plagge to accompany me on a hunting reconnaissance I had planned to the supposed shag colony.

Previously I had accomplished several exploratory forays around Mossman Hills going as far as the slopes that descended to Washington Strait. There the sea was not completely frozen due to the strong currents which impeded the formation of young ice that is a necessary precursor for forming a compact mass.

I already knew half of the route well, but we had to wait for the right opportunity to attempt the full journey.

At last the right moment arrived. With the weapon, a makeshift lunch, hunting knives, and ropes; wearing our windproofs, we left base to begin our journey over the ice of Scotia Bay.

Soon afterwards we climbed the chain of hills over Mossman Peninsula then descended on over the other side, until we reached the pack ice that was frozen to the shores of Washington Strait.

We began our reconnaissance as we walked, close to the shore, on sea ice which seemed firm; on no account could we risk a floe breaking off without our noticing and carrying us into the rapidly flowing open water from which we would have no hope of return.

We walked over five kilometres carefully scrutinizing all the islets. In none did we find any sign of life, although occasionally we saw bird droppings on some of the rocks that protruded from the ice.

Suddenly, towards the middle of the strait, over an islet with dark grey top, we saw several shags in flight, then they landed on it.

Taking a closer look, I could see that this islet marked the limit of the fast ice, as on its far side we could clearly see open water.

To reach this islet was not particularly dangerous; it would require only fifteen minutes walk across the level pack. The only thing which might present difficulties would be if some of this ice broke and detached, stranding us on the islet.

The weather was set fair; the wind was only a gentle breeze from the west and the sky was clear.

The Sun had no strength to warm us but, thanks to the clarity of the air, it was shining brightly and illuminating the entire landscape.

We resolved to brave the risky undertaking because we were determined to return with at least a few shags.

We continued on our way and, as we approached the islet I could see the birds sitting on the highest rocks.

When we were about one hundred metres from them, we took care not to scare them, and bent low as we drew closer, keeping cover behind the rocks.

36: Aboard a boat ready to hunt shags

At about twenty metres away there was a group of shags. Using the shotgun I decided to open fire as rapidly as I could. With Plagge's help in loading, that weapon turned into a veritable machine gun.

At the first shots a mass of shags took-off and began to circle overhead, but after several minutes they returned to the rocks until they scattered again after more firing. Most of them did not flee from the vicinity, only the more timid ones departed northwards out of sight.

We clambered onto the islet and started to collect those we had killed. In only a few minutes we had accounted for two dozen.

Among the higher rocks there were a few nests built from seaweed. The few shags remaining alive did not take off and remained perching on the rocks; they showed no fear of us and our untimely visit.

Each bird weighed more than three kilograms, and when I extended the wings of one specimen the span was rather more than a metre. Their plumage was beautiful, a very dark blue with a greenish sheen on the back of the body and wings, and completely white on the underside of the wings and on the breast. The colouring of their heads was magnificent, actually the general appearance and the lively colours of their plumage make them the most picturesque of all Antarctic birds.

We decided to return and stowed ten of these birds in our bags.

The return journey was extremely arduous; if it was a major effort to trudge over Mossman Hills without carrying any load, imagine what it would be with more than thirty kilograms of birds strapped to your back.

39: Life on Base

In each person there is an established pattern of behaviour. Personal habits which cannot be overlooked, even in circumstances most inimical to the trappings and comforts of civilization. When it becomes necessary to do without one of these civilized comforts, man ingeniously contrives to find the most practicable substitute. He will do anything to provide a degree of comfort to alleviate the harsh living conditions and frequent scarcity of supplies that an expedition to the South Orkneys is obliged to suffer.

Saturdays and Sundays are the days reserved for bath night. On

37: A colony of Blue-eyed Shags

Saturdays, after dinner, Valentiner and Bruhns had their baths. On Sunday at the same time, the rest of us bathed. The quotation 'A healthy mind in a healthy body' is not out of place, even somewhere as remote as in the Antarctic.

Personally I considered it necessary in the South Orkneys, even if only once a week, to have a good scrub; especially after all the physical work we had to do. The nature of our outside tasks; hauling heavy sledges, chipping ice, shovelling snow, etc., caused us to sweat profusely. That is the reason we were happy to queue every weekend to wash our bodies completely, and not just the top half which is all were able to do in the washbasin.

On Sunday, just after Otto had finished tidying the kitchen after dinner, he would put a galvanized bath-tub on the floor, of the hip bath type, while he boiled two fuel drums of fresh water on the stove.

I undressed in my bunk-room and walked to the improvised bathroom wrapped in a towel. The parlour stove blazed and the rings of the kitchen range still glowed red hot. I barely felt any cold.

There in that rudimentary manner I would bathe, yearning for the more comfortable bathrooms of modern houses in our large cities.

Occasionally, on Saturdays, Valentiner would call to me and say:

'Kleine, this afternoon may I trouble you for a few minutes of your time? I would like to give you the chance of learning how to cut hair.'

At the appointed time, he would sit near the stove, cloaked in a sheet, while I, armed with scissors, a comb, and a pair of clippers, would circle around him.

I will never be able to forget all the attempts I made to trim his mop of hair, nor the efforts I took to smooth the jagged steps left by my ruthless cutting.

Many times he would say to me:

'It is a shame we don't have any dogs this year ….'

'Why?' I asked him.

'They could have a meal from all the bits of ear you chop off.'

Half way through the year I found I had been unofficially voted barber-in-chief. All things considered I do not think I did such a bad job because, whenever a haircut was needed, my services were requested.

The real problem arose when I needed a haircut. I had to be brave and just let the machine pass over my head and hope for the best, without bothering about how I would end up looking. Otto was the best barber, though occasionally, in his enthusiasm, his hand slipped and I was left quite bald.

We made every effort to pass the evenings in friendly companionship. Plagge and I had occasional games of chess.

At other times we listened to music on the gramophone: the trouble was that by this stage of the year, I knew all the records we had by heart; whenever I hummed the tunes I could even mimic the scratches on the surface of the records.

Occasionally Valentiner appeared with a newspaper under his arm. For us this meant an issue of 'new' newspapers dating from the days before our departure. After reading and rereading them many times, we still kept them carefully; they were reminders of a distant civilization.

We were killing time while endeavouring to see that it did not kill us.

Everything was monotonous, always the same, but amidst that lack of variety in the details of our life, we did everything practicable not to upset or become a nuisance to each other. In spite of the care we attached to this common endeavour, minor sources of friction occasionally surfaced. Invariably they were caused by trivialities. For such a small group of men with limited distractions, such 'trivialities' acquired a disproportionate importance.

Except for the scientific work, the rest of the base jobs were divided fairly and equally among us all. Everyone took his turn in chipping ice to make water, hauling sacks of coal, continually shovelling the snow away from the building.

In the South there exists equality in this respect. When a sledge needed hauling or a shovel had to be grabbed, only in the most exceptional circumstances would someone be pardoned for an absence from such tasks.

We had our latrine. In summer we could use any of the beaches in the bays, but in winter ….

In an alcove in the store, formed by the partition wall of the pantry, was a smooth white seat with an ample hole in the middle, and below it was an empty paraffin drum. When this container was full, we had to dispose of it by putting in on a sledge and taking it down to the sea. Another empty paraffin drum replaced it.

Performing this operation was by strict roster, from the Leader to the Cook. That was the system and everyone was happy with it.

It was far preferable to do this, rather than undressing outside, at a temperature of thirty degrees below zero.

40: Pharmacy

Our polar life took its course without any major difficulties. Slowly, but surely, we were adapting to life in the South Orkneys. We had ample heat from the stoves, simple but sufficient meals, good clothing and warm beds. All this, supplemented by the healthy philosophical attitude we had adopted, made the days fly by with no significant alterations to the routines.

Occasionally one of us might awaken with slight indigestion, a headache or loss of appetite. The cure for this was well-known; a good dose of castor oil taken at the right time, or else a couple of aspirins. This quickly calmed and composed our digestions which struggled to cope with the abundance of condiments, spices, and strong sauces with which our meals were prepared.

If you were to subject our diet, and each meal served, to careful scrutiny, you could be sure that anyone who did not possess a cast iron stomach would experience some ailment resulting from eating penguin and shag flesh, or from tinned meat, or any of the associated ingredients.

Fruit and vegetables came in sealed containers of every shape and size, which generally displayed colourful eye-catching labels declaring *fresh spinach*, *fresh cauliflower*, *fresh asparagus*, *fresh fruit*, and so on.

Everything was declared *fresh* but after a few months, not to mention years, no one could face another meal with ingredients like that.

It is one thing to have tinned peaches in syrup three or four times a year, but quite another matter to have them regularly once a week throughout the expedition.

Merely one week of eating exclusively from tin cans is enough to put anyone off such food. Imagine what it would be like for those having to do this for months on end.

Ultimately he who is not dyspeptic to start with, eventually becomes so. He who prides himself on having a 'cast-iron stomach' will, when he gets home, have to go on a drastic diet to help restore his overworked digestion.

Commenting on this Valentiner observed:

'Nonetheless, we have nothing to fear from any Antarctic illness, as there are no germs here. Even if there were they could not survive this climate.'

'In fact,' I said, 'I have not had a single cold since I arrived.'

'You won't get one,' Valentiner replied; 'what you need to watch is not to acquire pneumonia, from that you could certainly die in the South Orkneys ….'

I have never thought of checking our medical kit. 'Is it complete?' I asked.

'Anything but', was the reply. 'A few things are missing, but you have to bear in mind that some medicines would be too difficult to use here. I do not think any of you' he added 'has medical training or knows what dose to give in cases that are not straightforward.'

We all nodded our heads in agreement.

'Just as well then,' he continued, 'I hope for everyone's sake that that we never need to administer laudanum[195] or morphine. That would mean we were in dire straits. The only things that I am allowed to prescribe are an amount of castor oil for you to take when the occasion demands and to supply the right dose of aspirins for headaches. However you are a fit and healthy lot; before you left Buenos Aires you had a complete medical examination. Therefore do not give any more thought to the possibility that you might get sick here.'

One afternoon I decided to have a look in the medical chest. Inside there were dozens of phials with various coloured liquids. Their labels were all in Latin or Norwegian.

Valentiner told me the skipper of a Norwegian whale-catcher had given him this medical kit as a gift. This was many years ago when the catcher visited the station to exchange some fresh whale meat for a bottle of whisky.

Ever since the contents of the phials had neither been used nor changed. Annually, with the change-over of expeditions, sufficient purgatives, of two or three varieties, were included for the winter party. These were supplemented by a supply of tincture of iodine[196], bandages, and several other items in common use and easy to apply.

Taking advantage of the inspection of the rudimentary pharmacy, Valentiner came to me with a little packet and said:

'I am going to put this in, although it arrived rather late.'

'What is it?' I asked.

'It is a box with a set of scalpels. It arrived this year and it has been sitting on my desk waiting to be put in the medicine chest.'

'Do they provide scalpels in case we need to remove someone's appendix?' – Bruhns enquired.

We all laughed at the thought and Valentiner said:

'No; it is not for that. If a bag of ice did not work there is nothing

195. Laudanum is Tincture of Opium; an alcoholic solution of opium alkaloids.

196. A 2% solution of elemental iodine in alcohol and water, applied as a local antiseptic.

more that anyone here could do for appendicitis. These instruments are in case we have a repeat of an operation similar to the one performed on Tapia in 1920.'

'I remember having heard about someone having to have his fingers amputated' I said, 'would you tell us the story?'

41: An Accident

We drew our chairs close to the stove as Valentiner began: 'in 1920 Wilhelm Kopelmann was the Leader. It was his second time in Orcadas, as he had been already a member of the 1914 expedition. The Assistant Leader was Bruno Kollasius, who, like Kopelmann, was German. The rest of the party were Augusto Tapia and Alejandro J. B. Boracchia, both Argentines, and the Cook was Georg Piper, also German.'

'On 6 July, deep into winter, Tapia was doing his usual morning shift. He completed the eight a.m. meteorological observations and then dressed to go outside again. Taking the sea temperature thermometer he went to Scotia Bay to make this observation.'

'These routine sorties did not attract his companions' attention, as everyone was accustomed to his brief absences; it was the established routine for Tapia to make this observation.'

'What struck Kollasius as the clock chimed nine, was that Tapia still had not returned, and this was unusual.'

'The alarm was raised immediately, and a search party of all base members available was promptly formed. At first they followed the footprints Tapia had left in the snow as he went towards Scotia Bay. There they found that he had completed his work. Retracing their steps they noticed that his footprints disappeared from view on the isthmus, and were leading to Uruguay Cove, where the sea surface had still not frozen.'

'They hastened towards the beach near the cemetery from where the sea temperatures were also measured. There too they found nothing to indicate that Tapia had been there. There were not even footprints which might have indicated that he had passed that way.'

'Then they began to search the entire length of the beach, when suddenly, from the top of the perpendicular face of the ice-foot[197] formed on the shore by the action of swell and tide, Kollasius noticed that several large blocks of ice had detached and were afloat, moving away with the wind which, at that time, was blowing a hurricane from the south-east.'

'Taking every precaution, Kollasius leaned over the edge which was about three metres high, and was amazed to see his comrade standing more or less upright on the beach below, leaning his body against the vertical face, grasping the cliff edge with both his hands, his fingers buried in that white wall of slush and ice. He was holding himself in that position so that he would not fall into the sea which was already washing around his knees.'

'Tapia must have been in that precarious position for close to an hour. The drenching he had received when the ice on which he was standing broke off; the loss of his glasses, and the shock at realizing his predicament, all contributed in the first few seconds to his struggling to scale the irregularly shaped wall, but he was unable to do this, and had remained in the position in which his rescuers found him.'

'At the moment Tapia glimpsed Kollasius, he lost consciousness. Kollasius immediately leaped down to the beach to keep Tapia's head clear of the water. Meanwhile the others, led by Kopelmann, fetched ropes and a ladder which they would use to get the injured man to the top of the ice-foot, and then, as an improvised stretcher, to carry him back to base.'

'It was not without great effort that they managed to hoist his unconscious body to the top of the ice-foot. There was the danger that his limbs, already stiff from the cold, would suffer from a jolt or any abrupt movement.'

'Taking him to the dwelling took a long time, although the distance was only two hundred metres. The rescuers were sinking in the snow beyond their knees, and when one man fell the others had to stop and help him to his feet.'

197. An ice-foot forms when sea ice is detached from land ice, it may be heightened by the action of swell and tide forcing ice landwards, heights may reach three metres thus it may form a precipitous, and unstable, cliff.

'The basic rule you are taught when bringing a frozen man back to life, is never bring him straight from the cold into a hot environment. With this in mind, Tapia was placed in the entrance hall of the dwelling, the temperature there was several degrees below zero as there was no heating in that place[198].'

'There they undressed him and rubbed his body with large handfuls of snow. His pale skin took on some colour by the minute, but from the beginning his companions all noticed that his arms, and especially his hands, were white as a sheet.'

'Gradually Tapia came back to life, his limbs were no longer as stiff and, while the rubbing with snow continued, his lips parted and he uttered a slight moan.'

'The acute pain that is felt in a frozen finger or toe as this thaws and comes back to life is well-known. The pain is intense. It feels as if a locomotive was running over your fingers on a railway line.'

'Judge for yourselves' Valentiner told us 'the intense pain that Tapia must have suffered after he regained consciousness while his companions, with solicitous care, gradually restored his limbs back to life'.

'What then?' I inquired.

After that, comes the most distressing part of that terrible accident; the amputation of eight of Tapia's fingers.

We all paid close attention while Valentiner continued the story.

42: Surgical Operation

'A few hours after the accident, Tapia's recovery was apparently making good progress. He was lying on the Leader's bunk. His companion noticed his fingers were still lifeless and very white, with no trace of blood reaching them. It was necessary to place and keep his hands in contact with a cold object, so they bandaged them in such a way that, holding his wrists above his head, he was able to grasp a large jar filled with snow.'

198. Modern medicine recommends very different, and more effective, treatments for hypothermia and frostbite.

'The following day huge blisters formed on his fingers, and when Kopelmann examined them he noticed several little black specks on the skin that was otherwise lifeless.'

'Something like an official meeting of the winterers took place. Everyone was there and it was decided that the patient needed an active treatment to disinfect the blisters. This was the only thing that could be done to try to save Tapia's hands, now that his symptoms indicated the presence of the dreaded gangrene.'

'Every day they tried different dressings and cleaned his hands; and this routine continued for two long months.'

'There was no possibility of sending Tapia to the mainland: it would have been nothing short of a miracle if any whaling vessel had appeared at that time. The sea surface was frozen solid to beyond the horizon.'

'Moreover,' Valentiner continued, 'in those days there was no radio link[199] on the base, so there was no way of calling for outside help'.

'Days went by and dates were torn off the calendar. Everyone watched with alarm and horror as the supplies of disinfectants used to treat Tapia's hands were rapidly diminishing.'

'Nobody wanted to think about what would happen after these were exhausted. The increasing anxiety was evident on all their faces as this eventuality approached.'

'The patient's hands continued to deteriorate. His fingers had swollen out of all proportion. Large blisters formed, one after another, on their surface. Shapeless mounds of blackish skin, indicated, without a doubt, the presence of the terrible evil.'

'Gangrene was present with all its unmistakable signs. As time went by it became a matter of urgency to do something to counteract the effects which were becoming relentlessly worse, with irreparable damage to the tissue. General blood poisoning was becoming a possibility, in which case there would be no hope ….'

'There remained one final course of action, but no one could muster sufficient courage to tell the others what it was. It was a heroic and desperate remedy demanded by the circumstances; amputation of

199. The radio link was established in 1927; this conversation, in 1923, is apparently anachronous.

the infected parts of the dead tissue of the fingers, the lifeless parts of Tapia's hands.'

'Eventually the patient himself, who from the start had followed the whole progress of his malady with great interest and calmness, called Kopelmann and said:

"I am convinced there is no cure for my hands; all the disinfectant has been used yet the gangrene is still spreading ···".'

'Something could still be done ····.'

'No, there is only one possible expedient. I have complete confidence in you and all the others'.

'However ····.'

'There is no other remedy, Kopelmann, no other procedure. Here are my hands, I ask you to do what has to be done ····.'

'It is a great responsibility,' argued Kopelmann, 'let me talk to the others about it and of course we will reach a decision'.

'As you like. I am ready to sacrifice my fingers, and you well know the sooner it is done the better.'

'By then it was well into September. More than two months had elapsed after the accident, and yet no one had mentioned the ultimate consequence. None of us had any experience of surgery, and, in truth, it required much courage to perform an operation of that kind.'

'At last they agreed. The Leader would perform the amputation. They composed a declaration stating that, as a last resort to save Tapia's life, and with his consent, they would amputate the four fingers of each hand. The thumbs were excepted as these were in a better condition.'

'Tapia took the pen in between the wad of bandages covering his hands, and initialled the declaration which legitimized this primitive surgery.'

'He was not given any anaesthetic, there was none to give, not even any local one to dull the pain.'

'One by one the second and third phalanges of his fingers were severed, Kopelmann skilfully separated them with the only scalpel available. To cut the tendons and the nerves he used a simple pair of nail scissors.'

'It is needless to describe the acute agony Tapia suffered. His companions had a bottle of brandy to hand. They held his arms and body, and when his determination began to flag, a good swig would revive him sufficiently to allow that terrible operation to continue until the eighth digit was amputated.'

'Truly, not only Tapia[200] but his companions also, needed unassailable courage to face a situation such as that.'

'One man's life had been snatched from gangrene.'

Only those five men could tell of the horrors they lived through during those two months, culminating in the dreadful moments of that crucial operation. They would have relied on the deepest reserves of human valour. They would have uttered, between tight lips, a supplication for divine mercy, which even the most sceptical agnostic on the South Orkney Islands has had occasion, more than once, to implore.

43: Mundane Recollections

Ultimately my mind had adjusted itself to all the requirements of that singular lifestyle: the freezing cold and unvarying whiteness outside, the strict hourly routine which we were obliged to follow, our moments of relaxation, and the mere business of living together harmoniously with four companions. We may have had similar wind-tanned faces but our temperaments were very different. All these things ultimately habituated me to an existence quite unlike any other, and to the surrounding infinite solitude.

The most trivial incident, of no significance whatever in any other part of the inhabited world, was reckoned here, in the circumscribed confines of our regimented life, to be an extraordinary event.

On one day it occurred to me to open one of my trunks in which I kept books, clothes, and other personal effects. Much to my astonishment I saw a moth fly out from between the folds of a woollen jersey.

My voice sounded like the alarm, the same tone that I would have used in a genuine emergency ….

'A moth …, a moth …!'

200. Tapia's son wrote a brief account of these events and his father's time on base: Tapia, 1971.

'Where? Where is it?' they all shouted.

All my companions wanted to see that minute lepidopteron whose larvae, concealed in some garment, had come from the mainland, and continued to survive sustained by the warmth from the stoves.

Their astonishment was well justified. It was aeons ago, countless weeks, innumerable days, since we had left Buenos Aires, and we had not seen a single insect, not even one fly, not one single fly …. Can you comprehend that, not a single fly!

In the South Orkneys it is impossible for an insect to survive outside[201], but inside our dwelling, in an ambient temperature of about twenty degrees above zero, and within someone's suitcase, it was quite feasible to come across such signs of alien life.

A moth inhabiting clothing that had not been used for a long time could well make a sporadic appearance; but the extraordinary thing was, it was considered such an unusual event that everyone wanted to look at it, follow its ephemeral flight, and experience as they observed its existence the strange sensation of finding, in that lost part of the world, a fresh topic of conversation, on the chances of survival of insect life in the South Orkney Islands.

Having opened my trunk, I began to rummage around and came across an immaculately tailored suit. It had been in storage seven months without being worn, nobody would think it could have any practical function for the life here. It was kept for the return journey, but it seemed a good idea to take it out and air it from time to time.

After seven months wearing heavy clothing; warm vests, thick shirts and trousers; and the endless putting on and taking off heavy work boots and snow boots, I had become accustomed to their weight. Then I had an urge to dress in that fancy gear belonging to 'civilization'. I had to squeeze myself into my trousers, but it was a relief to find the jacket still fitted me. Beneath it gleamed a white shirt, with a collar and tie. I put my shoes on and, thus attired, made my appearance in the parlour.

201. Of terrestrial invertebrates there are only four species of free-living insects in the South Orkney Islands. These are all springtails (Collembola), primitive, wingless arthropods barely 2 mm long. There are also about twelve species of smaller mites (Acari), numerous microscopic nematode worms, and several other species of micro-fauna.

Bruhns, who was sitting writing, turned when I called, and almost fell off his chair in amazement before this apparition from a distant world.

A South Orkneys man dressed in a suit like that, even in the smallest Argentine town, would be quite unexceptional!

Everyone gathered round and examined me. I also wanted to see what I looked like in a broken piece of mirror. When I stood in front of it I seemed to be gazing at a completely different person.

For my companions I must have seemed something like an apparition. They inspected me from all angles and when they were eventually satisfied, Plagge said to me:

'I appreciate the spectacle you have given us. I was already in danger of forgetting how strange we look when dressed in the clothes of civilization.'

It was true; none of us was interested in being elegant and wearing a tailored jacket, or having sharp creases in our trousers. We craved only warm clothing that was adequate to provide protection from the ferocity of the blizzards routinely sent in our direction from the South Pole.

Walking in those shoes 'from up north' felt quite weird; I felt as though my feet were being pushed up from below and lifted off the ground, It was a sensation resulting from my being so accustomed to wearing snow boots which had a thick sole of two and a half centimetres and weighed more than three kilograms.

To bring our fantasies back to Earth Valentiner said:

'You see this box here, it has something exceedingly rare inside, and I am sure none of you will have anything similar in your possession.'

We drew close to look inside and taking his hands off the top of the box, he showed us:

'Coins! Money from up there!'

I took one the larger coins and could not resist the temptation of examining it closely. On the head side it had an image of Liberty[202]

202. Liberty personified as a female face in profile derived from the symbolism of the French revolution.

and above her the words 'República Argentina'. I wanted to see the date but it was too dull. On the other side it said in numbers and words '20 - Veinte Centavos'.

After that a one peso note[203] appeared, and then another of five pesos; these circulated from hand to hand. Everyone examined every detail, even the water-marks.

We must have made an extraordinary sight, huddled in a group and scrutinizing those notes and coins. Was it any wonder? For a full seven months I had not held in my hand a single coin, nor bought a simple pack of cigarettes; I had not used my wallet, nor passed between my fingers any note worth even one Argentine peso.

Then Valentiner broke the silence, and said to me:

'Tonight Kleine, you can go to the cinema.'

We all laughed at the thought. We knew perfectly well that money had no value whatsoever in this part of the world. Anyone in possession of legitimate currency would not find a single person in a radius of a thousand kilometres who would accept it as payment, even for a packet of the cheapest cigarettes.

44: My Birthday

I awaited the day. Every night I tore a leaf off the calendar, crumpled it between my fingers and threw it into the blazing fire. On the first of July I calculated I had to tear off eight more leaves until 9 July[204] appeared, that was the day I eagerly awaited.

Many nights during my shift, while my companions were asleep and the wind-driven snow was freezing to the windows, I drew my chair close to the stove, lit my pipe and sitting close to the heat, I would begin to daydream.

Among coils of smoke my thoughts inevitably wandered; they unfailingly took a course due north, far far away, so far ….

203. In 1923 the value of one 'Peso Moneda Nacional' was approximately £(GBP) 0·08 or $(USD) 0·32; or 10^{13} Argentine pesos of 2017.

204. Moneta's birthday is on the same date as the Congress of Tucuman in 1816 when the independence of the Provincias Unidas del Río de la Plata, an antecedent of Argentina, was declared.

I would see my home city, Buenos Aires, my family, of whose fortune and even existence I was totally unaware.

I strolled amidst the infernal din of the metropolis, and as it was night I saw it illuminated by a profusion of lights from buildings and avenues.

Plaza de Congreso, Avenida de Mayo, Plaza de Mayo, Calle Corrientes, the heart of Buenos Aires, were conjured in my mind. I did not have eyes enough to admire the beautiful women of my native soil, my dreams and illusions were beginning to gyrate in my brain infusing me with a strange feeling of contentment.

Happy is the man who has the capacity to think, and, even if only in his imagination, to transport himself to distant, inaccessible, places where there are no barriers sufficient to limit his dreams or to check the velocity at which human thought can travel.

I loved to immerse myself in those dreams that always took me so far away .... It was no hardship to find I was actually in a wilderness of ice; on the contrary I was happy to be able to close my eyes and, for a few moments, dream I was back in the surroundings in which I was born.

My companions were not unaware that on 9 July I would have one more year to my tally. On the evening before they voluntarily stayed up until the clock chimed the last stroke of midnight and the new day began.

As with the 25 May celebrations, we discharged the usual volley of celebratory shots, and then, back inside the dwelling, we stood and listened to the National Anthem. Then followed a few drinks and the others retired to their bunks but not before offering me cordial congratulations.

I stayed alone in the parlour and, as always during my night shift, sat close to the stove and allowed my thoughts to carry me away to distant lands.

Only the chimes of the clock woke me from my dreams when I had to interrupt my peregrinations round the world to concentrate on the routine of hourly meteorological observations. I dressed in outside gear, went forth to record the data, and then returned frozen. I went back to the stove, lit my pipe and resumed the thread of my thoughts.

Previously, I recalled however far away I happened to be from my family, I always returned to Buenos Aires and spent 9 July at home. Very rarely had I missed the brilliant spectacle of the military parade. I was accustomed to seeing the lavish commemorative lights in the streets and the central buildings, with the display of flags draped on all façades. Even in the remote Antarctic, sitting alone so far away isolated from the world and from life, I could still count, in my imagination, all the electric lights that I once saw illuminating General Belgrano's statue[205] ….

My mind never tired of those imaginary forays; they distracted it, but sometimes plunged me into profound sadness. I soon resigned myself to this fleeting melancholy; it required no great effort to return to reality.

'Patience!' – I would counsel myself 'everything has a purpose, everything comes to an end ….'

At five in the morning of 9 July Plagge took over my shift, and I went to bed.

At midday I was awakened by the March of San Lorenzo and we took our seats at the table to be served by Otto who had decorated it for the occasion.

'Here is our substitute turkey', he said, as he put a large dish on the table in which was a golden-roasted shag.

After dessert a bottle of champagne appeared and I made a spectacular display of releasing its cork.

We drank a toast. Valentiner got up and returned a few minutes later with a small packet which he placed at my side, pointed to it, and said:

'Please Kleine, accept this gift ….' I had to accept it and opened it smilingly, when his arm stopped me and he added:

'Have a look at the address.'

205. General Manuel Belgrano (1770–1820) was much involved in the independence of southern South America and was a notable Argentine patriot who designed the Argentine flag. A prominent equestrian statue of him stands in front of the Presidential Palace in the Plaza de Mayo, central Buenos Aires.

I read:
Señor Polar Explorer, José Manuel Moneta.
National Magnetic and Meteorological Observatory
South Orkney Islands
Argentine Republic

I could tell there was something suspicious about this when I read 'South Orkneys, Argentine Republic'.

We had repeated discussions on the subject of the ownership of the South Orkney Islands. It was strange that the British government administered and taxed whaling by Norwegian vessels which, during the summer months, operated within the limits of territorial waters of the South Orkney Islands. I was even more astonished to learn that the government drew much revenue for granting those rights.

However, at that time, I was neither ready, nor willing, to address the question of territorial jurisdiction; I preferred to deal with it subsequently[206].

206. In the 12th, 1963, Argentine edition a footnote has been inserted. It reads: 'On repeated occasions, both in newspapers and in public debates to the present, Argentine rights in the Antarctic have been a subject for discussion. These rights have stood from precedence; they are seen to be affirmed by the course of our country's history and by her decisive actions in the Antarctic. She succeeded in occupying that remote territory, where no other country whatsoever has attempted to establish any permanent bases. That on the South Orkney Islands was the first to be built by any country in the polar region. In earlier chapters we described the origin of the current meteorological and geomagnetic observatory in the South Orkney Islands. On the question of sovereignty in the already distant past when the station was founded, we refer to the thoughts of the Scottish scientist Robert Neal Rudmose Brown, who, aboard the ship *Scotia*, wintered on the South Orkneys for eleven months during 1903 prior to the arrival of the first Argentine expedition to take charge of the observatory. In the book about the voyage of *Scotia* published in 1906, the matter of Antarctic sovereignty is referred to. Rudmose Brown wrote the following: *Often, among the varied topics brought forward in the cabin in the long winter evenings, arose the question of ownership of the South Orkneys. And after many long discussions we arrived at the pleasing conclusion that even in this age of imperialism the South Orkneys had escaped the grasp of any country, and that we enjoyed the privilege of living in No-man's Land. But I fear this is no longer so. Not that we claimed them for Britain, – for even had we been seized with a desire to widen the confines of our empire, we could not lay claim to new territory in our country's name without a Government mandate, and as for claiming them for Scotland, I feel that still less would have been recognized, though in Mossman they certainly had a Scotsman for their first governor. However, when the* Scotia *returned to the islands in February 1904, with an Argentine staff to take over the meteorological observatory at Omond House under the auspices of the Argentine Government, the Argentine naval flag was hoisted*

I untied the parcel and was delighted to find two pounds[207] of chocolate. I had a sweet tooth and my companions knew it; I knew how to appreciate properly this unforgettable present on my birthday.

45: Infinite Solitude

Every day, every moment that I had free from my duties, I would put on my ski and go out to glide over the endless plain of the frozen sea. Sometimes, even during the rawest storms in which temperatures of minus thirty degrees were registered, and with hurricane winds blowing, I still would go out to ski, but would hasten back after a few minutes, satisfied that I had completed one part of the daily routine I had set myself.

From the beginning I realized that one of the best ways for maintaining morale whilst living in the South Orkneys, was a good a night's sleep; calm and restorative.

During the winter months there were not many opportunities for keeping muscles in trim, or for taking enough physical exercise to tire the body so that it would need long hours of rest or repose. Therefore I made deliberate efforts to tire myself, so that when night came I was so exhausted as to be able to place my head on the pillow and fall into profound sleep. Seldom was I unable to sleep, as I took my daily exercise to ensure getting six or seven hours of unbroken slumber at night.

The skiing exercise gave me that satisfaction. I was determined to lead a methodical life in every sense of the word. I went outside daily,

---

on the cairn where formerly the Scottish Lion flew; and I presume the South Orkneys are looked upon as a possession of that power, – the nucleus of an empire, perhaps, they may even seem to ambitious Argentine expansionists. The South Orkneys are certainly the only spot in the Antarctic regions that have been inhabited without a break for a period of over three years, [this was written in 1906 and refers to the Argentine expeditions] and they bid fair to become a permanently inhabited meteorological observing-station of the Argentine Government; for that country has, for the present at least, agreed to keep a staff of six men there, to be relieved annually. That was the categorical position on the tenure of the South Orkney Islands by Argentina, before any other countries had imitated it by establishing a similar presence in the Antarctic.' [The quotation is directly from pages 78 and 79 of: Brown, Mossman, and Pirie, 1906.]

207. Almost a kilogram

whatever the weather, even if it was just to return after a few minutes, my body numb with cold.

It was in that way I progressively came to know a large part of the islands. The sea, frozen in winter, gave me this opportunity because, in summer, it would have been too dangerous to venture out in a boat through the surrounding channels and bays.

One of the first visits, across the ice-covered Scotia Bay, was to a place named Point Martin described as 'a large penguin rookery', the site where they would hatch and raise their chicks during the summer months.

According to the topographic maps, a distance of five kilometres separated me from the abandoned rookery, so one fine day I put on my ski, determined to visit it.

I dressed in light, but warm, clothing; took a spare pair of gloves, the indispensable camera, and my choice of weapon was a good hunting knife.

I made good time, about six kilometres an hour across the ice of Scotia Bay and once past Point Diebel, about one kilometre from home, a magnificent spectacle appeared before me.

An extensive plain, formed by the frozen sea, stretched before my eyes. To the left, it was limited by a wall of ice two hundred metres high, formed by enormous glaciers that terminated in the sea. To the right were the hills along Mossman Peninsula, between which were several glaciers extending to the fast bay ice[208].

There were no signs of life; I saw no trace of any living being anywhere; not a single petrel, gull, nor penguin. Everything was white: the mountains, the surface over which my ski sped, the ice of the glaciers, even the sky was obscured by the thick layer of cloud which allowed no trace of the blue of the firmament to show.

I skied non-stop across the middle of the bay, drawing near the point I had taken as a reference for my course which was to a high peninsula terminating in the frozen sea. I could distinguish that the ridge of the

208. Fast ice is a term defining sea ice which is attached, stuck fast, or frozen to the shore; it does not mean ice moving rapidly.

38: Rookery of Adélie Penguins at Point Martin

peninsula appeared to be covered in a faintly light brown colouration which became more apparent as I advanced closer.

After about an hour's walk I reached my destination, took off my ski, and began to climb the gentle slopes of the coast. In several places I saw rocks caked with penguin excrement and it was this that caused the discolouration of the snow I had noticed from afar; dirty melt water had frozen giving the snow an opaque brown muddy tint.

There was not a single penguin in sight. After spending about half an hour looking for one, I put my ski back on and headed slightly farther south.

I reached the extreme end of Mossman Peninsula and there I stopped to rest; I was fifteen kilometres away from base.

Again I took my ski off and ascended a high headland[209] from where I had a far more extensive view. I was at the southernmost point of Laurie Island and of the entire archipelago. All that lay spread before me was a limitless extent of frozen ocean reaching the horizon. In

209. Cape Hartree, 155 m elevation; while this is the southernmost point of the islands and rocks of Laurie Island; South Cape, on Atriceps Island off Coronation Island, is the most southern point of the South Orkney Islands, it lies at 60°47·9´S, only 1·7 km farther south of the rocks off Cape Hartree.

some places large icebergs emerged from it, which had become beset.

To the right I could distinguish the southern entrance to Washington Strait, and in the far distance to the west, the rugged outline of Coronation Island the largest of the archipelago.

Absent was any sign of life; not one plant, not even a single green leaf broke the monotony of that endless white.

I took a bar of chocolate from my pocket, it was hard to break with my teeth as it had frozen solid.

A freezing cold breeze coming from the south-west stung my face. To the eerie silence was added the frightful feeling of utter solitude.

I experienced the surreal sensation of finding myself totally alone in that remotest part of a lost world. Apart from those four men awaiting my return, five hours march from where I was, there was no other living person within a radius of one thousand and five hundred kilometres of where I was now standing[210].

46: Moods

Slowly the days turned into weeks, and the weeks into months. My survival technique was to find jobs that would interest and completely absorb me for a good part of the time.

Occasionally I would have time to read, but actually our library did not contain sufficiently interesting books to distract my mind or hold my attention. In those days most of the books in the library were by foreign authors, published in their original languages, English, German or French.

The few books we had in Spanish were mainly detective novels, which were of little interest to me.

With Valentiner and Plagge I would speak a few words of English, while I was also interested in learning some words of German from Bruhns. However I soon had to abandon this enterprise, as, after some initial attempts, learning the two languages concurrently caused Babylonian confusion.

210. This is approximately the distance to Tierra del Fuego or the Falkland Islands; South Georgia is closer (850 km). See also the distances listed in § 3.

On the other hand, Bruhns made some progress with the Spanish lessons I gave him. His own style of learning was to read any Spanish book he could lay his hands on. As most of these had already passed through mine I was able to tell him what was good and what was best left on the shelves.

There were old magazines in several languages, but in spite of their being well-thumbed none was thrown out. Whatever magazine happened to be issued in any particular month, that was the only one available. Regardless whether it was good or bad, I had to depend on it to satisfy all my visual cravings with its photographs and drawings until the next issue in the next month. Once read it was never discarded, but, on the contrary, carefully returned to the bottom of the stack from where it would resurface many months later as a 'new issue'. We had folders crammed with *porteño*[211] and foreign magazines. Even today, the illustrations which embellished them have remained familiar, imprinted in my memory all those years ago.

When any of us discovered something new in those old papers, his immediate reaction was to put it aside to show to the rest of us. Days when some antique magazine was exhumed became a cause for celebration, or at least provided a much-needed stimulus for our minds.

Sometimes, Plagge would appear with a United States style photograph and say:

'Look at this!'

I already knew that these would be bathing beauties; gorgeous women showing off their bodies squeezed into skimpy bathing suits.

Not even in the South Orkneys do you stop being a man. Undoubtedly there were days when the blood coursed more vigorously through our arteries. A memory, a picture, a conversation were enough to transport the imagination elsewhere.

During my four years in that part of the world, I saw many men come and go, and some of the more sensitive ones suffered. The whole secret lies in how you adapt to the circumstances. The important things are to maintain good morale, to take vigorous physical exercise, and

211. *Porteño* indicates someone, or something, from the port city of Buenos Aires.

to go to bed only when exhausted, when you are absolutely sure you will be able to sleep soundly.

It was still depressing to find oneself so far from home, to awake and always see the same weather-beaten faces around the table. For that reason, I put up a few photographs of female faces and bodies in my bunk-room. Every now and then I would change them for others. Like that I managed to delude myself that amidst that atmosphere of men, whose mask of toughness often hid an inner sensitivity, there wafted a faint perfume of femininity.

All that deprivation, mental and physical, was indeed difficult to bear. Inevitably there came days when we got up, still half asleep, and we would forget to greet our companions in the morning.

In itself that was sufficient to initiate a trivial personal resentment among colleagues.

Perhaps one of us would make an unfounded and involuntary criticism of the way one of the others was doing something at work. Perhaps the same fellow, without a glance in our direction, would help himself to tobacco from his private supply without inviting others to join him in a pipe.

On a day when an incident such as this happens, there might be a flare up of feelings of cumulative resentment. It is not difficult to comprehend the sort of petty things that may cause it. It entirely depends on how the offended individual reacts, it depends on his values, his education, his experience and philosophy of life; whether he chooses to forget about it, or whether with all the vehemence of an untutored mind, he magnifies everything out of any proportion.

Sometimes there is nothing to be done except to draw a deep breath, freeze the blood in your arteries, show yourself dispassionate, even though your blood is boiling and your nerves are frazzled. Only in that way can you triumph.

The days seem endless. To be sure there is companionship. From time to time there is a heated argument, but all passes and calm returns.

Nevertheless as the months go by, the general mood seems to be slowly deteriorating, the bitterness of the isolation is gradually undermining us. It is outside, well away from the dwelling, in total solitude,

39: Giant Petrel (photographed by David Wilton, from: Bruce, 1907)

as I investigate a geological formation or whatever, that I find the strength to endure the petty eccentricities of my companions. It is there that I learn to understand my own defects and to correct them, so that my imperfections may at least be tolerable to the others.

47: Spring Arrives

The most important dates in the year were clearly marked on the calendar, as if to avoid any possibility of their being overlooked. Needless precaution! Those dates are so fixed in everyone's mind, no calendar is necessary to remember the right day.

Thus on 21 June[212], someone would announce:

'Today is the shortest day in the year.'

That was not news for any of us. From the first month of our existence in the South Orkneys, we had known that date marked the ascent

212. The austral Winter Solstice is an important event traditionally celebrated on all Antarctic stations, ancient and modern. In the latitude of the South Orkneys mid-winter daylight duration is approximately 5 hours (compared to 19 hours in summer, with twilight in addition).

of the summit of a steep proverbial mountain, indicating that we were roughly half way through our time on base.

From that day forward, we could tell ourselves that we were 'on the downhill run', the shortest day of the year had been passed.

For my part, I realized only too well what that day meant. From the start I had understood that mid-winter did not mark the end of our uphill journey, but only rather less than half of it. There remained a long stretch to run; but it was at least some comfort to see that half the leaves of the calendar had already fallen and would not be going back on the wall.

The next eagerly awaited day was 21 September[213]; by then it would be possible to see the return of some of the wildlife. This would be a huge boost to morale, a kind of compensation for the boundless solitude of the winter.

As Valentiner explained: 'from early September Giant Petrels start to fly overhead again. We will see their numbers increasing every day. At the same time large flights of shags will arrive'.

'Which tells us,' I interrupted, 'that the sea ice is beginning to disperse from around the outer islands'.

'Yes, that's right, the ice starts to break up to the north and this is enough reason for the sea birds to resume residence on land, although they still need to seek areas of open water in which to feed'.

'What about the penguins? Will it be long before they are back?'

'By the middle of September we should begin to see them. In October they start taking possession of their breeding sites, and during the first days of November, we will be able to have fresh eggs for our *milanesas a caballo*[214].'

I asked another question; 'when will the seals[215] return?'

'Within a couple of weeks, then the first pups will appear.'

'In the first few days of September?' – I enquired with disbelief.

'Well that is the time when they have returned in previous years, I hope this year will not be the exception.'

213. The austral Vernal Equinox (daylight duration 12 hours).

214. *Milanesa a caballo* (on a horse) is a fried breaded steak topped with a fried egg.

215. Weddell Seals

After a pause, Valentiner asked me:

'When you made your previous visit to Cape Murdoch, did you see any seals by chance?'

'Not a single one', I replied.

'That is strange, as that is precisely the place where they congregate to bear their pups.'

'That may be so,' I pointed out, 'but what about the pack ice that was covering the sea in all directions?'

'That presents no obstacle for seals; presently they should be on the ice far from the shore but getting closer to the coast all the time.'

'What methods do they use to propel themselves?'

'Those mammals,' he explained, 'make breathing holes through thinner sea ice. To extend their time submerged, they look for small cracks or flaws in the ice which are usually found near icebergs. They enlarge these by scraping with their teeth, so that they can break through and come to the surface for air whenever they need it.'

Needless to say I could hardly wait to see all this. For me it would provide new interests and help the time to pass. Life spent confined within four walls would seem less monotonous.

Anticipating all the ski treks I was planning to make, I started preparing my equipment by fixing skins to the ski I used daily[216].

I added to my kit a pair of good ski sticks, strong and light. In my free time I made a canvas rucksack for carrying a few provisions for the journey. I prepared my best pair of ski boots, my windproof clothing and my camera. I sharpened my hunting knife giving it an edge which would have made even a dagger envious.

The sledges received a general overhaul; later we would have to use them to transport crates of penguin eggs and we had to be sure that they were strong enough to survive many journeys between the rookery and base without mishap.

The prospects were very encouraging; even heavyweight Otto, who had gained quite a paunch despite his best efforts to the contrary,

216. These are strips of tanned seal pelt so arranged that the direction of the fur fibres reduces backward motion of ski, somewhat like the effect of a ratchet, they also provide firmer footing on ice.

chose some ski on which to practise.

One afternoon, at the end of August, I climbed Mount Lola to observe the state of the sea ice. To the south everything was exactly the same as it had been in the depths of winter, but to the north, beyond Saddle Island and to the north-east, there appeared to be some indistinct patches of open water. This was a good sign, the ice was starting to break up. Wherever there was an opening it was now unlikely to freeze over again.

I came back downhill at top speed and told my companions the good news which they greeted with evident signs of satisfaction.

I recall that on that particular afternoon a strong wind came from the south-west. By nightfall it had turned into an unbridled hurricane of one hundred kilometres an hour, the outside temperature dropped to minus thirty-seven degrees. There was nothing to indicate the imminent arrival of warmer weather that September was supposed to bring. This was the time when the wildlife was due to return to these latitudes, a time when new impressions and changes of scene would help make our existence more bearable.

It was only that distant stretch of open sea that was significant, but first it was necessary for many days of strong southerly winds to disperse the ice far out to sea. With the present wind direction our hopes were waning; to me it appeared quite impossible for that scene of desolation to change in any other way than in our patiently awaiting the natural lengthening of the daylight.

Inevitably the daylight must lengthen; there was no remedy but to sit and wait.

48: Waves Under the Frozen Pack

September eventually arrived. A violent southerly storm had been battering the buildings for an entire week, making us suffer the full force of the harsh polar weather. On 2 September the wind began to abate; by sunset, only a light breeze wafted in the air. The sky gradually cleared and the clouds slowly thinned.

That night I was on duty until four thirty in the morning. When I went out at midnight to check the variometers, there was a brilliant moon shining in a cloudless sky.

Complete calm, the wind seemed to have stopped for ever. The snow, hardened by twenty-five degrees below zero temperatures, crunched underfoot. In a landscape rendered dimly blue by moonlight, the complete absence of movement seemed to enhance the feeling of overwhelming solitude.

While I was in the Variometer Hut busying myself with recording the instrument readings, I thought I heard a prolonged groan.

At first I thought it was caused by the pressure of the wind against the doors, or walls, or roof of the little hut, but I immediately realized it could not be that because there was no wind.

I listened carefully, and after a few seconds I heard another groan, even more prolonged than the first.

I was accustomed to going to the Variometer Hut several times a day and at all hours of the night, and I could never resist the instinctive temptation to cast a glance in the direction of the cemetery which was not far away.

I was startled when I heard the groans, which recurred at regular intervals; they seemed to be coming from the cemetery, or very close to it.

When I finished my work in the hut, I went out and glanced again towards the graves, but I could not see anything that might be untoward.

Nevertheless, the groaning continued and suddenly it dawned on me from which direction it came.

The sound was coming from the beach, next to the cemetery. I went there quickly, being very interested to find the cause of such a weird disturbance.

I clambered down the ice-foot which took me on to the fast ice of Uruguay Cove where I stopped to listen carefully.

Nothing could be seen to break the monotony of that white plain, and I was wondering what unknown force was causing that strange phenomenon, when I heard an even louder groan, startling me into reality.

I was struck by the fact that the surface of the ice on which I was standing seemed to be moving up and down very slightly.

I was over the sea, a few metres from the shore and the ice-foot, which connected the land to the fast ice of the sea. Along the entire foreshore, and parallel to it, ran a series of dark lines dividing the ice attached to the shore from the fast sea ice. Then I realized that it was the irresistible flux of the tide that had caused these cracks to appear as the solid white surface of the sea alternately rose and fell[217].

However the movement of the ice, and the groans which had made such a powerful impression on me, must have some basic cause. The ice rose and fell continuously, seemingly impelled by a force beneath the surface; it was this slow undulation that produced the tide-cracks. The friction as their vertical sides moved against each other caused the grinding sound that broke the silence of the polar night with groans.

The still night atmosphere was pervaded by eerie creaking sounds, sharp cracks echoed and multiplied by the hard crystalline walls of the glaciers.

The sea ice rose slowly and fell, impelled by waves passing beneath it; waves which had probably originated far out to sea, where the surface was not frozen. Then these travelled under the frozen surface and penetrated the bays, finally expiring on the foreshore or at the ice-foot.

The sea ice moved in an undulating motion similar to that of waves; with barely perceptible amplitude, but sufficient to generate frictional forces against the fast ice where the unnerving groans originated.

There was no doubt about it; that movement was a good sign.

If the waves could penetrate under the ice from such a distance, it demonstrated that its extent was not as vast as it had been in the depths of winter.

The fast ice was beginning to break up and it would not take long for ice floes to form and drift seaward.

Slowly I walked back to the base. The night continued to be splendid and the sky, littered with glittering stars, was cloudless.

217. Known as a tide-crack.

49: Blood on the Snow[218]

Prospects for making a trek on the following day could not have been better; I immediately resolved to make my planned visit to the seal colonies. According to the date, the female seals would already be taking possession of the best locations to await the birth of their pups.

While my companions were asleep I prepared my equipment and provisions: some sandwiches and a portion of chocolate. I checked the skins on my ski and the sticks. I remembered my indispensable hunting knife, and then waited for the end of my shift before starting my trek.

I had not slept, but thinking about the trek made sleep impossible, and I felt I had sufficient energy to accomplish it regardless.

I would rest after I returned, hoping to be back by four in the afternoon.

At four thirty in the morning I woke Bruhns, and as soon as he was up, I made ready to leave. I was putting on my ski when he asked:

'Are you not carrying anything to slake your thirst?'

'I don't think that is necessary,' I replied, 'there is enough water in the ice.'

'Well, it's best you take something to drink, I will prepare a flask of cordial for you and bring it straight away.'

A few moments later he returned and said:

'It is lime-juice mixed with red current cordial and water. I added some sugar that will make it much better than chewing ice.'

'Thank you! and see you later ….'

'Hope it all goes well', he replied. I then set off at a good pace heading south.

My object was to reach Acuña Island, in the extreme south-east of Scotia Bay. On the map a large seal colony was marked in the vicinity. Although the journey was about a dozen kilometres, I was in good physical condition, and able to reach the destination in less than three hours. I passed Point Davis and then continued towards the outer part of the ice-covered bay.

After an hour's skiing I stopped and opened the flask, I wanted to

218. In this chapter the first four paragraphs have been moved from the end of the previous one.

have a taste. The liquid had already frozen, even though it was in an insulated flask.

I continued walking and suddenly a black shadow soared close to my head. It was a Giant Petrel, the scavenger of the Antarctic which, according to the naturalists, feeds exclusively on remains of seals and penguins.

More evidence, I thought, of animals not too far away.

I was already at the outer limits of Scotia Bay, diametrically opposite Cape Murdoch, of which I could see the outline in the distance. As I turned to continue my trek, there, farther ahead, I saw half a dozen Giant Petrels circling over the same place.

That is where the seals must be, and as I was eager to see them, I hurried on.

I was about to reach Acuña Island. On approaching one of the offlying rocks, I suddenly stopped. There in front of me, only a few paces away, on the completely white surface, was a large red pool of blood which had frozen, and bits of something which, on first sight, appeared like the entrails of some animal.

That must have come from a seal, broad flipper tracks in the snow were all around it. I decided to follow the track and, a few paces further, came across another frozen object. It was a seal placenta, and it was covered in coagulated blood.

The red trail continued, so I followed it farther.

I reached a big block of ice, behind which the tracks disappeared. Suddenly I found myself, face to face with a seal which, big as it was, lay sound asleep on that frozen ground. She was not alone; pressed closely at her side, slept a tiny pup.

They were within a couple of metres from the tips of my ski, and they had not awakened. When I touched the mother's head with my ski-stick, she lifted it in alarm, poised erect on her fore-flippers and began to grunt angrily at me.

The mother's movements awoke the pup and as it turned round, I noticed its umbilical cord was still attached.

I stepped back a few places, took off my ski and turned back to observe that curious pair.

With my ski-stick I tapped the mother, who leaving her pup aside, dragged herself lumbering towards me. Her movements were very slow. I could see that her extraordinary weight and girth were not the cause of this, and that the poor beast was clearly in deep pain following the birth.

I wanted to reach the pup, but the only way I could do this was to separate it from its mother, but when I tried to approach it she had other ideas and opposed me. By distracting her with my ski stick I managed to separate them by about twenty paces; and then made a dash towards the pup.

The pup was about fifty centimetres long, and was covered in long thick hair. I took it in my arms and with this load scrambled on to a tall block of ice. The mother pursued me rapidly, but had to stop short as she was unable to climb to where I stood.

The desperation of the poor seal was pitiable, she bellowed in despair in her frustrated attempts to recover her pup. My intention was to hand it back, but I was secretly pleased to see her response to my brief abduction of which she was victim.

The pup made bleating sounds not very different from those made by lambs, and its tameness was similar to that of any domesticated animal. I could not believe that I was holding in my arms an authentic Antarctic seal evidently born on the night before. Those large red stains on the immaculate white snow announced, in the immense wilderness of the Antarctic, a new life.

50: Drowsiness

I returned the pup to its mother, who immediately started to pass her snout over its head while she grunted contentedly. I began to feel quite weary after more than three hours trek over the ice, and decided to rest there before continuing my discoveries.

There was sufficient time to explore an extensive area, my watch said it was only nine in the morning. The Sun was shining and time seemed to stand still, the weather looked as if it would stay settled. However although there was hardly any wind, the cold began to penetrate and chilled me to the bone. As the sweat from my exertions began to freeze,

40: Páez Montero with three Weddell Seals

I felt distinctly uncomfortable and longed to be somewhere warmer.

I looked for a nice sunny spot and lay down to rest on a piece of ice for a few minutes.

The seal and its pup were only a few paces distant, where, from my vantage point, I was able to observe the small family at my leisure.

The fright I had given the mother was still evident, she did not take her eyes off me, as if wondering what my intentions were. After a while, she slowly slid away from me, pushing her pup, until she was about ten metres distant, where she stopped and gazed at me again.

With that distance between us she now seemed more comfortable, and without further concern about my presence, lay down flat and placidly went back to sleep.

I watched her measured breathing, she held the inhaled air for some time before taking another gasp. The thing that most caught my attention, were the convulsions which continually shook her body.

She gave the impression of profound sleep, but every so often shook agitatedly, even violently.

Meanwhile the pup had moved closer to its mother, searched with its snout, and found a teat on her breast, suckled and then drifted off into gentle sleep.

I was pleased to have the chance to contemplate that scene. Few men, even those who have been as far as the South Orkneys, can have had the opportunity to observe, at close quarters, what I was watching at that moment. It was unparalleled; in everyday life some people may have seen seals, even if only in a circus, but to be able to watch them in their natural element, their home territory and environment, was indeed a pleasure denied to most people.

Furthermore, southern seals are not the same as the tamed ones that we are used to seeing, trained to perform in fairgrounds, which are usually from the Northern Hemisphere[219]. There are some similarities between the species, but the most noticeable differences are at their extremities. These are much smaller in Antarctic seals; the latter, however, have a larger and heavier body, as well as lighter and more varied fur than those from the north[220].

With regard to the question of whether these Antarctic seals have the same intelligence as the northern counterparts, I have serious doubts.

Recalling the tricks that those darker northern seals had been trained to accomplish; I doubt that any Antarctic seal could match the skills of the performing ones which may even appear on the theatre stage.

Down south, in the natural wilderness, desolate and primordial, it is very possible that Antarctic seals maintain their inherited natural wildness.

While I reflected on all this, I rummaged in my pockets for some chocolate which I chewed with relish.

I had rested sufficiently and decided I should put on my ski and continue to travel. However, a strange feeling seemed to force me to tarry.

My mind was lucid, but suddenly an overwhelming tiredness forced me to close my eyes. I was very cold, and although I knew I could counteract this by the exertion of walking back, an impulse made me stay there a few minutes longer.

For a few seconds, my head nodded heavily forwards, I suddenly

219. Performing circus seals are usually a Californian species, *Callorhinus ursinus;* one appears in Illustration 74.

220. This indicates the difference between the otarid and focid orders of seals, species of both occur in each hemisphere.

41 : Monti and Páez Montero returning from a visit to the seals

got a fright and leaped to my feet.

I was terrified that I might have remained asleep on the ice, and then I remembered how that intense cold acts on the body like a sedative. Many unfortunate accidents have occurred when polar explorers have fallen asleep in the open and remained at the mercy of the elements. Generally they never wake again.

On previous field trips I have frequently been aware of drowsiness coming over me when I stopped to rest for a while outside. Until that moment I had never quite experienced the immediate effect.

I continued on my way and reached the north side of Acuña Island. Between this and Laurie Island was a distance of five hundred metres. On the ice there were hundreds of seals, all in deep sleep.

A few of them had pups, while others looked as if they were about to bear them. The occasional male was visible among the females, but they were sparse and far between. They appeared to be standing guard over the sleeping mothers-to-be.

On my way I passed very close to these animals, most of them did not bother to open their eyes to see what I was; it was only the males which were alert.

Together they presented a rather odd sight; they slept soundly on their sides, but as they did they shook from the convulsions which constantly affected them.

51: More about Seals[221]

It took a while to ski my way between the seals. I felt hungry and decided to climb to the top of Acuña Island to take another rest and eat my provisions. The island was fifty metres high at its summit; its shape was circular and its area was approximately one hectare.

I chose a comfortable spot bathed in sunlight and settled to have another rest, but with the firm resolve to get moving again if any drowsiness overcame me.

I was extremely thirsty, but was unable to use the frozen contents of the flask. It did not take me long to notice that sucking snow and ice burned the tongue, making me even thirstier.

There was nothing to be done about this but to put up with the situation for another few hours. Better this than to miss a chance of observing the novel sights before my eyes.

My watch said midday, I ate a bit more chocolate then I went down to the sea ice to continue my travels and observations.

There were many Giant Petrels flying overhead. Looking at the seal colony, it was obvious why those birds were there. They could not do any harm to those huge lumbering adults, but perhaps their pups could fall victim to a surprise attack.

At the other side of another block of ice I could see a seal suckling its pup. Surrounding them were three Sheathbills, which were feeding on the remains of the frozen blood left on the snow. Neither of those two seals seemed to be aware of them, but the brazenness of those birds

221. It is notable that only Weddell Seals were described during Moneta's time in the 1920s although Fur Seals and Elephant Seals are mentioned in the historical summary, § 1, and other species in the accounts by Valette (1906) and Acuña (1982 and 2015). A Leopard Seal (*Hydrurga leptonyx)* also appeals briefly in the film. Sealing activities during the 1800s had so drastically depleted the fur and elephant population that they were very rare during the early 1900s. By the end of that century populations of both species had recovered; they had become abundant again in the South Orkney Islands and elsewhere.

astonished me. I saw one of them come close to the pup and start pecking the umbilical cord still attached its belly. The pup yelped in pain which woke the mother who lifted her head to look at the Sheathbills, but before she could let out another groan, the birds had already flown off.

I went close to another group of seals consisting of twenty-two females and one male.

They were clustered round a breathing hole kept open through the ice, the area of which was about one square metre.

Sea water was evident at the bottom of the hole and the ice around it was covered in organic detritus from the seals.

I went near the hole and estimated that the thickness of the ice was about two metres. Sea water reached about fifty centimetres below the ice surface.

Suddenly I noticed the water level rose rapidly and out popped a seal's head, its jaws opened and it expelled the stale air from its lungs.

It saw me, but it did not seem the slightest bit bothered by my presence. It took a deep breath, closed its mouth, and dived under again.

I remained immobile, thinking it might return. I was right, a few moments later the water rose forcefully and the same seal reappeared quicker and more vigorously.

Perhaps it had intentions of scrambling onto the ice, as it supported itself with its flippers pressed against the sides of the hole. With its jaws it was grasping a protrusion of ice that was hindering its free movement. By moving its head laterally the seal managed to scrape the ice clear with its teeth. Then, with its body and flippers, supporting itself on its chin, and using all its strength it eventually hauled its body on to the surface.

It was an average sized male, he immediately began to move away from the hole, but he did not get far, as I decided to see what his reaction would be on being harassed by a stranger.

I took my ski off and grabbing one of my sticks and ran after him and whacked him hard on the rump, which seemed to bother him, although I had caused him no harm.

I planted myself in front of him, but although finding himself cor-

42: Female Weddell Sea l with a pup on sea ice (from: Brown, Mossman, and Pirie, 1906)

nered, he made no effort to head back to the hole where he could find safety. After several more minutes of harassing him, without causing any injury to his pelt, he eventually decided to head back to the sea, where I let him go freely.

He reached the edge of the hole and stopped: he dipped his head in as if scrutinizing the depths, and then slid forward and, in a quick plunge, rapidly disappeared underwater.

The other seals continued sleeping, not one of them had noticed the commotion.

I decided it was time to return but, as I still had time to get back to base by a reasonable hour, I resolved to cross the bay to make another visit to the penguin rookery on Point Martin.

I put on my ski and, while sucking a handful of snow, went in that direction.

52: Meeting on the Ice

It had been eight hours from when I left base and, despite having moved around constantly, I did not feel particularly tired. From time to time I was troubled by thirst but was able to quench this, to some extent, with an occasional handful of snow.

When I found myself in the middle of the bay, I noticed Giant Petrels flying over Point Martin, where the penguin rookery was situated.

This small detail seemed to augur well, as undoubtedly there would be some penguins; otherwise those petrels would not have been around.

I continued on my way and after a few minutes, as I looked again in the direction of the rookery, along the coast I caught sight of some strange dot-like things moving on the snow. I had not brought my binoculars, so I increased my pace to investigate more closely what they might be.

There were approximately a dozen little black dots which were advancing in a straight line with a regular gap between each.

I diverted from my route and turned quickly towards the moving black dots, with the intention of catching up with them. I increased my speed and as I approached, discerned their shape and saw they were Adélie penguins, travelling north in the direction of open sea. Probably they were on their way to feed.

They all processed in their characteristic manner, lying flat with their white bellies on the snow and sliding along while propelling themselves with their feet. Their flippers were spread out horizontally at right angles to their bodies and served to maintain their balance as well as helping them along[222].

They slid along very smoothly, tobogganing many centimetres at every push, and gained considerable speed with little apparent effort.

When they saw me, they stopped, and stood upright. They had a moment of indecision, during which they appeared to be consulting each other on how they should react to such a strange apparition. The next thing I knew was that they were advancing resolutely

222. This type of penguin locomotion, sliding on the ventral side, is referred to as tobogganing; on a suitable surface it is much faster than walking (but slower than swimming).

towards me.

I remained absolutely still, and when they were about five metres from me, they stopped.

There was no doubt that my presence completely fascinated them. Possibly they thought I was also a penguin, covered in strange plumage and with a decidedly odd appearance. It was impossible to guess if they had any power of thought or discernment. I am sure I am right to maintain that the presence of a man, a biped, walking upright like penguins occasionally do, must have given them the idea that he is some sort of a giant cousin, of whose intentions they simply could not be sure.

I advanced a couple of steps, and far from fleeing from me, they swiftly formed a compact group as if to mount a joint defence.

They were a metre away, all looking up facing me and did not take their eyes off me. Their flippers were held wide and at right angles to their bodies. They flapped furiously with strong thrusts, as if showing the powerful blows they would be capable of delivering in a fight.

I extended a ski stick and gently tapped two of those in front. It was as if an electric current had passed through the creatures, they all began squawking simultaneously and beat their powerful flippers, even some of their own number fell victim to these blows.

The brawl had become general among this little pack of Antarctic hooligans, and I couldn't help but laugh at their naivety. For them it was more entertaining to fight each other rather than make a concerted defence against a common enemy.

My hilarity was short-lived. I was in the middle of that hullabaloo. Two of the penguins that seemed to be the most enraged by the fray, came up to my feet. I could not move freely as I was still wearing my ski.

Suddenly, one of the penguins seemed to realize that perhaps it was me who was responsible for all the fuss. He fastened on to my windproof trousers with his beak, and gave my legs a powerful beating with his flippers.

I had no idea that their flippers were so strong, it felt as if I were being whacked with a piece of wood.

After a while they calmed down slightly and separated.

Two of them started back to the rookery whence they had come, and then a third one joined them; then the whole gang followed suit.

As my intention was to reach their nesting site, I tried to keep up with them as they tobogganed along rapidly.

They travelled at quite a speed, in an 'Indian file' formation.

My description of the penguins' mental ingenuousness might possibly give rise to some sceptical comments; however anyone who has ever spent time among these flightless birds, even a single season, will be able to confirm that their normal docility and lack of fear are products of their instinctive curiosity.

53: It is not always Wise to be First

It must have been a very singular group we comprised. At times I quickened my pace to catch up with the penguins in the vanguard, which I reckoned were tobogganing on the sea ice at seven kilometres an hour. We reached the coast at Point Martin where the penguins stayed on the beach.

I ascended the ice-foot over hardened snow and took off my ski to continue my investigations on foot.

Over the entire crest of Point Martin, there would have been about a hundred Adélie penguins scattered around. They were engaged in the task of selecting appropriate sites to make their nests.

Giant Petrels, accompanied by a few Cape Petrels, were flying overhead constantly.

That long tongue of land which jutted into the frozen sea was still uninhabited. Its appearance made quite a contrast to what it must have looked like in summer when, according to Valentiner, this entire area is covered by penguins.

I noticed some of them lying down on the hard surface with their abdomens resting on the cold snow and ice. When I passed close to one, it stood up, leaving an oval indentation in the snow, formed by the heat of its body. This impression constituted its right to occupy that spot, where, at a later stage, it would build a nest.

That is why the few penguins which had arrived early at the rookery, were busy selecting the most desirable spots for their nests either on higher ground or places protruding above the ice. These were the best locations where they would be able to incubate their eggs away from the water from the coming spring thaw.

Nothing less than death would have made a penguin relinquish the spot it chose. As couples they took it in turns to look after the nest, each time the male, or the female, made the long trek to the open sea to feed.

There was not much else to see in the rookery and as it was already getting late, I went back to the shore, put on my ski and began the return trek following a direct course towards the head of the bay.

While I was skiing along I was thinking about the arrival of the first penguins at the rookery. The entire sea was ice-covered; to the south it extended to the horizon, to the north it covered Uruguay Cove and for many kilometres beyond.

In the vicinity of base, we had not observed a single penguin coming from the north. Consequently the only practicable route by which they were reaching the rookery would be from the south.

I halted, turned round to look southward, and could see that there was indeed another group of penguins arriving from Cape Murdoch making directly for Point Martin.

There remained no doubt: these creatures were coming from Washington Strait, where there were already areas of broken ice; then, from Cape Murdoch onwards they reached the frozen bay and arrived at the rookery by tobogganing over the fast ice.

That was an enormous traverse across the sea and the ice, but because penguins are better swimmers than hikers, it was the most appropriate route for them to take.

On my departure from the rookery a group of penguins broke off and came after me, tobogganing along my ski tracks. I was not surprised that they followed me; I knew that once the bay north of base was clear of ice it was where they would be heading because that would become the shortest route between the rookery and open sea.

At the moment, however, the route northwards from The Isthmus would still involve a traverse of many kilometres before the penguins would be able to dive to find their indispensable crustacean food. Nonetheless, these were the first penguins to pioneer this route; later it would become the one preferred by them all.

I continued advancing and several times looked back to see if they still followed me. Gradually I was getting close to my destination and although I was extremely thirsty and weary, I hurried on to have the pleasure of giving my companions the good news of the arrival of the first penguins at the rookery, and perhaps also the first in the general area of base.

At last I rounded Point Diebel, and the base buildings were in sight.

I stopped for a few seconds to look at my watch. It said five p.m. which meant I had been outside for nearly twelve hours, enduring temperatures of below minus twenty-five degrees. I was well-pleased with my trek; even the weather had been relatively good during that rewarding excursion.

I saw someone come out of the dwelling holding a meteorological notebook. He saw me and, as we were still five hundred metres apart, we waved to each other. He stood still for a few moments and was looking closely at the area around me; I guessed that it was the penguins that caught his attention, they were still following and had now caught up with me.

I continued on my way, reached the coast, and finally went slowly up the beach slope, still followed by the penguins. Finally, after walking on sea ice for twelve hours, I was stepping on solid ground again.

I was a few steps from the base, when I saw Valentiner appear, followed by Bruhns and Plagge; the latter was carrying a bludgeon. I immediately had a mental image of what was about to happen, and when I was about to say something, Valentiner said:

'Hello Kleine … thanks for bringing us fresh penguin meat.'

In fact that was exactly what happened. Plagge ran from one side to the other, and four penguins ended up in a pool of blood on the ground.

Those poor creatures had been up from dawn to take possession of the best sites in the rookery, but the unfortunate result proved an early start was not always for the best.

43: Collecting penguin eggs (photographed by Luciano Valette, from: Bruce, 1907)

54: Reflections

The next day, during lunch, we were discussing what I had found on my recent excursion and particularly the details of the arrival of the first penguins at the rookery.

'This is very important,' Valentiner said, 'We will have to prepare for the annual egg-collecting trips.'

'That's right,' I said, 'I have noticed that during the whole year since our arrival, we have been eating last year's penguin eggs from the larder.'

'Those are the ones left by the previous winter party. Last November, four thousand eggs were gathered. They had used approximately one thousand before we arrived, thus three thousand were left for us.'

'So now we should do the same for those who are coming to relieve us?' I asked.

Valentiner nodded, and while he filled his customary after-dinner pipe, he said:

'We will collect the eggs during the early days of November, preferably before the 10th of the month, as we must ensure we gather the freshest eggs. You two, Plagge and Kleine, must take it turns to make a daily visit to the rookery from 1 November, to monitor closely when the first eggs are laid.'

'It would be a good opportunity,' I suggested 'to start taking the empty fuel crates to the rookery'.

'Good idea', agreed Valentiner. 'Having the sea frozen makes that job easy, but there have been times when the ferrying of the crates had to be done by boat, because the fast ice broke out before November. The boat journeys can be quite hazardous, storms may take you by surprise in the middle of the bay.'

'Coming back to the egg-collecting,' I said, 'is there such a vast quantity of penguins in the rookery that we can, with impunity, loot four or five thousand eggs?'

'You will see,' Plagge replied 'there are thousands of penguins; what we take won't make the slightest dent in their numbers'[223].

'How many eggs do they lay in each nest?'

'Only two -.'

I inquired further, 'when a couple lose their eggs, do they lay again to replace them?'

Valentiner and Plagge, the two veterans who could have answered, looked at each other enquiringly, but seemed not to know the answer.

Valentiner said to me:

'In the four years that I have been here, it has never occurred to me to observe this detail. It would be interesting to know'[224].

'That's right,' I replied 'even though our survival takes precedence over any sentimentality, it would be interesting to know if Nature has endowed these creatures with the capacity to replace the eggs that we take so inconsiderately.'

223. Estimates in 1993 show more than 35 000 pairs of penguins bred at Point Martin: Woehler, 1993.

224. Later observations show the answer is yes; they usually lay another egg. A bad breeding season, with poor weather or scarce food, or both, might adversely affect this, as well as the general laying in the rookery.

I decided to investigate this as soon as I had the opportunity, but to make the study I still had to wait almost two months.

We continued our conversation and while we discussed the seals I had seen, I asked if their meat was suitable to eat.

'I have never tried it,' Valentiner told us, 'I imagine it must be disgusting.'

'However,' I pointed out, 'I have read accounts of all the Antarctic expeditions, and the explorers fed entirely on seal meat, eschewing penguin which they deemed unpalatable.'

Bruhns and Plagge agreed, but Valentiner said:

'I have also read those narratives, but I have never had the opportunity to acquire a taste for seal meat. If you like, you can do so at any time, as it would do no harm to know about this alternative food. However, we all agree that penguin meat, well-seasoned, is edible; you are not fed up with eating it yet, are you?'

'No,' I replied 'though I certainly miss a good juicy steak with a fresh green salad, but I can forget about this after eating one of Otto's superb penguin *milanesas*.'

'You are right,' said Plagge 'although I am not *criollo* like you, I like a good beef steak as much as you or your compatriots. As for the green salad, I have to confess I feel a huge nostalgia to be able to bite just a single fresh juicy lettuce leaf. How I miss the colour green!'

'Not just you,' I said 'I too miss it, I think we all do, and rightly so. Nine months without seeing a single plant, a single leaf. Yes, not even a single leaf of the commonest garden plant.'

In the Antarctic the plant kingdom does not exist, or rather its presence is barely perceptible. Occasionally, between the rocks, a stalk or two would appear, sickly green, more white than green, lacking chlorophyll, malnourished by the absence of sunlight during the long dark winter nights.

Those stalks, which never reach five centimetres long and a couple of millimetres thick, are very hard, and they do not have leaves of any sort. They are just primitive lichens; that with a few clumps of moss, dark-brown or grey-black, form the only vegetation in existence among the fragmented rocks in the hills[225].

55: Moral Turmoil

Days went by, but not with the desired speed. They passed slowly, one by one. The leaves were torn from the calendar daily, and its thickness was reduced, slowly, very slowly.

Many times I dreamt that, with a single tug, one movement, I was tearing off all the remaining leaves of the calendar, drawing back in a second the curtain of the four months still to go before we could leave. That was just a dream; I had to get back to reality and confine myself to summoning to my aid all the patience and philosophy needed to control this powerful impulse.

I realized that my work alone; elucidating certain aspects of the lives of penguins, seals, and other polar denizens; studying, writing and reading; could provide some distraction from brooding on our forthcoming liberation from that hostile land, of our return to life, to the world, to humanity ···.

However ··· would they come to fetch us? Would the ship coming to rescue us from that desolate prison manage to get through the ice to reach us? Would she ever leave Buenos Aires? What if there had been some catastrophe in the world, something which would condemn us to live here forever, to die here, imprisoned in the mute, white, polar ice?

Did the world still exist? Were there still people out there?

It was terrible to allow myself to drift into thoughts of such a nature, and invariably when I found myself brooding on such sentimental pessimism, I summoned all my will-power to drive myself outside. I would put on my ski, go out and hurtle dizzily down any of the slopes. Only with the help of ski would these weird thoughts be forgotten.

That total isolation was very difficult to endure, and more so, for not having any outlet for alleviating the grievances of the soul.

Only four men of very different temperaments were at my side. Work mates, yes, till the end ··· but friends? That was quite different.

225. See the footnote on vegetation in § 16.

I had to look for distractions. Invent them if possible. I had to busy myself with something, anything to avoid becoming prey to brooding thoughts. It was much better like that; but sometimes, despite myself, there was no remedy, I would still find my thoughts turning north, far north.

'The South Orkneys are for those who are strong-willed, and have greatness of spirit.' I would say to myself. To chasten my body, as opposed to my mind, I would deliberately face into the wind; for by now I was impervious to the physical pain. The sharp ice needles that stung and encrusted my face no longer hurt me.

The same must be happening to my companions. Our conversations around the table were dwindling and we seemed to have exhausted every topic of discussion.

From time to time we would fall out with one and other; one trivial comment, a trifling mannerism, would be enough to cause offence. Sometimes I too would take part in these arguments. After a while the tone of voice would rise. Then, someone would realize it was useless to flare up, and the conversation would return to normal.

However the other person would just carry on in the same vein.

Sometimes I would watch them and say to myself:

'The first one to get up and leave, will be the real winner, it is he who will have returned to reality.'

To argue in the South Orkneys was futile. What for? Everyone there was right.

Those scenes and arguments, even the most trivial, owed their origin not to any failure to comprehend the nature of the antagonism, but to the mere fact of that solitary lifestyle, so hard, so unusual.

Our feelings became exasperated; and sadly this could lead to aggressiveness. You had to be understanding, and then anybody with a modicum of decency would not fail to apologise, and the matter could be forgotten ··· well for a day or two anyway. Until the involuntary omission of a simple 'Good morning' would be sufficient to reignite a smouldering temper.

The South Orkneys no longer held any fear for us; the ice had been vanquished; even the fauna and flora no longer held any major secrets.

Now, after having run two thirds of the course here, came the hardest part: knowing how to wait.

Awaiting the moment, to be the first to sight the relief ship and shout:

'At last! It's here … they've arrived! They've arrived!'

It takes a huge strength of character to restrain your frustration.

It is impossible not to drift into thinking or saying: 'six, seven months still to go … who knows? Perhaps more!'

'Will they come? … Yes, they will, … but will they really?' and so on until the end.

56: The Saga of Four Bars of Soap

We were in the latter part of October. From the beginning of the month, Valentiner and Bruhns had been tabulating their magnetic observations to complete the annual programme. This took them a large part of the day; thus Plagge and I had to take turns to make most of the meteorological observations.

October is very important in the South Orkneys. Apart from completing the magnetic programme, it is the month in which one has to start thinking about things such as reports, samples, equipment, etc. that we will have to take back on the ship when we leave. Then the first fortnight of November will be needed for collecting penguin eggs. December is entirely taken up by repair and general maintenance of all the buildings and other installations.

We could expect the relief ship to arrive towards the end of January. Thus we had three and a half months left. Still plenty of time, but Bruhns and I did not see it that way, we spent every moment of our spare time preparing our baggage and packing personal effects.

I have mentioned in previous chapters the personal issue of clothing which each man received. Whether that clothing lasts or wears out depends on the care it has received during the year. In my case, I still had a few unworn clothes, absolutely brand new. I also had items other than clothing which, apart from occasional personal use and consumption, I had carefully kept, to allow for any unforeseen occurrences.

Among all the supplies issued to us was a small amount of toilet soap, of the better type generally used for washing face and hands.

Personally I had no interest in saving any of this perfumed soap. I had used it all during the first few months, then, when I had none left, I simply washed with yellow kitchen soap; the lack of perfume did not bother me.

One morning I was in my room arranging things in my trunk, Bruhns was watching me and asked:

'Have you used all your perfumed soap?'

'Yes, I finished it all some time ago and I have been using yellow soap.'

'I have some left,' Bruhns said 'and I can let you have some if you want it.'

'Thanks, but I do not think I will need any.'

'Perhaps so, but for the return voyage don't you think it might be more pleasant to have some decent soap to wash with, rather than the stuff you are using?'

'No doubt', I replied.

Actually I should have kept one bar in reserve, because if we were relieved by whale-catcher I would not be able to get any more until we reached South Georgia where there was the whalers' shop, the 'slop chest'[226].

I was beginning to get interested in the subject, so asked Bruhns:

'Do you have many bars of soap left?'

'Only two, but I can let you have one.'

I was reluctant to accept it, as I knew he had a predilection for using that soap, and disliked the ordinary one, whereas that did not bother me at all.

'Thank you again,' I replied 'but I think Otto and Plagge have their supply intact, so I may well be able to get a couple of bars from them.'

'Good!' – exclaimed Bruhns 'that's right, I had forgotten'. Offer Otto some whisky in exchange for a carton of soap, he won't mind at all making a swap.'

226. The 'slop chest' is a type of general shop at a whaling station, which sells anything a whaler might want or need (except alcohol).

I really liked the idea, and without losing any time, went straight to the kitchen, where I found Otto stirring the steaming hot saucepans on the stove.

'Otto,' I said 'have you still some of that nice toilet soap left, you know that perfumed stuff we were given with the rest of our kit?'

'Oh yes, I have it all, I have not used any.'

'What, none? …'

'That's right, I have not used any as I use the other one.'

'Well, I propose some business here. If you give me four bars of soap, a carton, I will give you half a bottle of whisky.'

Otto, who was at that moment busy mincing meat and getting a meal ready, looked at me in astonishment, and in disbelief at my offer, he asked:

'Really? Honest?'

'I give you my word', I replied.

To make sure he really understood I added:

'That's right, half a bottle of whisky in exchange for four bars of the good soap. I give you the whisky, you give me the soap.'

Otto was delighted and said:

'Very well, I accept … but you will have to forgive me I cannot do it today, I will have to do it tomorrow, as I have to look for the soap among my things, I cannot remember exactly where I put it.'

'That's fine,' I laughed 'when you find it let me know, and you will receive the agreed payment.'

'Done', I had just rounded the business deal, and was feeling pleased with myself. Overall, I had no need for any more whisky. A nice bar of perfumed soap to help soften my calloused hands was much more to the point.

I went back to my bunk-room and continued organizing my kit. As I went through the parlour, I passed Bruhns heading for the kitchen which I had just left.

Little did I imagine that those four insignificant bars of soap would later lead to a physically violent confrontation between us, which, because of the continual state of nervous tension in which we all lived, had the potential for a serious melodrama.

57: VIOLENCE ···

While I continued sorting out my clothes, putting socks, vests, and underwear in their proper places, I overhead an animated conversation going on in the kitchen between Bruhns and Otto. I listened instinctively, but when I paid closer attention I realized they were speaking in German.

At first I did not pay much attention, but, as I recollected their words, the significance began to strike me.

'··· *Seife ja, ja ··· mehr Whisky ··· ? was sagt der Kleine*? [··· soap, yes, yes ··· more whisky ? what did Kleine say?]'

'··· *eine Flasche* ··· [··· one bottle]'

'··· *ja, glaube ich habe vier Schachteln* ··· [··· yes, I believe I have four cartons]'

While I listened to these words, I thought my companion wanted to make the same business arrangement I had just made. Their voices, from the kitchen came falteringly, I listened more attentively while continuing what I was doing.

'··· *verstanden ··· eine Flasche* ··· [··· understood ··· one bottle ···]'

'··· *vier Schachteln* ··· [··· four cartons ···]'

'··· *aber sagen Sie nichts!* [··· but do not say anything!]'

After a while, Bruhns came back to our room and sat on the corner of the bunk and looked at me with a smile, a slightly mocking look.

I thought I understood. I only had one doubt.

'Possibly he wants to be a nuisance. He is capable of having arranged a deal with Otto to give him all the bars of soap.'

I was not sure, and could not confirm my suspicions.

I should wait a few days, and then I would know how far I should go. The soap didn't interest me in the slightest, but at that moment of wavering when I began to cast doubts upon the irreproachability of my companion's behaviour, they began to interest me much more.

I could not allow him to do such a thing, and treat me like a child.

The more I thought about it, the more indignant I felt. In one impulsive moment I thought of returning to the kitchen to question Otto, but then I looked at Bruhns who was calmly reading a book, so I refrained.

'It is not possible … I told myself, I must be wrong, I must wait and see ….'

Doubt was already assailing me. How was it possible he would do such a thing?

Even Bruhns himself had offered me a bar of his soap.

It made no difference, I continued to suspect that I was to be their victim and they were planning to cheat me. My thoughts were flying and my indignation increased, I was getting more worked up by the minute, my hands clenched tight in nervous agitation.

'Calm down,' I said, 'I must be mistaken. Ah well, we shall see ….'

From the room I could see Otto laying the table for lunch. Bruhns left the bunk-room and sitting in the parlour, near the table and the stove, continued reading his book.

I was watching him through the partly open door. Everything in his bearing denoted perfect tranquillity and while Otto passed close to him and laid the table, they did not exchange a single glance that might have indicated any collusion against me.

Despite all this, I was still seething within. I did not want my thoughts to betray me, and I made sure my movements indicated perfect composure.

Nevertheless, I had my doubts. Bruhns could no longer be my friend. I had lost all confidence in him. Despite this I resolved to feign complete ignorance of the overheard conversation between him and Otto, and particularly what he was expecting to get from it. After all, I could no longer continue to be nice to this man who, in my view, did not even deserve to be spoken to.

You need to be down in the Antarctic to experience these situations. If I had reasoned this through with a cool head, I should have realized that I was making a huge drama out of four insignificant little bars of soap. However in the South Orkneys, in that terrible loneliness, one's spirit, oppressed by the total lack of variation in that clockwork routine we led, gave way to angry indignation over small matters. I gave this episode a violent interpretation that it did not have, culminating in a frenzied outburst of righteous indignation.

Otto gave the call for lunch, and as I had finished the work I was doing, I washed my hands in the basin in the corner of the bunk-room. I was drying them when Bruhns came in to do the same. To help each other with our basic facilities, we had a routine of always emptying the dirty water from the basin after use, thus leaving it clean and ready for the next man. On this occasion I deliberately left my dirty water in the basin instead of throwing it into the bucket, and left the room without looking at Bruhns.

I sat at my place at the table and was soon joined by Valentiner and Plagge. Bruhns appeared and sat at his place opposite me.

He picked up his napkin and looked at me coldly. I held his gaze. He seemed to understand me.

My attitude in not emptying the basin seemed to make him reflect.

He ate slowly, without talking to anyone and from time to time, I felt his gaze on me; watching me.

Meanwhile I talked to Plagge and Valentiner; Bruhns took part in the conversation with them, but not with me. We knew each other so well, we only had to look at each other to understand what the other was feeling, and we did not say one word to each other.

Hostilities had commenced! War had been declared! ….

58: Fight! …

Three days went by, and Bruhns and I did not exchange a single word. Without looking at each other, we understood what the other was thinking. In the bunk-room our life in common continued as normal, everything followed the formal established routine, as well as the informal one agreed among ourselves. In the scientific programmes, where we were duty-bound to work together, we also had very little to do with each other. There too our labours conformed to the set routine which had to be followed throughout the year. Any deviation from this routine would, if necessary, be properly sanctioned by our Leader, from whom we received direct orders.

The rancour between us did not upset the communal routine of the Observatory, it merely translated itself into a reciprocal silence, indicative to all outward appearances, of our mutual indifference to each other's company.

For my part I lived in a state of increasing nervous tension. My fury had increased considerably and the inevitable chance encounters with my companion, inside our shared bunk-room, became intolerable.

The third night, after dinner, Bruhns retired to bed, as he had to get up as usual, at four thirty in the morning. He closed the bunk-room door and I thought he was undressing.

I remained with Plagge in the parlour, and we played a game of chess next to the warm stove.

A few minutes later I saw Otto go past, he was heading towards my bunk-room from the kitchen with several packets in his hands.

I jumped out of my seat and said:

'Wait a minute! ··· What have you got there? ··· Where are you going? ···'

'They are four bars of soap for Sr Bruhns.'

'But,' I said: 'what about mine?'

'I am sorry *Herr Kleine*, but I do not have any more.'

He pointed at the bunk-room door and said:

'He asked me for them ··· and you know ···.'

I was furious and had to dig my nails into the palms of my hands, I could hardly contain myself.

Plagge asked:

'What is wrong with you, what on Earth is happening?'

'No ··· no, nothing, nothing. Let's carry on.'

'Look,' said Plagge, 'if it's about soap, I have several boxes, and I can give you whatever you need.'

'Thank you, I am grateful. No it's not that.'

'Yes, man, I can give you two boxes ···.'

'I repeat, I will not accept them. Please let's carry on with the game.'

'As you like.'

I could not carry on playing. My brain was going round in giddy circles, and it was impossible to concentrate on the game.

Plagge won, and then he went to bed.

I remained alone.

I had to record observations until four in the morning, then at a quarter past four I had to make that inevitable visit to the Variometer

Hut. On the way back, at four thirty I had to wake Bruhns, and then I was free to go to bed and sleep until midday.

I could not do any work that night. Every hour I automatically got up to make the meteorological observations; then I returned to my chair and to my thoughts.

No! ··· things cannot go on like this. At the same time, I asked myself is it worth making all this fuss over four miserable bars of soap ? No ; it was so petty. Was I still so childish, that I did not realize how pathetic it was to get so worked up over something so trivial. In effect it was just so stupid. Let's not give it another thought. What! - not think any more about it? What about all my companions? What would they say when they found out. They could hardly not find out, that Bruhns had diddled me out of the bars of soap. Just because he had the advantage of their own language to persuade Otto to make a better deal.

All night long these thoughts went through my head.

Four a.m. came. Then four thirty. I should wake Bruhns ··· but first I had to prepare myself. I had a plan ···.

Stealthily I opened the door and went into the bunk-room. There was a dim light inside, and on the long table facing the bunks, I could see the carton of soap.

I took out my sleeping clothes, a warm vest and long johns, and changed next to the warmth of the stove.

When I finished, I went back into the bunk-room I put my clothes on a hanger, lit the paraffin lamp, went over to Bruhns and shook him to wake him up.

'Thanks,' he said 'what's the time?'

'It is twenty to five', I replied.

I had pulled up my sleeves. I was tempted to do nothing at all, I felt like jumping under my blankets. However I could not contain myself.

I stood in front of his bunk and, in an imperious tone, said :

'Tell me, those bars of soap! ···'

'They are mine!'

'You liar!' I shouted at him 'You have no right to them!'

'That is not true!' he exclaimed as he sat up in bed.

The argument increased in tone, and I wedged myself against his bunk. I was seeing red and the shouting became infernal.

Bruhns wanted to get out of bed, but I would not let him, he could not move freely in that restricted space.

He gave me a shove, but I sprang back next to him using my fists to advantage.

My antagonist struggled to get out of bed, but I would not allow him.

It was an unequal struggle, but in our present positions our strengths were well-matched.

Suddenly the door opened and Valentiner appeared behind me, grabbed me brusquely and threw me outside.

Great was my surprise when I heard him burst into a loud guffaw. Splitting his sides laughing he told me:

'Calm down, let's have a couple of drinks and then you can go to bed. Don't make an issue of it. Things like that are always happening here.'

'Yes, but Bruhns?' I asked.

'He will get over it. More whisky? Fine, cheers!'

59: Reconciliation

I felt a prisoner of extreme nervous tension. I was sitting with Valentiner in front of the fire, expecting Bruhns to come out of the bunk-room at any moment. I was afraid neither of his bulk nor the hardness of his fists. I only longed for the moment he would appear through the door. I was just curious to see what attitude he would adopt. If he started any trouble, I had no qualms about hitting back.

Despite Valentiner's conciliatory attitude, I wanted, and was inclined, to go on fighting, although I knew that I would be worse off, we did not have equal strength.

Valentiner well understood my attitude of just waiting for a confrontation; he put his hand on my shoulder and said:

'Calm down lad for God's sake. Just forget these fleeting grudges, and when Bruhns appears, go to bed and sleep.'

'Truly it is ridiculous,' I said, 'over four bars of soap ….'

44: The dwelling in the silence of a winter night

'Ah!' – Valentiner said: 'it is because of the soap ?' and he let out another peal of laughter.

At the same time the door opened and Bruhns appeared, all dressed and ready for work. He did not look at me and went straight over to the desk where the notebooks for the meteorological observations were kept.

'You go off to bed,' Valentiner said, 'as far as I am concerned the matter is forgotten ....'

I stood, went into the bunk-room and climbed into bed. It took me ages to get to sleep, my head was spinning thinking about all that had happened, yet I tried to reflect on it calmly. After a while I must have dropped off.

The next morning I was awakened by a hand shaking my body. I sat up in bed, and caught sight of whom else but Bruhns. Before I could recover from my surprise and speak, he said:

'Good morning!' – 'Good morning', I replied.

'Kleine,' he said, 'I need to speak to you. I have awakened you a little earlier, it is eleven a.m., as we are alone in the dwelling; everyone else has gone out.'

I sat up in bed, while Bruhns, standing next to me, looked at me attentively.

'Fine,' I said; 'but first of all I want to apologise for last night.'

'That is exactly what I want to talk to you about. It is I who needs to apologise.'

'No Bruhns, let's forget all that happened. It is painful and regrettable that things like that should happen amongst us ….'

'That's right,' he interrupted 'but you know that here in the South Orkneys many things happen that make us lose our tempers, even the strongest of us. Let's be friends again, companions, and again I ask you not to harbour any ill feelings.'

'Delighted; believe me, I am truly sorry about the whole thing. I too am well aware of the constraints of this way of life of ours, I can't forgive myself for a momentary loss of self-control.'

'It was my fault, believe me Kleine, for not having told you that I wanted to give you a surprise present of two cartons of soap!'

I was completely amazed and only managed to utter:

'I believe you … I know you are really generous.'

Bruhns continued, 'I realize I am a bit rough, I'm a bit older than you, but, although I out do you in years, my temper often gets the better of me. When you left the dirty water in the basin, I realized you were annoyed; although I had no intention of becoming your enemy, I couldn't help fanning the flames of mutual antagonism, and I in turn became angry.'

'Well, well,' I replied 'I do not need any explanations, I understand everything: let's forget it and pretend nothing happened.'

'Delighted, give me your hand.'

We continued chatting amicably and suddenly he asked me:

'Tell me Kleine, when Valentiner came into the room last night and went back with you to the parlour, what did you talk about? Did he say anything?'

'Nothing important.'

'But did he not make any comment about our fight?'

'None, absolutely nothing. We had a couple of drinks and then he sent me to bed.'

'Yes, possibly … but, afterwards he laughed … didn't he?'

'That yes,' I replied 'He laughed a lot and I think all the others would have laughed as well.'

'Certainly, they would have all laughed at us.'

'Surely ··· but, what is wrong with that?'

'That's the point,' Bruhns exclaimed uncomfortably, 'they did laugh at us!'

'Tut tut!' – I said 'you should not take it so much to heart. You have to realize we behaved like children; let them have their laugh. You know the secret of living here, is to be able to laugh at someone without his realizing that we are laughing at him.'

He reflected on this and seemed to calm down, and, while I dressed, we continued having a cordial conversation.

Suddenly he seemed to remember something; he went to the table and handed me two cartons of soap saying.

'Take these Kleine, please accept them.'

'Ah! – that no! No way.'

'I beg you, please accept them.'

'No. You keep them and when we leave here, then I'd be grateful if you would let me have just one bar.'

'Very well then, but remember you may have as many as you like at any time. I will leave them here.'

'Accepted. Now that we are friends again, do me a favour.'

'Of course! ··· What?'

'Bring me a jug of water from the kitchen ···.'

He left and came back a few minutes later with a jug of warm water. I put some in the basin and began washing with my modest yellow soap. When Bruhns saw me, he declared:

'No, please don't use that one ··· take this one, it is better.'

Then he handed me a bar of perfumed soap.

'Well, thank you!' I said laughing while taking the proffered bar.

He also laughed uproariously and, to close the matter, said:

'Use all you want, here are the boxes, but do not waste it ···. Otto has no more left!'

60: Break-up of the Sea Ice

October drew to an end. The programme of magnetic observations to determine the coefficients and calibrate the instruments was the next to complete. The temperature had risen slowly and the snow which had covered everything during the winter, according to the poles we used to measure it, had decreased to half its depth.

From my periodic excursions to Mount Lola to observe the state of the sea ice and its movements influenced by winds and marine currents, I could confirm that the area of open water to the north was enlarging.

Several times, however, winds from the east, strengthened and reached hurricane force, lashing us furiously for a whole week. Then, when I went back up the hill after the blow, it was a bitter disappointment to see that the pack had again returned, almost unbroken from one horizon to the other.

Later the winds would blow from the south-west or the west, and clear all this pack away, leaving the sea completely open to the north beyond Saddle Island.

Enormous swells and violent waves from the open ocean would penetrate beneath the pack ice, reaching Uruguay Cove, and causing an undulating motion in the whole ice field. This would eventually cause it all to split and break.

In the last few days of October, the swell from a severe north-westerly gale broke the last of the fast ice of Uruguay Cove, and a breeze from the south cleared it all seawards, leaving the bay completely free of ice.

Summer was returning.

The thermometers outside in the meteorological screen did not drop below minus fifteen degrees, and daily we noticed a slight increase in temperature.

Despite the prevailing cold, the snow continued to diminish in depth and this phenomenon, inexplicable at first sight, was due to sublimation as a consequence of the dry Antarctic winds, which blow continuously. That is why, even though the temperature did not go above zero degrees centigrade, the snow and the ice diminished in volume, and therefore in depth.

Everything indicated a change in season, but occasionally we had a few fleeting reversals, which usually occurred with winds from the south. The temperature would drop steeply; it would snow heavily and, from higher ground, you could see the pack ice coming in and covering the sea again.

'If this carries on,' I said to Valentiner, 'the ship will not be able to get in when they try to relieve us'.

'This is looking quite bad,' he replied, 'but don't get too upset, there is still plenty of time left'.

'Have they always been able to relieve the expeditions?' I asked.

'They have never failed to do so, those who come south for that purpose are well aware of the responsibility of their mission. Sometimes they have encountered a lot of ice to the north and then it is a matter of waiting a few days, doing a few circles at sea until the ice disperses and they can get closer to the land. In the event that the approach from the north is blocked, it becomes a question of seeking an alternative passage through either of the straits, or making an eastbound or westbound circumnavigation of the archipelago attempting to find a way into Scotia Bay.'

'What if this is also frozen?'

'In that case, there is no other choice than to disembark on the sea ice and transport everything ashore with the sledges.'

Indeed, that idea did not look very agreeable, and, after thinking about it, he would say to me:

'What a prospect! – but if there is no other solution! ...'

Days went by, the south wind dropped and veered, and then it began to blow from the north. The temperature went up and Uruguay Cove filled with pack ice to the horizon.

To the south, dark reflections appeared on the underside of the white cloud ceiling; and when we saw them we exclaimed happily:

'Open water in the south, very distant, but open water at last!'

The huge area of sea ice was on the move and that was what interested us. Soon the wind would do the rest, sweeping the ice floes far from land, whether north or south, it did not matter, the only thing we longed for was for it to go, clear off, go as far away as possible from the archipelago on which we were trapped.

One day, at the end of October, the wind blew from the north at

45 : Ice stalactites on the eaves of the dwelling from snow melting from the roof

ninety kilometres an hour. The temperature had been increasing from the early hours in the morning, and while we were sitting around the table after dinner, Plagge came in with a big smile on his face.

'At last it's summer!' – he exclaimed, 'the temperature has risen above zero!'

'What's the air temperature?' Valentiner inquired.

'It is one degree above zero and the snow is melting rapidly.'

'That's very good', the Leader said, and looking at me continued, 'How many days since your last visit to the penguin rookery?'

'I was there five days ago, but I still did not find anything ….'

'Yes, that was still a bit early, but now we have to go every day. If you can go tomorrow, do not forget to have a good look at the penguins, especially the ones sitting on their nests.'

'Are the crates all over there?' Bruhns asked.

'Kleine and I took twenty-five crates over, I reckon that should be enough', Plagge said.

'Good,' said Valentiner 'I can't wait for some delicious fresh fried eggs.'

46: The grave of Otto Diebel with two Chinstrap Penguins

61: They were Five Men

The South Orkneys cemetery[227] …. Forlorn aspect of four rough wooden crosses painted black, some names and dates carved with a knife. Anyone who knows this cemetery will have experienced those first impressions of fear and anguish before the solemn presence of the unknown, of death pervading the white wilderness, evoked by the presence of those isolated crosses.

Whichever way you look, from the windows of the dwelling, from the door, from the meteorological screen, from the beach, the bays, the hills, from everywhere those crosses are visible. They emerge from the immaculate snow, like solitary perennial sentinels, indicating the presence of beings who once breathed life, and are now doomed to rest there for eternity.

It is a melancholy sight that instils deep respect. The urge to escape from their presence is futile, you cannot avoid looking at them as you pass. With time you become accustomed to their presence and towards the end of the year we went to read the names of those who lay between the rocks and ice, covered by a white shroud of eternal snow.

There are four crosses, Bache-Wiig's one is absent; it was placed on top of Mount Lola, in the vicinity of where he fell to his death.

The first fatality was Allan George Ramsay, on 3 August 1903. He was Chief Engineer of *Scotia* which wintered during that year. It was this expedition which led to the establishment of the current observatory.

He was followed by Otto Diebel, the officer in charge of the 1905 winter party. He died as a result of an illness contracted after spending a night in a blizzard at the summit of Mount Mossman making observations of some sort.

On 20 August 1910 Jon Eliesen died, victim of a heart attack while getting over an illness of several weeks duration. It was only when that expedition was relieved and had returned to Buenos Aires that it emerged that Eliesen had suffered heart symptoms before his

227. The graves are oriented meridionally; this began with Ramsay's interment in 1903, his 'head to the southward, his face looking to the sun, northward and homeward' (Bruce, 1992, page 134.).

departure, which were quickly exacerbated during six months on that faraway archipelago.

In 1913, on 7 May, Harald Wiström died[228]. For a month he had suffered an acute attack of appendicitis which all the ice and snow of the polar world could not have alleviated.

Following the chronological order of the men who lay there I mention here the date 30 April 1915 on which Bache-Wiig was last seen, as related in chapter 36.

Five men, whose lives were extinguished, far away from civilization and far from a normal way of life. The last glimpse they had was of that desolate world of whiteness, stretching to infinity, which could not have seemed so very different from the world of eternity they were entering[229].

Deep emotion and sadness overcame me every time I went near the cemetery, and I could never stop thinking of their mothers, of their wives, of the children of those brave men who, in the name of science or looking for an ideal, had the fortitude to face the wild Antarctic seas.

★ ★ ★ [230]

It was indeed a youthful bunch resting in that place. Eliesen, Wiström, Bache-Wiig, perished at 27, 30, and 29 years of age, testimony to the fleeting passage of their life on Earth. At that age you are always young, the whole world lies before you.

Perhaps you are tempted to ask yourself 'Will I survive this year … ?'

When you are young you do not reflect, you tend to go after the adventure. You do not gauge the possible consequences of a burst of boldness, and sometimes the inevitable happens; nature wins.

Five stayed on ….

228. Obituary in: Andersson, 1914

229. After Moneta's time other men from Orcadas have been lost, the total in 2017 was ten on Laurie Island and six who are buried on Signy Island; all are listed in § 5 and footnotes.

230. Three asterisks appear several times in the text of every edition of the book, these do not indicate an ellipsis of any text between editions.

47: The cemetery in summer, 1929; left to right, graves of Escobar, Wiström, Eliesen, Ramsay, and Diebel

62: All Souls' Day[231]

A date always remembered throughout the world. I too wanted to pay my humble respects to those who rested under the ice. I took advantage of some free time that afternoon to go to the cemetery. I took some tools and began walking in silence between the graves. With love I lost myself in the task of repainting them black to improve their preservation. I then replaced some rocks which had fallen from the graves. I wandered round, putting them back in place one by one, buttressing rocks that were loose and finding others to hold them securely in place.

I walked to the neighbouring hills and looked between the rocks for some of the diminutive lichens, the ones with some tiny blue flowers[232].

I placed the plants on the graves and while I was doing this, some

231. All Souls' Day is 2 November; the traditional day for remembering the dead and tending their graves.

232. Lichens are primitive vegetation, they do not produce flowers but some genera, such as *Cladonia*, have green-blue podetia that might look a little similar to a flower bud.

penguins appeared wandering up from the beach.

They stood next to me as if inquiring the reason for my presence, then two of them climbed on to the grave and lay down to rest near the base of the black immutable cross.

I felt no desire to chase them off; to my way of thinking they were not guilty of the least desecration.

Valiant men, those Antarctic explorers entombed beneath the polar ice; white mountains of rock-hard ice, frozen seas, penguins under their crosses ….

It made a solemn picture as, balaclava in hand, in the immensity of that polar wilderness, I looked at the crosses; tears came to my eyes, and a prayer learnt in my childhood came to my lips, already tanned and weather-beaten by south winds.

63: Something about Penguins

The next day, 3 November, after lunch, I put on my ski and went with Plagge in the direction of the penguin rookeries. Our intention was to note the date when the first eggs were laid. If we found any I intended to bring some back, as evidence that the annual reproductive cycle was underway and laying had started.

We arrived at Point Diebel. There was an extensive view over the whole expanse of Scotia Bay, which was still frozen. I observed an unusually large number of penguins on the move. They were coming from the south and heading towards The Isthmus with the intention of reaching the patch of open water to the north of Saddle Island. Long columns were tobogganing over the white plain. When we met one of these columns, the penguins halted, stood upright, and stared at us brazenly as we passed. Then they lay down again and rapidly continued their journey, that had been interrupted by our presence.

While we skied on I asked Plagge:

'Do you think they are searching for open water where they can feed?'

'Yes, that's right. It is a shorter distance to the open sea beyond Uruguay Cove by crossing the isthmus rather than passing south around Mossman Peninsula to go through Washington Strait.'

'These are enormous journeys for such small creatures!' – I exclaimed.

'Yes, indeed, the penguins will have to travel at least ten kilometres over sea ice before they can have any hope of reaching open sea.'

'Once they find open sea they won't stay there, right?'

'They may have to go even farther swimming many kilometres from the coast before reaching areas rich in food.'

'What is their food?' I asked again.

'Crustaceans, and specially one which must be considered a delicacy for them, called krill[233]. It is a little whitish shrimp, very abundant in these Southern Seas.'

'Is that all they eat?'

'They also eat small fish and some algae, often I have found remains of these in their stomachs.'

'There is another matter that interests me, and as an old hand, you might know about it.'

'Please ask,' said Plagge 'I will tell you if I know.'

'What I would like to know is when one of the pair leaves its nest in search of food, does this adventure take it a long time?'

'Just imagine it yourself', he replied. 'Bear in mind that he or she has to go a long distance over the ice, then swim from the ice edge to at least thirty or forty kilometres from land, or perhaps even more. Then they fish or eat, whatever you want to call it. To the time taken for all this you must add reasonable resting periods, then the time for the return journey.'

'That would take days!' – I exclaimed in astonishment.

'Bearing in mind the present extent of the ice pack, I reckon that the return journey would take a minimum of three days - and even more if Uruguay Cove happens to be full of pack ice. In this case they would have to travel all the way to the edge of the ice.'

'so, when they return to the nest,' I said 'they are exhausted and have empty stomachs?'

'You would think it would be like that, but that's not the end of the

233. *Euphausia superba*

story. When they get to the rookery and reach their nest, their mate who has been custodian of the eggs, is relieved. It is now the turn of the other to make a similar journey to the sea.'

I thought it would be a fascinating project to investigate in detail the entire life-cycle of these animals, whose social organization appears to be so perfect.

We were arriving at our destination; vast columns of penguins were crossing the frozen bay going to, or returning from, the sea.

As we came close to land, I could see that the whole of Point Martin and adjacent slopes were occupied by penguins in their thousands.

It was difficult to say exactly how many; the entire terrain, rocks, snow and ice, was covered in penguins, and their combined squawks were deafening.

It was a superb sight to see that mass of penguins stretching up the hillsides, each next to another, with only a few centimetres between them.

The fringe of rocks along the shore, and the higher ground beyond, right to the top, over two hundred metres above sea level, was completely occupied; to the naked eye it seemed impossible that there would be enough space for a man to get through them.

On the beach, surrounded by penguins that looked at us from all sides, we took off our ski and I watched Plagge wade into the thick of that conglomeration.

Looking carefully I noticed they were in pairs, but admiration had no limits when I saw there was nothing there that resembled a bird's nest.

Some of the penguins had half a dozen pebbles under their feet, but the majority had none.

By contrast, others had formed the pebbles into a kind of truncated cone-like structure, concave on the inside. They seemed to be taking considerable pride in their handiwork for those structured mounds of pebbles were the most perfect nests that could be found in the entire rookery.

The most amazing nest building demonstration was given by the penguins which had arrived too late for the general division of territory. They had not been able to find twenty square centimetres of unoccupied space on which they could instinctively place their half-dozen pebbles to mark their personal territory. Thus they had to be content with merely establishing themselves on the snow. This, of course, melted with the heat from the male penguin or the broody female. Gradually this formed a deep hole, into which the penguins disappeared until they were below the level of the snow surface.

I approached one of the holes, the back of the penguin was twenty centimetres below the snow surface. I poked and levered it out with my ski stick, but this took some doing as the penguin fiercely defended its right to be there. At the base of this cavity in the snow, there were no more than five pebbles, each about the size of a hazelnut. The male was keeping a close eye on the movements of the female; as soon as he saw her emerge, he immediately climbed into the hole himself, intent on guarding such a precious treasure with his life.

I laughed, for to me it seemed impossible that this could be a nest intended for a long incubation.

64: The First Egg

It was exactly like that. Every couple had its own pebbles which they defended tenaciously whenever some other opportunist penguin, on the scrounge, passed close to them, and attempted to steal pebbles for its own benefit.

That intuitive resistance, to being deprived of the pebbles which form their nests, is one of the most deeply rooted instincts in penguins of all species.

The pebbles are essential to raise their eggs off the damp rocky ground, even if it is only by half a centimetre. This avoids water from thawing snow, mixed with faeces and other detritus, which constantly runs through the rookery. So the more little pebbles a nest has, the better incubated the eggs will be, as they will have some protection against running water, and indeed against the snow itself.

During this visit to the rookery, I noticed they had fierce fights over the possession of one small pebble, causing a tremendous hullabaloo.

The neighbours joined in, pecking and beating all and sundry with their beaks and flippers.

On other occasions, the pair guarding a nest composed of only a few pebbles would abandon it in the heat of the fight, the neighbours would turn that moment to their own advantage and, in a few seconds, all the abandoned pebbles would be appropriated to the neighbours' nests.

Another time, I spent ages watching a pair that was more pacific. The female was lying on the rocky ground, but around her body she had built a compact ring of pebbles. The male, standing by her side, seemed to be watching vigilantly for the slightest sign of movement among the neighbours. Another penguin, looking for its nest, and perhaps intent on relieving its mate and giving it the opportunity to feed at sea, waddled within range.

When it passed close to the other nest, the already tense situation ignited and turned into a general affray. As they fought it out among themselves, all nests were abandoned. The first to come within range of warring beaks and flippers would receive powerful pecks and beatings. The racket and uproar were deafening.

Suddenly, the tranquil penguin, which was watching over his female and her nest, saw that there was another nest momentarily abandoned. Without caring one iota about the brawl continuing around him, he ran in the direction of the loose pebbles and, with rapid pecking movements, rolled them over to the female who remained horizontal on her nest, quite oblivious to the chaos all around. She picked the pebbles up gently and placed them next to her body; then she stood up, and with the male, that had so efficiently done his job, started reorganizing her pretty and comfortable nest. She seemed so proud of her companion! In penguin country it is clearly not deemed immoral to build your nest with another's pebbles.

While all this was going on, Plagge had managed to get into the middle of that frenzied and excited mob. From the spot he had reached he beckoned for me to join him. For my part I wasn't particularly keen to wander into the thick of that savage and closely-packed throng, but I was determined to follow my companion's lead. On taking the first

48: A rookery of Adélie Penguins (note the ciné camera on a tripod)

few steps in his direction, I soon discovered that just my boots brushing that imaginary demarcation line between two nests was enough to start another free-for-all.

Eventually I reached Plagge, and taking a seat next to him, on a tall rock, I asked;

'Have you found or seen anything?'

'Up to now, nothing.'

'There are many penguins lying down', I observed.

'Although the nests are empty', Plagge added.

'Unfortunately, our journey is wasted.'

'I think not,' Plagge told me, 'let's go farther up, that's where the first arrivals will be, and it is possible they may have laid some eggs.'

The idea seemed to be a good one. We had advanced about one hundred metres when I saw Plagge approaching a female and standing next to her. He picked her up by the tail, checked the nest, and put her back.

'They are broody', he said. 'Do the same and you will see they don't even react'.

I turned round and went close to another penguin that was also lying down. My boot was not far from her beak and she pecked it, surprised at its hardness. I bent down, and picked her up by her tail.

'Plagge!' – I shouted, 'the first - look at it!'

It was a penguin's egg.

65: Novel Food

We worked for over an hour, during which we lifted hundreds of penguins of both sexes that were resting on their nests, but we managed to find only fourteen eggs, which we stuffed into our coat pockets.

We began our return journey, and once we moved away from the coast, I asked my companion:

'Are you hungry?'

'Rather, and you?'

'At this moment I feel quite ravenous', I replied.

We continued and I suddenly had an idea.

'Tell me,' I asked Plagge, 'how does a good *milanesa* with chips and a fried egg sound?'

'Why don't you add a salad as well?' – he replied.

'Because a salad would be impossible to find here, while what I propose could be found.'

'I know that', said Plagge 'penguin eggs, boiled tinned potatoes and meat, penguin meat ….'

'No, not penguin meat ….'

'What then?' he interrupted. 'Do you want to make *milanesas* with salted meat, or corned beef, or any other meat you can find in a tin can?'

'No,' I replied; 'I too am fed up with tinned food. As a change from penguin meat, what about seal meat?'

'Man, I have never eaten that,' he replied 'I remember your conversation with our Leader about this matter. Frankly I cannot explain why it has never been tried, especially as it has been one of the principal sources of nutrition for all other Antarctic expeditions.'

I realized the idea was beginning to enthuse him.

'Do you have a knife?' I asked.

'I have not brought one. Do you have yours?'

'Yes, I do and it has a very sharp edge. Shall we go to Acuña Island ? In three quarters of an hour we will be there.'

We stopped and looked at each other, each understanding what the other was thinking.

'Let's go!' – he said.

'That's the direction', I indicated.

We had changed our course, the snow was in ideal condition for skiing and we advanced at a good speed.

I asked Plagge:

'What do you think about keeping the discovery of the eggs secret until the meat we are about to get has had time to marinate?'

'Fine; on the day after tomorrow we can give everyone a nice surprise.'

Slowly we arrived at Acuña Island, looking back to the left, we could see The Isthmus and beyond it, some little dots were standing out which were the Observatory buildings.

In front of the coast of our destination, there were also some black dots; they were the seals we were after.

At last we reached the colony which I had already visited numerous times and while I undid my ski, I asked Plagge:

'Which do you think is best?'

'Any, they are all the same.'

'Perhaps,' I said, 'but it would be better not to kill a female so as not to orphan her pup. Also we must bear in mind that the meat of some of these animals might be tainted by the process of calving.'

'The most practical,' I proposed, 'is to kill a male, a young male.'

'What do you think of that one?' I asked Plagge.

'Very good, let's go for it.'

It was a beautiful animal, about a metre and a half long, its skin did not have any of the scars frequently seen in older seals. It slept on the snow and when we were half a metre away, it awakened and raised itself, as if inquiring the reason for our presence.

I gripped the knife tightly in my right hand and said to Plagge:

'Get as close as you can, and distract him with your stick. I will attempt to reach the carotid artery in a single blow.'

It was easier said than done, as the seal moved constantly, and advanced towards us with the intention of biting our legs.

We skinned it and then removed a good piece of tenderloin; very dark red meat in which we could see a few veins, arteries and tendons as they moved with their natural contractions.

It looked just as good as penguin meat, but there was much more of it.

The tenderloin I took out must have weighed about ten kilograms. I divided it into two parts, and we each carried one and set out to return to base.

66: A Local Meal

At dusk we reached the base, where our companions were at the door; they guessed from the joy expressed in our faces, that we had found penguin eggs.

As we were ascending the slope from the beach, I said to Plagge:

'What do you think if we keep the seal *milanesas* as a surprise?'

'Good idea ··· but look, they have already seen us coming.'

'That is the least of our worries. We can depot the seal meat here on the beach, and I will come and get it later.'

That said, we deposited our load on the white ground, and I covered it with snow to stop the petrels making off with it.

We reached the base, and Valentiner who came to meet us, asked:

'Did you get many eggs?'

'Only fourteen,' Plagge replied 'I think in three or four days, laying will have become general.'

'Well, that's something ···.'

While we took off our ski, I called Valentiner to one side and said:

'I have a favour to ask.'

'You say ···.'

'It's to do with the penguin eggs.'

'I am listening', he said.

'I suppose all the egg collecting is for the benefit of everyone here.'

'At least that is how it is specified in the rules governing our communal life' Valentiner responded.

'Good,' I said 'If we had brought only ten eggs instead of fourteen, the same would apply?'

'Certainly, but where is this leading?'

'It is a simple matter. I personally would like to take four of the fourteen we have just brought back.'

'For a port cocktail?' asked Valentiner, a smile coming to his lips.

'I beg you not to ask. I will give you my assurance that they will be used for our common use, for every one of us.'

'Something new for dinner? Another dessert speciality of yours?'

It may seem odd, given the fact that in three or four days, or very soon, of having thousands of eggs available, that I should consider seeking permission from the Leader to dispose of four of the fourteen that I personally had lifted from the nests. A spirit of discipline was, however, necessary for our lives to function normally, the fact that I was planning to use a communal item which was, in effect, the property of all of us, was what prompted me to make such a strange request.

After dinner I made an excuse, went back to the beach, and dug up the pieces of seal meat buried there.

When I returned, every man was in his room resting. I had asked Otto to wait for me and found him by the stove in the parlour, carefully reading the label of my whisky bottle.

'This is what it is about', I said, handing him the pieces of meat.

He took it in his hands and examined it, saying:

'It looks very good.'

'I am of the same opinion,' I said 'prepare a good strong marinade, and serve it tomorrow night as I suggested.'

He nodded and left to prepare it.

Before he left the room, I said:

'Afterwards come back and have a drink or two.'

'Danke schön, mein Herr Kleine! [Many thanks Mr Kleine] I will be right back.'

'You will see ··· and it will be good!'

That night, while on my shift, I was reading narratives of Antarctic expeditions and what they had to say on the subject of expedition food, and found the evidence confirming that their meals were based on seal meat.

The most experienced Antarctic explorers; Shackleton, Nordenskjöld, Scott, Amundsen, and Charcot, had all eaten seal, and preferred it to penguin which they found smelly. Therefore I saw no reason why seal meat, clean and tender and the same colour as penguin

49: Returning to base with a sledge loaded with fresh meat

meat, should not be included in our usual diet.

The next day we made preparations to go to the penguin rookery and that night, we sat down to dinner, expecting the same old rice and stew. Otto appeared with a great steaming dish in which enormous golden *milanesas* were crowned with penguin eggs and adorned with chips.

'What is this?' all asked in chorus.

'A local dish,' I replied, 'seal *milanesas* with penguin eggs.'

'Yes, of course!' said Bruhns, with a tinge of mockery, and asked:

'Are the potatoes also local?'

'Not exactly from here or cultivated in the South Orkneys, actually they are tinned ... but that's good enough, isn't it?'

Before us was genuine seal meat, though somewhat disguised by the eggs and golden crust of breadcrumbs. We looked at each other, not daring to try it, until Valentiner said laughing:

'Ah-ha! That was why you asked me for the four extra eggs, wasn't it? Go ahead and tuck in!'

What else could I do? I took an enormous *milanesa* which stretched well over the edge of my plate.

I was about to regret my audacity, when I saw Otto urging me to start eating without a second thought.

I took a mouthful, and instinctively my throat tightened, but with considerable effort I started chewing.

They all looked at me attentively, while a smile with a touch of mockery appeared on their faces.

It was all right! Yes, not bad at all, in fact it was actually quite nice.

It seemed like an ordinary common veal *milanesa*, well-marinated.

I started to eat with relish. Not tough at all! Excellent!

My companions were still in two minds about it, then Bruhns, who seemed determined not to go to bed with an empty stomach, served himself and started eating. In a short while everyone did the same.

'It's seal !'

'Yes, genuine seal.'

'It is really good!'

Now we had a new variation in our daily menu. Previously I had dreamt of an enormous loin steak, golden-roasted in the oven, and garnished with chips, now, two days later, the dream became reality.

67: Merciless Looting

On 7 November, at six in the morning, Valentiner, Plagge and I went to the penguin rookery on Point Martin. Between us we dragged the long sledge on which Otto had loaded several bottles of cordial and a box containing bread, ham, cheese and chocolate. That way we carried supplies for the whole day, plus a paraffin stove[234], and a suitable pan.

The weather prospects could not have been better. Although the sky was rather cloudy and the thermometer indicated minus seven degrees, the conditions for spending the entire day outside were quite good.

To help with the work of egg gathering we also carried several buckets on the back of the sledge.

234. A pressure stove of the *Primus* variety

It was agreed that we would be back by nightfall, and Otto would have a substantial meal ready for us.

During the day, Bruhns would make the observations and record each man's data; that way we had no worries about our absence from base.

The three of us hauled the sledge comfortably and barely felt its weight. On the surface of the snow that covered the sea ice a thin sheet of freshwater ice had formed. This cracked as the tips of the ski passed over, making it easier for us to slide along.

There were penguin tracks all over the sea ice. On the level ice close to the coast marking the rookery limits, you could see long lines of penguins marching to the sea in search of food.

After an hour we reached our destination, halting at a spot overlooking the foreshore; we depoted the sledge in a safe place, took off our ski, and went to look for the empty crates that, with Plagge's help, I had brought over a few days previously. When all was ready, and each man had a bucket, we lit our pipes and began our task.

There was a tremendous number of penguins, thousands in every direction; they extended over a vast area of the foreshore and up the hillsides beyond. Most were lying down, like hens on their nests. Where the male or the female had not gone to feed, the other stayed close to its mate on the nest in an expectant pose.

Many skuas were flying very low, over the penguin's heads, as if looking for something. Suddenly one of them folded its wings back and then dived, almost vertically, swooping over the penguins, which milled around the predator, which then lifted off with an egg in its vicious beak.

Plagge who was at my side watching the scene said to me:

'That is nothing. You should see them when they grab a chick which they carry off to the hilltops to devour!'

Time to start gathering the eggs, and with my bucket hanging on my arm, I sidled in among the penguins.

I wasn't going to waste time exhausting myself for nothing; all the penguins that were horizontal on their nests had eggs concealed beneath their abdomens.

I filled my bucket in a few minutes and went back to where the stack of empty crates had been left. I unloaded the eggs, then returned to the rookery to continue gathering.

Despite the passivity of the penguins every time I came near to their nests, I had to suffer innumerable pecks on my legs from those that I had disturbed on the way. Several of them held on to my trousers with their beaks, while striking me with their powerful flippers to the accompaniment of a tremendous squawking.

It was impossible to hear what Plagge was saying right next to me. I was utterly deafened by their extraordinary racket.

After working for an hour I was quite convinced it was a most gruelling and exhausting task, you had to bend over and pick up each penguin by its tail, and with the other hand you had to hold the bucket up high enough so that it would not be overturned when the birds attacked, all this on rough terrain made slippery by detritus and thawing snow.

However the crates were filling. At midday we had six full, and we decided to have a rest. Plagge and I prepared some lunch, the main course being a tasty omelette of fresh eggs.

After lunch we continued the exhausting work which demanded much patience; my hands were bleeding where they had been pecked and my legs covered in bruises caused by blows from their flippers.

It was only a matter of time before one of us would fall over, dragging the bucket with him. The bucket crashed to the ground breaking its contents and making a filthy mess which dirtied one's clothes; but the unfortunate casualty picked himself up and went back to the job as if nothing had happened.

On one of our forays I found myself next to Valentiner, with whom, between shouts, I managed to maintain the following conversation:

'How many should we get?' I asked.

'Four thousand five hundred, or five thousand if we can ....'

'How many fit in each crate?'

'Two hundred and fifty', he replied.

'Then we will not finish today', I said.

'We will do half, and tomorrow we will carry on.'

50: A sledge-load of penguin eggs (photographed by Luciano Valette, from: Brown, Mossman, and Pirie, 1906)

At dusk we stopped gathering. We had twelve crates full and we decided to take back four of them on the sledge.

I was absolutely exhausted, my body ached and was somewhat the worse for wear from being pecked and struck by penguins, enraged by the looting of which they were the unfortunate victims.

As we hauled the sledge back, and stopped for a moment to rest, Plagge began bewailing about what we did to the penguins ….

'Poor little things … what disgraceful robbery! Talk about looting!'

However we were satisfied with our valiant efforts and the valuable contribution made to replenish our much-depleted pantry.

68: Life Returns

December … summer … but no trees, no leaves, no green, no suffocating heat …. The last drifts of snow around the dwelling, the rocks, and all that was an obstacle to the wind, rapidly dissipated as the temperature rose above zero.

During the day, especially when the wind was from the north, I could feel the warmth. There was nothing for it but to believe that summer was here, even though the thermometers were only showing one or two degrees above zero.

Sometimes the temperature went up to four or five degrees, and Bruhns could be seen arriving at the dwelling carrying his jacket under his arm, mopping his brow with a handkerchief.

'Uf! ··· isn't it hot!' he would exclaim.

'It certainly is ··· we actually have five degrees above zero!'

In truth, I felt the effect of that humid hot, saturated sea air.

However it wasn't just the outdoor temperature which on its own caused that feeling of stifling heat and suffocation; four or five degrees above zero was actually quite cold, even in the raw winters in Buenos Aires.

That heat we felt came from our routine of dressing in thick clothing, and our habit of never shedding our bulky woollen vests and essential jerseys. The latter had a weft the thickness of a finger. These tended to be worn throughout the bitterest part of the winter when we had temperatures more than forty degrees below zero.

The extreme temperature range between winter and summer is more than fifty degrees. That is why, when the thermometer indicated five degrees above the freezing point of water, we felt hot.

Outside on these occasions all the steep slopes had water cascading down them which gushed from rifts between the rocks, forming multiple torrents that reached the sea. Only the high glaciers remained immovable reserves of ice, utterly unaffected by the heat of the Orcadian summer.

Sometimes, on calm nights, while on duty, I would hear something like the sound of distant thunder, and then a muffled sound of an enormous mass of ice crashing into the sea, shaking the ground and the structure of the dwelling.

I would quickly put my head out of the door, or dash to the window. In the moonlight over Uruguay Cove, beneath the ice wall of

Mount Ramsay, I caught a glimpse of the column of spray raised by a monstrous mass of ice that had broken off from the glacier, and from a height of five hundred metres had crashed vertically into the sea.

At first those calvings filled me with anxiety, but after hearing them daily, I became accustomed to them and even strove to be the first to reach the door and witness such spectacles with my own eyes.

The opportunity wasn't slow in arriving; one afternoon when coming back from the Variometer Hut, I heard the sound of a detonation caused by ice when it cracks, and was in time to behold the magnificent spectacle of huge masses of ice plummeting into the sea.

Summer progressed, and the surroundings became populated with other wildlife.

Several species of petrel, gull, and shag, hastened across the sky, giving that Antarctic solitude a new infusion of life.

Vast convoys of penguins waddled across the isthmus, and sometimes at dusk, they stopped to rest and spend the night on the beach of Uruguay Cove.

Everything took on an aspect of new life. As if to cap this sensation, so radically different from the depths of winter, I discovered something which, even if I had been told about it, I should have deemed impossible to happen.

After the gathering of penguin eggs was finished, I became accustomed to making excursions high in the hills, in search of evidence of movements of the pack ice, which I had to record in the relevant observation book.

One afternoon I had climbed, as usual, one of the hills of Mossman Peninsula. I was exhausted from the ascent and had lain down on the rocks on the summit. There were still a few hollows filled with snow which the north wind had not yet melted, but other depressions in the rock had water in them which the wind gently rippled.

I lay face down, my hands cupping my head, while my thoughts lingered on the time still left until our liberation; I looked at that little pool of water, no bigger than my fist, just a few centimetres away from my face.

I saw that the ripples were not caused by the wind, but by some interior force. I peered to see better, and what I saw seemed incredible. Moving, swimming in that tiny pool, were minute insects no bigger than a pin head[235]. I put my finger in the water and had no doubts; the presence of a foreign body in the water caused all the beasties to flee.

I was perplexed. To find something with life up there, in a place where in winter the water turned into a solid frozen layer at forty degrees below zero and where the wind could blow for days at a time at more than one hundred kilometres an hour, seemed absolutely impossible, but I had to give way to the evidence before my eyes.

I went down to the base, preoccupied with my discovery, and finding the Leader, I asked him what they were.

Valentiner listened to me and said:

'What you say is very possible. The snow is pervaded by live micro-organisms. Think of the huge surface of a glacier, for example, with its greenish discolouration in some parts and red in others, formed by vast colonies of microscopic algae that can endure that environment.'

That was true. Life in the ice; not on the surface, but inside the ice.

Organisms which, with the feeble warmth of polar summer, came to life again, moved, and continued Nature's inexorable cycle of regeneration, in one of the most remote places on Earth, and the most inimical to human life. In it lived millions of diminutive beings which were defended from the inclemency of the weather and the low temperatures, by becoming encased in solid ice at forty degrees below zero!

69: Possibilities

We were already in the latter part of December which was the official time when we had to start preparing for the decisive day of relief. Our own personal conviction was to believe that the new expedition would not arrive before the end of January, or more likely, February; possibly even March.

235. These are copepods, known vernacularly as 'water fleas', of which eight fresh water species are recorded from the South Orkney Islands. One genus, *Pseudoboeckella,* occurs in some of the most remote polar and high altitude water bodies being able to withstand freezing in winter.

Will they come?

This question was in all our thoughts, and guessing the exact date when the relief expedition would arrive to liberate us was an obligatory topic of conversation every day.

'What date,' I asked Valentiner one day, 'is the most usual date for our relief?'

'February, possibly March, it depends on many factors.'

'such as? For example ….'

'Of primary importance,' he replied, 'is what has been happening up north in Buenos Aires, and whether they have found a suitable vessel to dispatch and relieve us.'

'That's true,' I reflected 'the corvette *Uruguay* no longer makes the voyage, it was withdrawn from service a few years ago.'

'That is no great problem,' continued Valentiner 'but it has to be dealt with in Buenos Aires. You would think that of all our war-ships they could find a transport vessel strong enough to face the fury of the Southern Ocean.'

'What if they cannot find a suitable ship?'

'In that case, the relief will be done as it was done when we made our voyage via South Georgia. However this is not the only thing that the date of our departure depends on; there is another important factor, the sea ice ….'

'Certainly,' I replied 'it could be tricky.'

'Very tricky,' said Valentiner, 'you have to remember we are in the middle of December and Scotia Bay is still closed, blocked by pack ice. To the north there are extensive fields of pack and drift ice which could blow into the entrance of Uruguay Cove blocking access to that as well.'

'What would happen then?'

'Nothing, just wait.'

'Just wait …' I echoed mechanically.

'It's the only answer,' continued Valentiner 'and in truth, this year is unusually bad for navigation due to the exceptional amount of pack ice around.'

'Has it happened,' I asked, 'that an expedition has remained unrelieved because of the ice?'

'Here in the South Orkneys, I have never heard of a case, though almost all polar expeditions, especially Antarctic ones, have had serious difficulties, in some cases even losing their ships, which have been crushed by ice pressure.'

'However in our case, is there a way they can reach us overland and rescue us?'

'Sometimes what has happened is that ships have had to sail around the islands for weeks waiting for good weather or for the right moment to approach land. It is just a matter of time and patience, waiting for the ice to retreat from one or the other of the bays ….'

'Then the ship can get in', I interrupted.

'There is no other way.'

I wanted to get to the crux of the matter so I asked again:

'Let's imagine the worst scenario, that they cannot reach us; the ship they have dispatched has been damaged or sunk by ice and cannot approach even to within thirty kilometres of our position.'

'If that happened,' Valentiner said 'we would have to resign ourselves to remain here another year. Could you survive?'

'Perhaps,' I replied 'but what about the provisions ? The coal ?'

'As far as provisions go, we could remedy that. We have an abundance of fresh meat; it's all a question of logistics; we would have to collect enough eggs; the limited stock of tinned food, and a few really essential items would have to be rationed. Heating would be a problem as we only have four or five tonnes of coal left, but we could supplement this by using seal blubber for heating and cooking.'

'A year under those conditions, I observed, would indubitably be quite dreadful.'

'Grim,' said Valentiner, 'but there is no need to despair, it is still ages to go till February arrives and there is no way we can expect to be relieved before then.'

'That means when February comes round we'll be living in a state of constant expectation and apprehension won't we?'

Valentiner did not reply, he knew more than he was letting on.

Our only option was to wait patiently, holding out day by day for that special moment when one of us would shout with all the power of his lungs:

'They are here! They are coming!'

We would all run to the beach and try to locate that almost invisible sign, a smudge of smoke, on the horizon, far away in the north.

70: End of the Year

The 23 of December. In the afternoon I went into the kitchen and asked Otto:

'What do we have for tomorrow night?'

'Up to now nothing special', he replied.

'Why not? Tomorrow is Christmas Eve, isn't it?'

'Yes, yes, but the Leader has not yet given me instructions.'

'Never mind. We should do something nice.'

'Yes, we must do something'. Christmas Eve and Christmas Day, the real end of the year celebration of the northern Europeans. Even if my companions preferred to eschew it and celebrate on 31 December instead, it would be my duty and moral obligation to show we were not oblivious to the traditions of others.

'Well, Otto. Tomorrow night we expect a good meal ....'

'Dessert?' – he asked.

'Plum pudding with brandy sauce.'

'Strong sauce?'

'As strong as possible.'

'Then after?'

'After the meal, a punch with rum and whisky.'

'Hm, this is a brilliant idea', said Otto and his little round face lit up with pleasure.

With so much brandy sauce and punch to hand, there was a good chance that we would end up without a cook, so I added:

'Tomorrow night when you need the brandy and whisky, ask me for it.'

'Very good, *Herr Kleine*!'

The 24 of December arrived. Already in the afternoon, at teatime, we experienced the first nice surprise; Otto brought to the table an enormous cake exquisitely decorated with an inscription in chocolate saying:

Orcadas, Fröhliche Weihnachten, 1923

The most memorable thing about that celebration took place in the evening, during dinner.

We had no other lights in the base except for the paraffin lamps, but on this occasion, Bruhns appeared with some beautifully carved wooden candelabras, that he had painstakingly made, and inserted stubs of candles which were alight.

The paraffin lamps had been extinguished, and that primitive candle light was enough to illuminate the farthest corners of the room perfectly, suffusing all the familiar objects around us with its gentle glow.

After dinner, once Otto had finished his work in the kitchen and come to join us, my companions sang a few German Christmas carols, while I filled our glasses with steaming punch.

It was an extraordinary spectacle that those men made, as they sat round the table by the glowing hot stove. Isolated in a base surrounded by a world of ice, and dimly lit with that most primitive of lights, they sang carols to the baby Jesus, carols learnt in childhood and, in my case, engraved on my memory ever since.

It was Christmas Eve in the South Orkneys ….

Even that too passed. Days came and went, until one morning, Plagge woke me and handed me a carefully folded piece of paper.

'What is this?' I asked rubbing my eyes.

'A telegram, read it ….'

At once I realized it was a joke. It could not be otherwise, in those days we did not even dream of having a radio station in the South Orkneys.

I unfolded the paper and laughed as I read:

José Manuel Moneta,
South Orkney Islands.
We will be there in five months.
Best regards,
Relief Expedition.

I could not stop laughing as I involuntarily glanced at the calendar to see the date:

28 December, Holy Innocents' Day[236]!

Three days later we found ourselves celebrating New Year's Day, 1924, a date that was a decisive mark in the annual motion of the globe and in the passing of a man's life.

At midnight on 31 December, we went to the beach of Uruguay Cove, where the flag pole stood. We each took a gun, and when Bruhns gave us the exact second we marked the passing of one year to another by firing a simultaneous salvo.

We then made our way back to the base where bottles of whisky awaited our return.

I waited until everyone was seated, then went into my bunk-room and returned carrying a heavy bottle with a thick neck. Without anyone noticing, I took off the wires and the cork shot out making a muffled bang.

They turned around surprised and I just managed to shout:

'Glasses! Quick!'

I served champagne, and toasted:

'To the health and happiness of all of you!'

'Prosit Neujahr!'

'Happy New Year! …'

'Skāle! …'

'¡Salud! … ¡Feliz Año Nuevo!'

71: Preparations

Personally we had nothing to prepare. For months our luggage had been packed ready to leave at a moment's notice. We did have a few jobs to get done refurbishing things which we would be leaving for the next group.

On 1 January, the day after the convivial New Year's Eve party, the Leader said to us, at lunch:

'Tomorrow we will start painting and cleaning everything in order

236. Holy Innocents' Day, or Childermass, commemorates the Biblical slaughter of the infants, ordered by Herod after the birth of Christ. It is also a day for a 'Feast of Fools' in the Spanish tradition, similar to All Fools' Day on 1 April.

to hand over the buildings in a state as good as new.'

'Very well,' I replied 'so we have to paint everything?'

'From top to bottom. We will begin with this building. We will paint the parlour, then the bunk-rooms, and finally the kitchen. We also have to do some carpentry repairs; I have inspected some of the windows that have suffered from the pounding they received during last winter's weather.'

'Is this work done every year?' – I asked.

'Every year,' Valentiner affirmed 'that is how the buildings have been maintained intact from their construction in 1906.'

The next day I made some paper hats, similar to those used by decorators; we wore our oldest clothes and began to wash the walls and ceilings with soda.

There was a huge amount of dust from the stoves. Every time we moved the large piles of books, instruments, jars, papers and bottles, thick clouds of dust rose from the shelves.

Everything needed to be moved from one side to the other of each room to clear enough space to paint.

While painting the parlour ceiling, helped by Plagge, streams of oily blue paint ran down my arms and face, splattering my clothes.

I worked hard, I knew that every brush-stroke was bringing the all-important moment of departure closer. I couldn't help laughing when I thought how odd it was, this mixture of scientific and manual work; how mistaken people were who believed that the job of a painter and decorator was at odds with the essential scientific work which was the principal object of all involved with polar expeditions.

It did not end there. After painting, we had to wash the floor, repair the stoves, clean and even patch the chimney pipes.

When the indoor work was finished, we went outside to renew the Ruberoid cladding of the walls.

One afternoon, Valentiner asked while nailing some tacks into the wall:

'What colour would you paint the outside of the building if you could choose one?'

'I would leave the roof as it is; black.'

'Yes we can give it a few good coats of pitch.'

'The walls, I would paint them a bright colour ….'

'For example?' asked the Leader.

'Red, dark red … possibly green ….'

'I like red better,' Valentiner said, 'it would stand out well against the white of the snow.'

'Red walls and white windows and a black roof. It is not a bad combination for the South Orkneys', I replied.

We decided to paint it like that. However it took us much longer than we anticipated, because the wind was too strong for us to be able to paint every day.

On other occasions, when the temperature dropped substantially below freezing, the paint became too stiff to apply to the wall, and the work had to be abandoned.

Later we dedicated ourselves to refurbishing the magnetic instrument hut.

Hammers, nails, Ruberoid, paint, we put in a long month on this work; daily routine was disrupted by all the mess caused by trying to move things around in such a confined space, not even the bunks escaped the regulation coat of grey paint.

Amidst all this work, we still made enough time to mend our outside clothes, our underwear, and boots. We also prepared personal books and any special objects that we wanted, or had, to take back with us when we returned to civilized life.

Often I came across Plagge, Otto, Bruhns and even the Leader himself, struggling to thread a needle or sewing on a trouser button.

I also found a serious problem. From wearing the thick soled boots we had in the South Orkneys, I no longer had socks without holes, and could not darn them as there was no suitable wool. I resorted to using string, weaving the ends patiently, until I made a thick, very strong patch, which had to last until the date of my departure.

72: Waiting and Hoping

January came to an end and the sea continued to be unfavourable for navigation. Scotia Bay remained ice-bound to beyond Point Martin.

Although Uruguay Cove was ice-free, the pack could be seen farther out to sea; there was a white line on the horizon denoting the edge of an extensive field of ice floes.

Daily we trekked to the top of the local peaks to check the movement of ice. From that elevation we could observe, to the north and south, that the ice was considerably diminished, compared with what it had been during the worst of the winter.

However, sometimes I would see Valentiner gazing abstractedly out to Scotia Bay.

'It's strange,' he said, 'to see Scotia Bay still frozen. I have never seen it like that in January.'

'Sometimes it must have remained fast like that all summer?'

'In the known history of this area this is the first time.'

That was cause for alarm. February arrived and Scotia Bay remained ice-bound beyond Point Martin. To the north, very far off, some ice could be discerned, and it blocked the entire the horizon.

It was at that precise moment that an atmosphere of nervousness, of expectation, of alarm set in …

I would find myself working on the table, near the stove; at any moment Plagge would burst into that small room like a whirlwind. When I saw him going straight for the binoculars my heart would give a leap. He then dashed outside again, with the binoculars.

I jumped past the table and chairs, and in two bounds was right on his heels.

We ran outside and down to the beach.

'What's happening, what's happening?' I would ask panting.

'Nothing, let me see … over there … to the right, pointing at a black dot on the horizon, while he focused the binoculars in that direction.'

'I can't see anything. Lend me the binoculars', I begged.

'Hang on … seems not … but yes! … no, it's nothing, merely ice.'

That happened often; five, ten, twenty alarms daily, and it was always the same old thing, ice. A fragment of pack ice in a weird deceptive shape, adrift on the sea, and in which we thought we discerned the shape of our long-awaited relief ship.

At night the same thing would happen. When I had to go to the

Variometer Hut, at two or four in the morning, I would stare north.

I had never made such a concentrated effort with my eyes, to penetrate that dark Antarctic night.

Sometimes, while I walked along scrutinizing the horizon, I would remain in one spot as if petrified …. I would look into the distance and there was a light! Once I recovered from this overwhelming emotion, I would dash to find the binoculars and look at that bright light again.

Yes, another deception; it turned out to be a star low on the horizon; if not it was a piece of crystalline ice which shimmered like a mirror, reflecting the moonlight.

During the day I had to control my nerves, and look calm, as I noticed my companions unhurriedly continuing their regular tasks. I did not envy them; I knew that every one of them, even though he did not show it, inwardly suffered the same false alarms as I did, to the point of not being able to sleep.

It was difficult. By now it no longer mattered to me if they did not manage to reach us. Perhaps I could resign myself to spending another year here with complete peace of mind, if I knew in advance that nobody was setting out to look for us.

In my own mind I had already told myself that we would be going back to Buenos Aires; it was an illusory return, and my morale suffered severely from the anguish of not knowing. Only the appearance of a ship could end this.

I once mentioned to Valentiner when we were discussing this subject:

'Now would be the ideal moment to profit from radio communications. That way we would definitely know, one way or another, what our fate was to be.'

Valentiner replied; 'South Orkneys with a radio! I have thought about it a lot, Kleine, and frankly I don't think it would make a huge difference to our life here.'

'Think,' I retorted, 'with a radio we would be in contact with so many delightful things.'

'Or disagreeable ones', he interrupted.

'In effect, yes, but at least by now we would have received some definite news to ameliorate this cruel period of waiting.'

'Nevertheless, I personally prefer to remain ignorant of what is happening in the world beyond. I know they will come, they have to come. It would be unthinkable that, because of some trifling difficulty, the authorities would abandon five men to their fate.'

'Yes … but come February …, and more ice … and they do not appear.'

'They will come,' Valentiner affirmed 'I have seen the same scenes, nerves and atmosphere of expectancy towards the end of every one of the four years I have spent here.'

'But with a radio link …' I repeated.

'Yes, I have no doubt it would be better. You are still young and, as you have shown on several occasions, have set your mind on coming back; you can dedicate a future winter to the project to getting a radio link established.'

'It will be one of the first things I do when I get back to Buenos Aires.'

'Bear in mind that down here the receipt of a piece of bad news could have a disastrous effect on an individual, if not on the base', said Valentiner.

'Yes, I know that, and I realize your sentiments on the subject. In a case like that, there is absolutely nothing that could be done, that is what you are thinking, isn't it?'

'Certainly.'

'Forgive me for disagreeing,' I said, 'in the likely event of someone in the South Orkneys receiving bad news that might depress his morale, surely the effect of this could be counteracted by the same means, by the radio link. Surely this would be preferable to having to endure a whole year in that lost world and not to have any news of life outside, especially without news of things affecting our own lives, our future.'

73: A Bet

February too drew to an end. It was apparent that autumn was already arriving in the South Orkneys. The daylight, which had reached its maximum length towards the end of December when the Sun was above the horizon for twenty-two hours a day was now unmistakably getting briefer.

The fast ice persisted on Scotia Bay, while on the northern horizon beyond Uruguay Cove the distant white line of the edge of the pack could still be discerned.

Every day, tirelessly, I would ascend one of the neighbouring hills and spend hours scanning the sea, scrutinizing every piece of distant ice; in silence I would gaze at the vast floating icebergs, anxiously expecting our relief ship to emerge from behind one at any moment.

Nothing different or abnormal was to be seen. Nothing that would break the monotony of that desolate wilderness.

Thousands of conjectures passed among us and it was frequent to hear dialogues such as:

'They are taking their time, aren't they?' said Plagge as he accompanied me on a trek.

'I don't know what to make of the delay.'

'For my part, I am not particularly bothered yet … but there is so much sea ice around ….'

'All the same,' I replied 'I don't believe there is much ice floating about to the north of the islands. Some days ago I observed that it was only toward the east that there is any significant amount of pack ice. To the north, beyond Saddle Island, there are large expanses of open water in which several strips of brash ice may be seen emerging from Washington Strait and Lewthwaite Strait.'

'It could be that the ice is not an obstacle and will not prevent the ship reaching its destination, but you must remember that, during the previous few days, there have been some strong blows from the south-west. It is also possible to imagine that one of those storms could have caused the ship to sink ….'

'That is alarmist talk', I commented.

'We will continue hoping ….'

'For my part, I am quite calm about it', I said, in a tone that I hoped sounded sincere.

'Yes,' replied Plagge, 'you say that, but in reality you are as nervous as the rest of us.'

'Don't believe it. It seems that we are all experiencing similar feelings.'

No one let a day pass without glancing a thousand times towards every detail of the horizon. Every instant we thought we saw the long-awaited ship emerging between icebergs. How many times did I think I saw smoke rising from the funnel, and even heard from inside the dwelling, on a calm day, the distinctive sound of a ship's whistle announcing its arrival!

The races to find the binoculars had multiplied and no one took offence at the jokes made by his companions when the object he spotted proved something very different from a ship.

Instances of men being deceived multiplied. A factor which contributed to this was the ocean currents that cut across the archipelago from the south-west to the north-east. Running strongly through the principal straits between the islands they swept oddly-shaped icebergs with them which, from a distance, caused the erroneous impression of ships sailing at high speed.

I had already begun taking fresh stock of the situation:

'Spend another year in the South Orkneys!' I think I could survive it, perhaps ….

There would be some things I would have to do without, but in that sense I could be considered an optimist; physically I was quite capable of doing it. The one serious doubt I had, hoping for a relief which did not happen, was about the possibility that it could seriously undermine my morale and that of my companions.

I did not care to dwell on the consequences of such a possibility.

Daily, at four thirty in the morning, I handed my shift over to Bruhns and I went to sleep. At midday Bruhns would wake me.

In the morning of 4 March, when Bruhns had awakened me, I asked him:

'What do you reckon? Will it be today?'

'It will have to be soon', he replied.

As I wished to remain an optimist until the last, I added:

'I propose a bet ….'

'About the arrival?'

'Let's do it like this: when do you think they will come?'

'I suppose before the 15th of this month.'

'Well, I say they will arrive after the 15th. Accepted?'

'Accepted, and I think I will win,' he said 'bear in mind I am an optimist.'

'So am I,' I replied 'but I don't think they will come before that date. Valentiner says it could be April or May before they have favourable conditions.'

'What shall we bet?' – Bruhns asked.

'Whatever you like. Cigarettes?'

'Cigarettes?' – he asked stunned 'You don't have any!'

'Precisely, that is why I bet cigarettes,' I said, 'three cartons of cigarettes in exchange for half a bottle of eau-de-cologne, the one I have which you like ….'

'Done, and remember, I will win.'

'We shall see', I replied.

While I climbed into of my bunk I asked:

'Could you let me have a packet of cigarettes on credit? I have nothing to smoke.'

Bruhns smiled and handed me some and said:

'See you later!'

'Good luck!' – I replied and then added, 'I hope you win!'

74: Awakening

During my sleep I dreamed; I found myself in a strange country; in a big city. Everything spun around me. I saw long straight streets, flanked by very tall buildings which reached the clouds. There was movement all around me. Trams[237] with their drivers desperately clanging the bells; cars sped past me; a continuous line of pedestrians; lights - lights everywhere …. I was enveloped in a mælström of humanity.

A mist descended before my eyes and I could not see nor could I think ….

237. Buenos Aires had the largest tramway system in the world in the early 1900s; it closed in 1963.

51 : José Moneta, Ernst Bruhns, and Otto Zeiger

After that several graceful female figures appeared, draped in silks and fine chiffons. They smiled as they looked at me, while I contemplated them in silent ecstasy. I extended my arms towards them and they too disappeared into that unfathomable darkness, now surrounded by profound silence ⋯.

Something green appeared; green until infinity ⋯ and suddenly I found myself entering a green paddock where it was impossible to see the horizon.

The ground was carpeted with luxuriant green grass ⋯ green? ⋯ but was it not white, the ground that all these people walked over?

Then, something distracted my attention and, without noticing, I found myself beside a torrent of crystal-clear water. It was very clear, transparent, and ran at my feet cascading amongst the rocks of a riverbed with leafy banks.

A forest then appeared ⋯.

Later I found myself in another remote world. There was nothing; absolutely nothing! Whiteness ⋯ water ⋯ snow everywhere; wind of a terrible force ⋯ just a little red hut with a black roof emerged in that immaculate whiteness ⋯.

There were several men, their faces tanned by the icy wind; they had long beards and were dressed in thick, heavy coats; while an icy blizzard and driving snow prostrated their figures.

They returned from the beach, saddened, their heads bowed, and when we faced each other, they looked at me in silence, but continued on their way without looking back ⋯.

A feeling of anguish overwhelmed me. I wanted to shout! To follow them ⋯ but I was powerless ⋯.

I was petrified on the spot and, as if unconscious; neither the cold, nor the snow, nor the wind tearing my face, excited me sufficiently; I was unable to utter a single cry.

I tried in vain to make a sign for them to see me.

I just could not ⋯.

They were leaving, abandoning me ⋯. Alone! Condemned once again to the South Orkneys!

Abandon everything, absolutely everything for another entire year of my life, to find myself once more in the Antarctic ….

Whilst this anxiety was choking me, God took pity on me and caused me to awaken.

I looked at my watch. It was a few minutes after eight in the morning.

I turned over in bed, made my head comfortable on the pillow, and tried to sleep ….

I was plunged into a mist, with my eyes closed and my thoughts spinning round my head, without my being able to put them in order.

Gradually my mind cleared. I lay on my back and stared at the boards of the ceiling. Then I stretched an arm and reached for my pipe, putting it between my lips I began to think:

'Overall … another year here is not so desperate. It has some good points, not many, but some. Afterwards I would organize my daily life to make it easier to survive a second year.'

'Philosophy ! Bah! It was simply a new adventure.'

'A dangerous adventure if you like, … but that's what it is.'

'Up there? What will my parents think? If only there was a radio link, at least … for myself … all things considered … here or there, it's all the same ….'

'What an adventure!'

Then a doubt assailed me:

'Does the world still exist? … has there been a cataclysm?'

Afterwards, resigned, I thought:

'Very well, we are staying … but I will fight my whole life to get out of here ….'

'Resignation! Oh my God, help me! …'

Then everything ended.

I slept.

Suddenly I heard strange sounds that were not part of my dreams.

Loud knocks … rapid footsteps … a man running … shouts!

Then the door of my bunk-room burst open suddenly and Bruhns appears with an excited expression on his face, and shouts:

'They are coming! … they are coming … from the north!'

I leaped from my bunk, and while I dressed, I had to wipe a few tears from my eyes.

Straight away I left the bunk-room, buttoning my leather jacket, but returned to look out from the window to the beach: the flag pole stood empty.

I seized the blue and white flag from its shelf and ran to the beach to rejoin my companions. They were all looking north, directing my gaze to the point where they were looking, on the horizon, far away, I could distinguish a thin column of smoke rising over the sea.

There was no doubt: it was a ship!

I jumped for joy, while I raised the flag of my country on that solitary pole set in the pristine Antarctic snows. I saw my companions following my movements attentively; they respectfully uncovered their heads before the bicolour emblem as it blew in the winds coming from the South Pole, while tears of emotion ran down their weather-tanned cheeks.

75: The Relief

Two hours later we could distinguish the ship perfectly, it was a whale-catcher again[238].

'The funnel is painted sky-blue with a white band', said Plagge while looking through the binoculars.

'It must belong to Compañía Argentina de Pesca whaling from South Georgia', Valentiner commented.

'Does that mean that they are coming to relieve us with the same type of vessel and by same route as we took when we arrived?' – I asked.

'For sure. Possibly in Buenos Aires, they did not find a vessel strong enough to make this voyage. Whale-catchers are built to navigate on these seas.'

It appeared to be a slight improvement on *Rosita,* the vessel which had brought us. It was off Saddle Island and I could see it was making directly for *Manchón Austral,* as if looking for the entrance to Uruguay Cove.

238. On this occasion the whale-catcher was *Karl,* built in Norway in 1907 and similar to *Rosita.*

'It will not be here for at least an hour', Valentiner assured us, as he started to walk back to the base, followed by Bruhns.

Otto caught up with them, but Plagge and I preferred to wait on the beach.

At times, the silhouette of the tiny vessel appeared bigger. It had two masts and between them, you could see a tall funnel letting out a thick plume of black smoke.

While the ship approached at full speed, Plagge and I got the boat ready in which we decided to row to meet her as soon as she was close inshore.

'It might be a good idea to take Valentiner with us', I told my companion.

'I'll run and tell him', Plagge said, hastening off towards the dwelling.

A few minutes later, we were dragging the boat to the water's edge, but we still had to wait a few minutes, as the catcher following her course, was out of sight behind Point Lola.

Impatience now got the better of us; by mutual consent, with barely a word spoken, we resolved to launch the boat and row out to meet our new companions.

With Bruhns' help, we launched it and, in a few seconds, were floating on the waves. Valentiner was at the helm, Plagge and I grasped the oars.

At the same time the catcher came into view from behind Point Lola, we could discern a group of men crowded on the bridge.

'I wonder who the relief crew will be', I asked Valentiner.

'I haven't the slightest idea', he replied.

The whale-catcher had reduced speed and was turning in our direction. We stopped rowing as we were some two hundred metres from the beach, and fewer than a hundred metres from the ship. I could see an Argentine flag on the stern pole, but it didn't have the emblem of the Sun on it, it must be the flag of the Merchant Navy[239].

Instinctively I looked towards the beach, I wanted to make sure the

239. Flags of the Argentine Navy, and other governmental organizations, have a heraldic golden Sun emblazoned on the centre (as in Illustration 70). Civil organizations usually use a flag without this adornment.

flag I had left on the pole was in place, and I could see Bruhns standing next to it.

The catcher came nearer. Valentiner, without letting go of the binoculars was watching every movement and suddenly exclaimed:

'Who do I see? Yes, it's Berg! Yes, Berg is one of them.'

'And the others?' – I asked anxiously.

'I don't know any of them ··· I can't see them very well ··· Look – look!'

That was an unnecessary exhortation; Plagge and I did not take our eyes off the ship.

I could see two men busily working on the foc'sle preparing lines, as if they were getting something ready.

The catcher had stopped making way, barely a hundred metres from us. Aboard you could see several men moving around, but it was still too distant to be able to distinguish their features.

As the wind was coming from the north, it was easy to hear the rough command; immediately the heavy anchor dropped to the sea bed dragging a length of heavy chain with it.

At the same time a puff of steam escaped from the ship's whistle and its strident sound was followed by multiple echoes around the solitary rocks and ice of the cove.

Then came another whistle, and a third more pronounced, three blasts altogether.

On the beach, Bruhns answered the salute by slowly raising and lowering the flag of our homeland. That signal took us out of our silent and emotional contemplation.

We grasped the oars and plying them vigorously steered towards the whale-catcher now at anchor in the bay.

A few minutes sufficed to reach her side, but we were rowing with such determination, so forcefully, that we misjudged the speed, and ended up crashing into her steel hull.

They threw us a rope-ladder, and we climbed aboard.

Standing there was a group of five men, dressed in full winter clothing; while a bit farther away, some Scandinavian sailors were watching us attentively.

I did not know, I had never seen, any of those faces, but that didn't stop me from grasping the arms and embracing the man who was closest.

We had never seen each other, but you would have gained the impression that we had known each other all our lives!

76: Tidings from Afar ···

When I got over my emotion, I saw Valentiner talking animatedly to one of the new team. He was of average height, very white and blonde, clean shaven, his face lit with a nice smile.

'Carl Berg', Valentiner introduced him. 'The Assistant Leader of the new wintering party, on his fourth expedition to the South Orkneys!'

We embraced warmly and I then went on to meet those who would be relieving us: Vladimir Dobrovolski, the Leader; Alfredo Witt, the Chief Observer; Horst Irrgang, Second Observer, and Otto Hess, who replaced Otto Zeiger as Cook.

They all impressed me in different ways. As soon as I met Berg, I discerned in his features and gestures, and in his easy good natured conversation, that he was, heart and soul, a veteran of previous South Orkney expeditions, and an expert in everything related to snow and ice. I later discovered that he was Swedish.

The Leader, Vladimir Dobrovolski, of Russian origin, was making his first voyage to the South Orkneys. Being from a country of snow and ice, he was clearly experienced in all the problems that he would have to deal with here.

Witt's tall figure stood out among his colleagues. He was also very thin. Immediately a current of mutual bonding was established between us, as, in the entire company, we were the only Argentines.

Irrgang, a fine young German lad who had not yet turned nineteen, and Otto Hess, also a German and a veteran of the 1922 expedition, completed the small group of men who had arrived to isolate themselves in the South Orkneys.

Witt was the first one to address me in the familiar form:

'You are Moneta?'

The familiar form of address came as no surprise: I realized that if we could not instantly be friends there, we would not be friends anywhere, and I replied in the same style.

'The same. Have you any news of ....'

Without letting me finish, he took my arm and said:

'Come with me, I have letters from your parents.'

Before I could say anything else, he added, pointing north 'up there they are all fine'.

While we went below to the ship's mess, Witt continued:

'Your father was with me several times before I embarked, and when he knew I was to be the only Argentine, he wanted me to give you the first items of news.'

'Thank you. I hope one day to be able to repay this favour.'

He handed me a large bundle of letters and a packet with newspapers and magazines.

'The other package was sent by your mother, it has warm clothing and a few delicacies which you will have been missing. Here, separately, is the last letter your father handed me aboard, as we were leaving. You can read it ... it has good news.'

You have to have lived in the South Orkneys for a year to experience the unequalled emotions and the overwhelming desire to devour in a glance these few lines hastily written by the ones you love.

Witt understood this and he left me alone for a few minutes, while he sorted his luggage. I sat on a cushion in the bridge and began reading.

There was not much time for reading. Valentiner and Plagge came in a few minutes later, followed by the rest.

Everyone was talking at the same time and I felt as if I was in the Tower of Babel, with snatches of conversation in English, German, Scandinavian, and Spanish; all mixed.

Then through a door another figure appeared whom I already knew; the newcomer embraced Valentiner and exclaimed:

'*Hullo Lads!* ...'

It was Larsen, the same Skipper of the whale-catcher who had brought us to the South Orkneys in the first place and who now came to fetch us.

His athletic figure had not changed and after greeting us effusively,

I heard him telling us:

'Well lads, now to work. We have few hands and there is a lot of ice farther north .... I don't want to end up in this place with the vessel trapped ... come on ... let's get on with it!'

Then the friendly reunion broke up and we began unloading the expedition provisions.

77: Until We Return

Accompanied by Larsen, we went ashore in two boats, our own, and another boat which the whaler's crew had launched. We walked on the track to the dwelling to show the new arrivals the Observatory. Bruhns and Otto were waiting for us, and Otto had prepared some tasty morsels which were offered to the new arrivals.

Afterwards, Valentiner said to me:

'Kleine, you show Witt and Irrgang around, while I do likewise with the new Leader and Berg. Otto will deal with the new cook, who will need no instruction as we know he is experienced.'

I began my commission by showing the new recruits their respective sleeping-quarters. Then we toured the other Observatory buildings, and I couldn't help noticing that they looked at it all with some amazement.

They asked millions of different questions, to which I replied in the confident tone of someone who has had the experience of wintering.

'What's it like living here?' – asked Witt.

'Well it depends on several factors', I replied, while Irrgang listened attentively.

'You will have already seen that, above all, order reigns here. Every tool, every object, has its place. You will soon realize that it has to be like that to enable you to live with a modicum of comfort.'

'What about the isolation, the loneliness?'

'That also depends,' I replied, 'on the amount of philosophy that each of you brings to the job. Down here you have to have sufficient tact not to be a nuisance to your companions; it takes all the altruism you can muster to think always of others, and not of yourself.'

While I was making these observations to the new recruits, I was showing them round the buildings and attempting to inculcate in them the best philosophy to bear the difficult life they would soon begin.

'There is no need to be afraid of anything', I told to them. 'You have to live prepared for anything and everything, but, at the same time, it is important to live in perfect harmony with your comrades. Remember, my counsel is based on the arduous experience of an interminable year. The man who manages to get on well with his companions, without irking them, is the moral winner: it is he who will show himself to have mastered everything I have been expressing.'

Time was passing quickly and Larsen was pressing us to hurry with the unloading. For quite some time already the catcher's crew had been working on the task of loading and taking ashore boats stacked high with sacks of coal and boxes of provisions. They deposited everything on the beach, close to the water's edge. Whenever a boat reached the beach those of us ashore had to carry all this farther up, beyond the high tide level.

By the night time, we were all exhausted, we had worked so hard that, in only a few hours, the entire cargo, including the personal luggage of the relief team, had been landed.

The new men installed themselves in the dwelling, and we took our belongings aboard the catcher.

Although it was dark no one thought of sleeping. I returned ashore to spend the last few hours in the company of those who would remain. Neither did the new men sleep, and when I entered the place where I had spent the last thirteen months, I went into the parlour and saw them busily writing their last letters to friends and relatives. I offered to take their latest news to Buenos Aires and hand it over personally, which they all accepted.

I spent the night with the new team; while they filled reams of paper with writing, I made them coffee, alternating with glasses of whisky and brandy with which they tried to distance the raw remembrance of the farewell at Buenos Aires, which, even here, occasionally filled their eyes with tears.

We hurriedly gathered the last of our personal baggage, the expressions on our faces made a striking contrast with theirs. They did not take their eyes off what we were doing; occasionally a fleeting look of misgiving seemed to cloud their brows.

They hardly spoke, and when they did, it was only to ask questions about the way of life on base.

On the contrary, we were joking and loquacious. Our whole attitude showed the self-confidence that comes with experience, or rather, the feeling of competence, of knowing that a job has been well done.

We spent the night in continual comings and goings, sidestepping all the gear cluttering the small room which we dignified with the grandiose term of 'parlour', until the darkness gave way to the first light of dawn.

At five I heard the front door opening and saw Larsen arrive out of breath.

'*Hullo Lads*! Let's go! Let's get moving! Wind's from the north. Ice and fog on their way ....'

I was in no rush to leave, the fact that they had come to relieve us was quite enough for me.

Larsen's warning was worth listening to: fog and ice combined! This was the most dangerous combination of all for impeding polar navigation.

The new Leader handed me a thick bundle of letters. We then had a drink, toasted the happiness and well-being of those who would be staying; and they responded by wishing us *bon voyage*. Then, *en masse*, winterers mingling with those going home, the whole gang made its way to the beach where a boat awaited.

At the water's edge we embraced each other in amiable confusion and then rowed towards the waiting whale-catcher.

When we arrived, the boat was hoisted aboard and, at the same time, we heard the clanking of the chain on the capstan as the anchor was raised.

Five men remained on the beach, of whom Witt's tall silhouette stood out above the others.

They walked in the direction of the flag pole, and formed a line near it.

Suddenly I heard the bell ring in the catcher's engine room. The water churned and eddied under the stern like a mill race. We were off!

The blasts from the ship's steam whistle deafened the air around us. In the distance ashore, I watched the blue and white flag slowly descend the pole only to be raised again to the top.

Another whistle blew, followed by a more pronounced one, longer ··· and then twice more the flag went down and up again where it remained. The men standing at its base waved in farewell, and remained motionless, as if petrified. Their gaze was fixed on our vessel rapidly departing northwards. Left behind, in that remote place, were five beings who would work for the scientific progress of humanity, their souls guarded and protected by the flag of our homeland.

78: *Orwell*[240]

I am not sure how long I remained on deck gazing at the outline of those lands I may never see again in my life. It was only when we reached the open sea and Saddle Island interposed itself between the vessel and the primitive abode where I had lived for a whole year, that I went below to the cabin that had been reserved for our quarters.

There I heard another piece of news, when Valentiner told me:

'We are going to *Orwell* ····'

'Where?' – I asked, not fully understanding.

'To *Orwell,* the floating whaling factory that has been operating since November off Signy Island. Larsen has just given me the news.'

'You mean,' I said 'the Norwegians have been hunting whales since November in the South Orkneys and we knew nothing about it?'

'I had guessed they might have been,' replied Valentiner; 'in years gone by, before the First World War and in fact during it, every summer, they used to come and hunt whales in the seas around the South Orkneys[241].'

240. *Orwell* was a floating whale factory built in Britain and owned by Tønsbergs Hvalfangeri of Norway which operated at the South Orkney Islands from 1922 to 1925 with two whale-catchers. Some whale processing was at a land station, in Factory Cove on Signy Island, during several summers. A new ship replaced *Orwell* in 1925, also named *Orwell* and built in Britain (indexed as *Orwell II*).

241. Whalers worked around the South Orkney Islands from 1907 to 1915 and 1920 to 1931:

'According to what you are saying, it looks as if they are anchored off Signy Island, south of Coronation Island ?'

'Exactly, in the same archipelago through which we are sailing. To be more precise, forty nautical miles [70 km] from the Observatory, towards the west.'

'How can it be that we never spotted any of the whalers, or the auxiliary ships, that they use in their work?'

'Larsen has also just informed me that the ice, the pack ice, has closed in at this eastern end of the islands, but that the west is ice free, allowing ships of all kinds to navigate freely.'

'So that within five hours we will be with them?'

'Approximately at noon we will be there. The catcher has bags of mail from South Georgia and from Europe, mail that was left in Grytviken, waiting to be sent to the South Orkneys at the first opportunity.'

'This is it, right?'

'That's correct', replied Valentiner.

I was interested in visiting this ship and learning about the operations of a floating whale factory.

That is how I found that it was a large vessel, registered in Norway and operated by a crew of the same nationality of about four hundred men. In addition *Orwell* was accompanied by four small vessels, whale-catcher's, similar to the one we were aboard. These hunted the whales which were processed aboard *Orwell*, the oil being extracted from their blubber.

We had rounded Saddle Island and were proceeding towards Washington Strait, but an extensive ice-field prevented further southward progress.

Larsen put the ship about, and I guessed his intention was to attempt to find a passage through the next strait, Lewthwaite Strait; but we could not enter there either due to the pack ice. There was no choice but to continue along the north coast of Coronation Island to its western end.

Hart, 2006.

Only at first light the next day were we able to catch a glimpse of Signy Island, but it was 6 p.m. before we arrived.

As we drew closer, the catcher's speed slackened, and when directly in line with the small harbour situated on the east side of the island, I caught sight of an enormous ship at anchor in the middle of the bay. Moored alongside were a couple of whale-catchers and half a dozen recently caught whales.

At our appearance, a large number of men gathered on *Orwell's* deck. I noticed they were all wearing thick leather coats as well as the classical sou'westers and long thigh-length boots.

In the calm water of the bay there was floating whale waste: intestines, stomachs, and bits of meat which, despite the prevailing cold, permeated the air with a stench of decomposition.

The air was thick with circling petrels and gulls; every now and then they would alight on the water and take a lump of offal in their beaks and greedily swallow it.

On the coast of the island were two wooden sheds; Valentiner told me what they held:

'One is the powder magazine[242] where explosives were stored; gunpowder used for the harpoons and cannons of the whale-catchers. The second is a store for miscellaneous ship's equipment which would only occupy space in *Orwell,* that could be more profitably used for loading whale oil.'

We came alongside *Orwell,* a rope ladder was thrown down to us, and Larsen clambered aboard carrying a large sack of mail on his back. I followed next.

I found myself on the deck of a true floating factory. It was a novel type of ship, the deck was cluttered with machinery of all kinds; a quick glance was sufficient to tell me what each machine was used for.

Standing on the whales that were floating alongside were several men, the heels of their boots had long claw-like spikes that stopped them slipping and falling into the water.

242. In 2017 the powder magazine remained on Signy Island adopted, in 1947, by the precursor of the British Antarctic Survey as an emergency depot for their scientific station.

These men were equipped with the traditional long-handled knives to flense the whales. With these they cut long strips of blubber, which were winched aboard.

The next stage was to feed the strips of blubber into a large machine that sliced them into thin strips. From there they were conveyed into a cooker, where the oil was extracted by steam under high pressure.

On another part of the deck, steam powered saws were operating. These, were for cutting the bones to a suitable size to fit into another cooker for production of bone meal.

While Larsen dealt with the formal introductions to the Captain[243], officers and scientific personnel of the floating factory, Plagge gave me a quick commentary on who they were and what they did.

Then came my turn to be introduced. I was taken first to the Captain of *Orwell* who extended his hand cordially, and then to the Officers who had formed a circle around us. I shook hands with each in turn until I reached someone who stood alone. At first sight he looked like another officer, but then I noticed that his uniform was different. He was looking at me and smiling. Instinctively, my gaze fell on his cap badge and, straight away, I discerned the Royal coat of arms of Britain. On the sleeves of his jacket were several stripes.

I stretched out my hand, which he grasped genially, and I managed to have the following short conversation with him in English:

'I am Moneta, Assistant on the expedition.'

'My name is Rumbolds[244]. Customs Officer ….'

'… Customs Officer?' I interrupted.

'Yes, of His Majesty's Government', he replied while saluting.

A slight smile lit up his face, guessing the reason for my confusion, he put his hand on my arm and said:

'Let's go and have a drink. I was never in Buenos Aires, but I like Argentines.'

Plagge followed us, while Bruhns accompanied Valentiner and the

243. Captain Ingvar Thom, with Søren Berntsen Manager of the whaling operation

244. William Charles Rumbolds (1892–1969); appointed Falkland Islands Dependencies government representative for the South Orkney Islands from 1922 to 1930. He had visited the Argentine meteorological observatory during his first summer (1922–23) while aboard *Orwell.*

Captain. Otto had also found some compatriots among the crew, and decided to join them.

* * *

We dined that night aboard the floating whale factory, and the next morning, very early, we said our goodbyes, and re-boarded the whale-catcher that would take us north.

I had struck up a friendship with Rumbolds, the Customs Officer of His Britannic Majesty's Government, and through him I learnt that the British government monitored all whaling activities in the Antarctic seas.

I was leaving the South Orkney Islands behind, destined for South Georgia aboard the whale-catcher ably commanded and navigated by Larsen. Huddled in a corner of the poop deck, I looked with nostalgia at the fading silhouette of the islands of ice and rock.

Plagge came to me and we had the following conversation:

'What are your thoughts about the year we have just experienced?'

'Heavens! ··· I don't know ··· deep down I think it has had some minor influence.'

'Would you like to go back?'

'Why not?' – I replied 'As far as morale goes I may have suffered a bit, but I think I have come through it well. All in all, whether I live here or somewhere else makes no difference to me.'

Then, expressing an idea that had been germinating in my mind for some time, I said:

'Yes ··· I am thinking of going back, but it would have to be with an expedition committed to some new project.'

'Such as?' – Plagge interrupted.

'A radio station link. That way the solitude would be easier to bear.'

In fact, the idea of returning has been in my mind for some time, but it depended on many things 'up there', in the north, and on what would happen in the meantime to affect my future.

79: Second Expedition

Ten months have passed, and I am on my way back to the South Orkneys, as part of the expedition of 1925. We are sailing aboard the transport-vessel *1º de Mayo*[245], of the Argentine Navy, which is heading to Isla de los Estados to refuel from a bunker ship waiting there for us.

The Leader is Bruhns. The Assistants are Alejandro Guerra Boneo and Albert de Tolnay. The Cook is Georg Piper. I will introduce them:

Guerra Boneo is a 27 year old Argentine lad, clearly *porteño,* not very tall, but well-built, fair, ample forehead and lively eyes. We have struck up an easy friendship and had used the familiar terms to each other since our first meeting many months before departure from Buenos Aires. He is a distinguished chess player[246] and has promised me a few lessons during the interminable polar nights that we will have to while away in the South Orkneys. He is modest and affable. I dedicate to his memory these affectionate recollections of the ups and downs we suffered together in the south ….

Albert de Tolnay, 30 year old, Hungarian, quite tall and thin, dark-skinned; his face shows all the characteristics of his race. It is interesting to listen to his colourful conversation, as he has travelled round much of the world and has had a thousand different jobs. All his activities, gestures, outlook, and behaviour reveal the personality of an incorrigible adventurer.

Georg Piper formed part of the expedition of 1920, taking an active part in the surgery performed on Tapia when he had to have his fingers amputated. German, 32 years old, medium height, a genial blonde with the physical fitness and the robust good health of men brought up on rural work. He had a fine crop of fair hair and lively eyes, like two sparks in his small face.

These were the men with whom I would be thrown together for another year in the South Orkneys. This time they all seemed to be

245. Captain Ramón Augustín Poch

246. Guerra Boneo was an internationally distinguished chess expert who would probably have become a Grand Master had he not died in an automobile collision at the end of 1926.

diametrically opposed in character and temperament, but the thought of being isolated with them for a year held no fear for me. I had my previous experience to thank for that.

Amongst our equipment was a complete radio station, transmitter, receivers and a dynamo powered by a petrol engine. This itself would be an incentive for much of the year to pass by without our even noticing, thanks to the amount of toil and effort that would be required to set it up and establish communications with Argentina.

For something else to fill the time, I had bought ciné equipment, intending to bring to the screen the highlights of our way of life and the scenery, which at that time was completely unknown to Argentines.

A firm in this business had provided me with the photographic equipment[247]. I had signed a contract with them for the commercial use of the film. Although I had no intention of making a fortune from this venture, I felt that the task would be an enjoyable hobby and, at the same time, would enable me to make a little money on the side.

The experience gained on the first expedition had been so instructive that I was able to bring with me now the basic essentials to enable me to be usefully occupied for the whole time.

The voyage was not as rough as the first on the whale-catcher, although *1° de Mayo* still rolled violently, and was not unworthy of the reputation, conferred by the officers, of being a lively ship.

80: In the South Orkneys Again

Again I found myself in the little wooden dwelling that was our living quarters. Nothing had changed from the previous year. Everything the same, the personal objects, the rudimentary furniture, the instruments and utensils, the walls, the doors and windows; everything was in the same place.

This time I knew two of the men who accompanied me, but the others were new.

There was a small change in the daily work routine as well; the

247. This company was Cinematografía Valle founded in Buenos Aires by Federico Valle in 1916 and closed in 1958 after another serious fire. See § 105 and footnote.

installation of the radio equipment. This was an addition to the work-load, needing every spare moment.

We erected a four metres square hut near the glacier on Mount Mossman, a few paces from the cemetery, almost on the shore of Uruguay Cove[248].

Tolnay, who was in charge of the radio equipment, directed the work of that simple construction and later, with the help of Guerra Boneo, we put together a steel pylon that would support the weight of the antenna.

A long month went by on this work, and gradually it began to dawn on me, that despite all our efforts, we were not going to be successful; our work was doomed from the beginning in its fundamentals.

The technician, or rather the man who had been put in charge of the technical side, didn't strike me as being a technician at all. The first serious difficulties arose when he installed the electrical equipment.

My uneasiness was evident, as I realized only too well that we were doomed to total failure. When I tried to persuade my Leader to give me authority to personally supervise those installations, it was already too late, as the armatures of the main dynamo had been burnt out by badly made connections.

The only things remaining intact were the petrol engine which drove a supplementary dynamo on its shaft, the bank of storage batteries, and a receiver which sporadically allowed us to listen to radio transmissions, although rather unsatisfactorily.

At the beginning of the winter, without any possibility of being able to repair the damage caused to the apparatus, I proposed we make use of what remained intact; the petrol engine and batteries, to install electric light in the dwelling. Everyone accepted this; Guerra Boneo and I immediately started work on laying an electric cable between the radio hut and the dwelling.

In May this was complete, and from that memorable event when we first lit the parlour with modest electric light, we stopped using the primitive paraffin lamps.

248. This was later used as a store for flammable materials, see Plan 1 and Illustration 52.

Otherwise, I applied myself to filming the most outstanding features of South Orkney life, in all its shapes and forms, which had the advantage of considerably increasing the amount of time that I was fully occupied. I wasted no opportunities to record on celluloid the most outstanding action shots of our daily activities. Thus, as time passed, I filled the circular containers with film exposed and developed, for the first time[249], in this part of the world.

Moreover, my previous experience had given me much knowledge which I now tried to inculcate in my new companions, Guerra Boneo and Tolnay, who, in a short time, became familiar with the underlying secrets of coping with that singular way of life on the islands.

Despite all my earlier experience I was unable to deal with the personal conflicts that afflicted the spirits of my companions. Many times they were depressed by the disparity in character among themselves and those with whom they had to live. This was one of the major difficulties of that year.

The disaster with the radio aggravated the arguments between the Leader and Tolnay, quarrels which redounded to the immediate detriment of the general harmony which should have prevailed among everyone.

On my part, I had to continually strive to bring everyone to agreement, and avoid those disputes ending painfully and irreparably.

Nevertheless, time still went by ··· slowly. The long evenings had their focal point in Guerra Boneo, who displayed his prowess as an expert in chess by playing simultaneous games with Piper, Tolnay and me.

The most interesting part was that 'our professor' as we affectionately called him, would go to bed after dinner, put his light out, and while he dozed, we moved our pieces on the board which was on the parlour table, and at the same time, noted the moves. However we still kept losing! Our adversary's rare ability was so manifestly clear, with his extraordinary memory, that even when he gave us the advantage

249. William Bruce's Scottish National Antarctic Expedition (1902–04) had previously made brief ciné films which included aspects of the South Orkney Islands.

52: First radio-telegraphy station (1925), later the flammables store

of a few more pieces, he still managed to beat us.

81: Second Winter

The sea had frozen over completely and the propitious season for going on excursions had begun. In those journeys I found the Cook, Georg Piper, who shared a bunk-room with Guerra Boneo, was a welcome companion. That was how I made innumerable visits to the neighbouring islands, and reached places even farther off than during the previous year.

Piper had a predilection for those adventures and he was always prepared to go on every occasion I invited him, even when the fiercest storms lashed us.

Sometimes Guerra Boneo would accompany me, but his sharp mind relished the chess board far more than the grand horizons of polar solitude.

Hundreds of metres of celluloid were running through my ciné camera. I was filming details of the lives of seals and a few petrels that had not yet abandoned the islands for the open sea before the pack ice arrived.

By that time I had become a confirmed smoker and despite the extra provisions of cigarettes and tobacco I had brought with me, I had to reduce my consumption to make my ration last to the end of the year. Guerra Boneo was one of those who suffered the most; all the tobacco supplied was not enough for him, but he managed to take his mind off this scarcity by exploiting his ability at playing chess.

Tolnay, who delighted in playing chess with Guerra Boneo, though he always took the part of the vanquished, was often in dialogues with him such as:

'Mister Professor, my dear friend,' Tolnay would say, 'I will challenge you to a game in exchange for two packets of cigarettes.'

The proposer of the challenge knew beforehand that his proposition would be accepted; but you could hear Guerra Boneo saying indecisively:

'No ··· it's not possible ··· I will beat you.'

'That does not matter, but ··· give me a castle as advantage.'

'I will give you two castles!' Guerra Boneo would reply 'and I still think I will beat you.'

'Bet?' – 'you're on!'

'Accepted ··· two packets!' Guerra Boneo replied.

Once he told me:

'I don't know why I play this game with you fellows! Even though I take you both on at once, in two simultaneous games, it doesn't really interest me, it's still too easy ···.'

That was how it was; I could see his obvious superiority as 'professor', and it was no match for our rudimentary knowledge of that enthralling game.

In August the only radio receiver that we had fell silent: the last two valves had burnt out. However, this accident was no great loss, as we had heard little or nothing to that date. Tolnay had received one or two fragmented bits of news, and these were from a broadcasting station. With our radio receiver it was impossible to understand anything else, as he did not know the Morse alphabet.

Life continued as normal amongst us, apart from a few altercations between Bruhns and Tolnay. Calm would reign again, there was a constant need to invent all sorts of work to make the time go by.

Excursions, trips, photographs, studies ··· these were the usual pastimes, apart from the routine work of the Observatory.

As a result of my excursions with Piper, when we roamed the snow-covered hills around us, and where we occasionally became lost, the bond of friendship between us strengthened and we came to use the familiar terms. This helped to increase our self-confidence as well as enlarge our circle of friendship. Already Guerra Boneo and I treated each other as brothers.

On one of those treks with Piper, I said:

'I have noticed there are many places which still have no geographical names on the maps, which leads to much confusion. I am prepared to baptise anything new that we find and bestow an appropriate name on it for a new map of the region'.

We had gone out with the intention of exploring in detail a group of islets in Scotia Bay, located near the foot of the glacier at Point Diebel, about five hundred metres from it.

That is where Piper went over a cliff edge, ten metres high, while still wearing his ski. I remember how it scared me to see him lying there unconscious. I ran to rescue him and when he came to I asked:

'Have you done yourself any damage?'

'What a fall!' – he said, while trying to sit up: and added 'no ··· something hurts ··· my leg ··· here!'

'Let me feel', I said.

'It's nothing, it's gone ···.'

He got up slowly and carefully, and in a few minutes had quite recovered from the accident.

I had an idea: I took a chunk of snow, made it into a ball, then moving towards the piece of rock that was jutting out of the ice a few paces from where we stood, I shouted:

'Deserted rocks! Antarctic ice! ··· I baptize you in the name of the islands of Georg Piper!'

I then hurled the snowball as hard as I could and it shattered into a thousand pieces as it struck the rock.

I ran towards Piper, who was watching me with a smile on his face; I embraced him and told him:

'At least there will be a reminder of your landing on Antarctic islands!'

82: Nothing Without Excitement

Weeks turned into months; winter went by and spring arrived.

We had a new project to look forward to in spring: boat excursions in the open sea. As well as the pram[250] brought down in 1923, we had another boat, six metres long with a cedar hull to which we had mounted a petrol outboard motor.

My inseparable companion, Piper, followed me in all I suggested, and he had no hesitation in accompanying me out to sea.

To make these journeys I had to wait for calm days, with a very slight swell, because in those waters it is foolhardy to go far offshore, even on days when the sea seems set to remain flat calm. A gale may begin without warning and, in no time at all, the mirror-like surface is transformed into huge waves.

With this novel form of transport we could reach the remotest parts of the coast of Laurie Island. This permitted me to study some of the sea birds, especially the shags which nested on inaccessible stacks surrounded by sea. In this way I learnt that these swiftest of birds had no fear of man's presence, and we were able to film scenes with Piper catching them alive with his bare hands, by leaping on them when they were on their nests.

On one particular occasion we were returning from our first hunting trip to the Shag Rock, close to Cape Mabel, the northern headland of Jessie Bay. While we were in the middle of the bay, a strong wind blew in from the north, which quickly developed a furious sea and huge swell.

Piper was crouched in the bow, holding up a waterproof tarpaulin to reduce the sea washing in, but it was to no avail. We bailed furiously, but all our efforts could not keep pace with the seas filling the boat, and the waves threatened to sink us.

250. A pram is a Norwegian wooden, flat-bottomed, rowing boat adapted for landings on rocky shores in difficult seas.

53: Georg Piper's islands and the sea frozen to the horizon

I had no other alternative but to jettison some of our cargo to lighten the load of the fragile craft, and it was truly providential that we did not founder there. We were unable to return to the rocky stacks or to run into the shore, which, in reality, was nothing but the vertical ice cliffs of a glacier front, thus we passed anguished moments. Our only chance was to continue ahead to try to reach Uruguay Cove before we sank.

Eventually, after a constant battle of bailing the water as rapidly as it came flooding in, we managed to get a glimpse of the base, and I gave the motor full throttle to get maximum power.

Our companions ashore, being alarmed and anxious, were waiting for us on the beach. As I leapt ashore I made a firm resolution never to attempt another escapade like that. My wishes were, however, in vain: on the next week I was out exploring by sea the rocks off Cape Robertson, at the entrance to Washington Strait.

* * *

The months went by and another year drew to a close. Then came January, February ··· and one day a whale-catcher[251] appeared, its presence indicating that once again we would be relieved via South Georgia, as in Buenos Aires it was not considered sufficiently safe to

251. This time the whale-catcher was *Foca;* built in Norway in 1921 and now a museum ship in Horten; 210 gross register tons, 32·3 m long.

commit the transport ship *1° de Mayo*[252] to another voyage to South Orkneys.

The relief party was led by Berg, who was now returning as Leader; Witt was Assistant Leader; Irrgang was First Assistant and Orlovski, the Second. The Cook was Edmund Müller, a German[253].

We embarked for the return voyage and now, more than ever, I was determined to return, but this next time it would be as Leader, so that I could recruit a winter party consisting entirely of Argentines.

★ ★ ★

The Skipper[254] of the whale-catcher, a good-natured Norwegian, was not unlike Larsen. He informed us that we would have to wait for a full month in South Georgia until one of the auxiliary vessels of the Compañía Argentina de Pesca of Grytviken was ready to leave on a voyage to Uruguay. We arrived, at last, at the whalers' island, and were greeted by the station manager Mr Jacobsen[255], with the utmost courtesy and attention. He put his house at our disposition and, on asking if there was anything we would like straight away or urgently, did not show the slightest surprise when we all replied in unison:

'A bath!'

A prolonged month of waiting went by in the placid surroundings of Cumberland Bay, and then we boarded *Harpon*[256] which took just over a week to take us to Montevideo. From there we took the commodious overnight steamer to Buenos Aires.

252. Regardless of this comment *1° de Mayo* was used for two later relief voyages to Orcadas in 1928 and 1929, and to other Antarctic stations in 1942 and 1943.

253. Extracts from Müller's diaries are published in: Krause, 2007.

254. Captain Johan A. Johanssen

255. Fritjof Jacobsen; however from 1921 to 1930 Captain Wictor Esbensen was Manager of Grytviken, see § 8.

256. *Harpon* was a transport ship, built in Germany in 1897, used by the Compañía Argentina de Pesca from 1923 until scrapped in 1957; 4950 gross register tons, 122·4 m long. Captain Anders M. Abrahamsen.

83: First Entirely Argentine Expedition

During the months when I lived in Buenos Aires after returning from my second expedition, I devoted my waking hours to organizing a third expedition, in which I would have the satisfaction of being Leader.

In truth I desired to return once more to the South Orkneys to enjoy the somewhat vain satisfaction of being the first native-born Argentine to command a wintering expedition. This would be the first wintering party to be composed entirely of men of the same nationality as myself. A further incentive to attempt this new venture was that it might possibly be me who would successfully inaugurate a radio link to those faraway polar lands.

One further desire; unfortunately, the ciné film I had made during my previous year, had been destroyed by fire at the headquarters of the company that had financed it[257]. Consequently I planned to make another film, benefiting from the experience I had already acquired.

That was how, about the middle of 1926, I came to be designated Leader of the 1927 expedition. I took advantage of the authority that was vested in leaders in those days and I was able to appoint my team personally. I chose them from a small pool of men whom I thought were endowed with a natural ability to tolerate that kind of life.

As regards the future installation of a radio-telegraph service in the South Orkneys, its construction was entrusted to the Argentine Naval Communications Service, an esteemed organization that would guarantee the quality of the instruments supplied to the expedition. As they embarked on this initial preparatory work, they were fully aware of the previous team's failure in 1925, and the lessons to be learnt.

In October 1926, the complement of the expedition was:

As Assistant Leader, my lifelong companion and friend, Miguel Ángel Jaramillo, who, during the preceding few months was perfecting his professional training at the magnetic observatory in Pilar,

257. Ciné films in that time had a cellulose nitrate base which is hazardously flammable, almost explosive. See footnotes in § 79 and 105.

54: The dwelling in winter snow

Córdoba[258]. This great friend, who died prematurely, was an efficient collaborator in my ideas during our stay on those distant islands. On repeated occasions he demonstrated strength of character and a tireless energy in overcoming the trials and tribulations of our basic and difficult life. Jaramillo's decision to accompany me on this venture was all the more admirable when one bears in mind that he abandoned well-paid work to follow me to a part of the world where any remunerative potential was limited to a very modest salary, and where dreams of glory and a brilliant future would depend on personal references and recommendations after return to Buenos Aires.

As First Assistant I had proposed Pedro Martín Casariego; and for the Second Assistant, Luís Fállico, both Argentines.

To fill the post of Radio Operator, a key man in the circumstances, I appointed Emilio Baldoni. His selection received official approval from the Naval Communications Service and the Director of the Meteorological Office. The young non-commissioned officer of

258. This is a geophysical observatory about 40 km south-east of the city of Córdoba, central Argentina.

the Argentine Navy had an impeccable background, he had been an outstanding pupil. Subsequently he won a scholarship to Germany to gain further qualifications.

Konrad Becker, appointed as Cook, was of German origin but had taken Argentine nationality. He was experienced in the art of preparing good meals from tinned and dehydrated ingredients, as he had worked on countless merchant ships on all the oceans of the world.

Thus the expedition was complete with native-born Argentines, with the exception of the Cook, but even he was under the protection of our flag by virtue of his Argentine citizenship.

For the first time, an expedition would be departing for the South Orkneys consisting of six men instead of five. Originally it had always been five but it must be remembered the sixth post was the new position of Radio Operator which did not exist previously.

The problems of supplying clothing, provisions, apparatus and instruments for several different studies, radio equipment, fuel, and building materials for extending the dwelling to accommodate the radio station were resolved quickly. By the end of October we were ready to sail as soon as instructed.

Bearing in mind the numerous structures that would have to be erected for the new radio station, I alerted my superiors of the need for us to leave Buenos Aires before the end of the year. The success of the project to install a radio link between South Orkneys and the South American continent would entirely depend on having the time available for the construction of buildings and ancillary installations, especially the thirty metres high galvanized steel pylon. All this would have to be accomplished before the beginning of winter, especially if that bleak season were to arrive prematurely. In such an eventuality, all the essential outside construction would inevitably be scuppered.

My suggestion and fears in this matter received a sympathetic hearing from my superiors. They arranged for us to make the voyage to South Georgia aboard the vessel *Tijuca*[259], an ancient sailing ship

259. *Tijuca* was built in Sweden in 1866 and plied to South Georgia from Buenos Aires from 1907 to 1942; wrecked 1946 off southern Brasil; auxiliary oil internal combustion engine fitted in 1924; 846 gross register tons, 54·2 m long.

belonging to the Compañía Argentina de Pesca of Grytviken, South Georgia.

84: Some Comments

On 26 November 1926 the entire deck of this ancient vessel, moored at a wharf in the South Dock, was swarming with people. Most were relatives and friends who had come to say farewell to us. The crew worked hectically to stow the last crates of our provisions for Orcadas as they were lowered into the hold. On the foc'sle a group of people was engaged in animated conversation about the four dogs I had been responsible for taking aboard *Tijuca*. It was my idea that they would help us haul the sledges on the island[260].

It was a hot day, and in the midst of the comings and goings of all those people, I noticed my new companions were besieged by inquirers wanting to know their latest impressions.

At the hour of departure, fixed at four p.m., someone approached and drew my attention to the arrival of several vehicles that stopped near the gangway between the wharf and the ship.

I went forward to receive the newly arrived dignitaries, for they were no less than the Minister for the Navy, Admiral Manuel Domecq García, and the Minister for Agriculture, Engineer Emilio Mihura. They were accompanied by the Director of the Meteorological Department, Captain Enrique G. Plate, and other worthies whose names it is impossible for me to recall.

After I greeted these gentlemen and made the conventional introductions, we went to the vessel's wardroom, where I remember Captain Plate, in his capacity of Director of the Meteorological Department, asked me in the presence of all the others:

'So you think the most likely date you will be able to communicate with us by radio will be only in the middle of next year?'

'I hope it will be earlier,' I replied 'It entirely depends on the prevailing weather in the Southern Ocean.'

260. Sledge, and other, dogs had been on Laurie Island sporadically since 1903 with the arrival of *Scotia*.

'More or less, can you give a date?'

'Let's say ··· well, March', I said.

'March?' chorused several voices.

'Yes,' I affirmed, 'if we have not established contact before 31 March, this silence could be taken as another failure.'

'Why do you think March will be the most appropriate date?'

'Because, after that date, it will become practically impossible to do any outdoor construction work because winter will have begun. Of course it might happen that this time the winter arrives a bit later than usual; which will make the completion of building work more feasible, but in the South Orkneys we cannot count on that ···'.

The conversation turned to other topics, and a bit later my presence was claimed by some friends. Among them, in a group which included my father and others, was Don Alberto Desimone, Director of a popular Buenos Aires newspaper. We greeted each other warmly and I said:

'It won't be long before you are receiving enough material to fill whole pages of your newspaper.'

'I know that already,' he replied laughing 'Jaramillo is will be sending me dispatches; I have appointed him the *South Pole Correspondent*.'

After this, he patted my friend and future companion genially on the shoulder.

'That's not the point,' I replied 'mail can only be dispatched once with the relief ship. After that we become icebound ···.'

'Then what?' he asked.

'What I am referring to is the radio communications which you will be able to have with us, assuming that we get the equipment installed successfully, then you will be able to contact us here. Then, we can reply by the same method. What do you think?'

'Brilliant!' was the unanimous verdict.

'I will take care of that,' said my father 'you can be sure we will not forget you.'

'Neither will we ···' added Desimone.

We continued chatting, when my attention was diverted by a conversation Jaramillo was having with a friend. I heard them saying:

'I saw in the newspapers that you were leaving for the South Orkneys, and I ran to say goodbye and wish you a safe journey and, better still, a speedy return.'

'Thank you,' replied Jaramillo, while he embraced his friend.

'You must be happy …' his friend continued 'You have always liked adventure, but I think this one will be a bit different from the boat trips we used to have on the Río de la Plata when we competed in regattas. Do you remember those adventures when we were far offshore and the *pampero*[261] began to blow?'

Jaramillo's friend continued.

'I suppose you will have fifteen to twenty days sailing to South Georgia.'

'Twenty, according to the Captain'[262], replied Jaramillo.

'Excellent. Twenty days confinement aboard a sailing ship, … congratulations, you will be fulfilling one of your dreams to make a voyage aboard a genuine sailing ship!'

'Yes, it's a barque. You can see … look!' said Jaramillo, while at the same time he gazed aloft and, making a sweeping gesture, pointed to the tops of the masts. He then added:

'A ship with three masts, rigged like a frigate. Look at the rigging, can you imagine anything more complicated.'

'She's a good size, but to me she appears a bit old …' observed my comrade's friend.

'Here, between the two of us, shall I tell you something?' said Jaramillo.

'I am all ears.'

'Do you know how old this hull is?'

'One can see it is a good age, although it has been well-maintained. No; I would not care to hazard a guess how many years she has been in service.'

'All right, but don't be astonished when I tell you. She is sixty-seven years old!'

261. The *pampero* is a cold squally, often dust-laden, wind in the pampas region of southern and central Argentina. At sea it may quickly reach gale-force and reduce visibility.
262. Captain Gotfred Thorsen

55 : *At sea aboard* Tijuca *under sail*

'How many?' asked Jaramillo's friend, thinking he had misheard, and added 'is she really sixty-seven years old?'

'Yes! she is; and still sailing', reflected my companion.

'Man!' said his friend, very slowly and looking around suspiciously 'then this iron hull must be completely rusted through ··· that is very dangerous! I would not go aboard. Think of what might happen if it were to collide hard with some ice at sea. The hull plating, I am sure would not just dent. With all the rusting it must have suffered over the years, it would break like crystal ··· and then what? So far from land, and on the high seas!'

'Don't be alarmed', said Jaramillo.

'Was there no other vessel to use but this one? I have heard of newer ships being withdrawn from service because they are too old ··· yet they are sending you aboard on this old rust bucket?'

'Come, let's have something to drink ···' said Jaramillo, taking his friend by the arm and dragging him off. He added:

'Not a word about this, it's really not worth it. Going to the South Orkneys is bound to involve taking risks ··· and this is just one of them ··· let's go!'

85: Voyage via Grytviken

Aboard *Tijuca,* the facilities we found were distinctly basic. It is true that the vessel had a much larger displacement than the ordinary whale-catcher's on which I had made several previous voyages. However aboard *Tijuca* the space allotted for passengers was confined to the poop house. It comprised a saloon with two tables and several wooden benches to sit on. A bit farther aft was another murky space, with an unpleasant smell and no ventilation; here were six bunks where my team would sleep.

My companions had already taken to these bunks where they were now suffering the cruel anguish of that form of confinement. It wasn't long before *Tijuca's* rolling obliged them to stay in their bunks, their bodies completely overcome by the misery of seasickness.

With regard to myself, I was in no better state than any of my comrades, as I had been allocated a tiny cabin next to that of the Master, an old Norwegian sea dog, made in the same mould as the unforgettable Larsen.

My companions, having little or no appetite for the food aboard, preferred to abstain for much of the time. The cheerful spirits and camaraderie they displayed kept up the morale and good humour of all.

After twenty-one days at sea under sail, we sighted South Georgia. We had set course for Cumberland Bay where we would transfer to a whale-catcher. When we were all gathered on deck contemplating the bleak coast of the island as it kept disappearing behind fog banks, Casariego suddenly asked:

'Are there any women there?'

We all looked at him in amazement. I laughed and replied:

'I don't imagine there are very many. On previous voyages I have only seen the wives of company managers and perhaps one or two others. All together I don't think there would be more than four or five, all ladies of good social position.'

'You shouldn't be thinking of things like that,' Fállico told him 'remember you have a daughter of marriageable age, and besides you are over forty. Anyway, with that Apollo-like body of yours, a big fat belly and rolls of flesh, I have no idea whom you are trying to impress ….'

We had a good laugh at these observations, but the unflappable Casariego was not to divert from his idea:

'The reason why I ask about ladies is because I have an idea … and don't think I am being vain', he added, laughing.

'What are you on about?' asked Jaramillo.

'Just a minor detail of hair ….'

'What do you mean, hair?'

'Yes that's right, I want to have a close cut and shave my scalp,' said Casariego 'we are still in summer, and they say if you do this it grows back with renewed vigour.'

'*El gordo*[263] is right', said Jaramillo. 'I propose we go one better; let's

263. A nickname for Casariego meaning 'the fat one' in Spanish, best translated as 'tubby' (see Illustration 11).

56: Shaven heads aboard Tijuca

all shave our heads!'

'Oops! I think you are getting a bit carried away!' – observed Fállico.

'Be convinced,' Jaramillo replied 'there are no women here …. Absolutely no one to impress! Let's shave our heads! Come on let's get on with it! I'll go first … Becker !'

'At your service!'

Jaramillo added:

'Start with the clippers, and then the cut-throat razor!'

That was how, within a few minutes and to the astonishment of the Norwegian sailors, we all had shiny bald heads.

In the early hours of the next morning, *Tijuca* was anchoring in the middle of King Edward Cove.

At the head of the cove, the extensive installations of the Compañía Argentina de Pesca bustled with activity.

Soon afterwards a motor-boat cast off from the whaling station pier and tied up alongside *Tijuca.* Later I noticed another launch leaving the jetty at the point[264] where the British administrative buildings were situated and the Magistrate was stationed. This boat flew a

264. King Edward Point, the site of the British administration and the Argentine meteorological station.

British colonial flag[265] at its stern.

I went to meet the Compañía Argentina de Pesca manager and staff, who came aboard at that moment. A few minutes later the second boat came alongside with the Magistrate, who also clambered aboard.

In my previous voyages I had made the acquaintance of Mr Binnie[266], the Administrator or Magistrate. It is effectively he who governs this productive British colony, and is responsible to the Governor of the Falkland Islands in Stanley.

After the initial greeting we went to *Tijuca's* wardroom, where I was present merely as a simple spectator of the following scene:

The Magistrate asked the Master of the vessel for the cargo manifest.

The Customs Documents had already been passed to the Master of *Tijuca* by the directors of the company in Buenos Aires; it was they who had chartered the ship for our transport.

Mr Binnie examined the papers, and suddenly asked me in English:

'If this radio-telegraph station is installed and running in the South Orkneys, by what call-sign will you identify yourselves?'

'I do not know yet what call-sign we will use.'

'Nevertheless,' he added with a smile, 'you must know the call-sign … have you any instructions … orders?'

Without another word he put the documents into a wide pocket, bid farewell to us, boarded the motor-boat and sped back to the point where the tall radio communication pylons of the British government stood.

We went ashore aboard the whaling manager's motor-boat.

That afternoon, as we were having a social gathering in the Manager's villa, I saw the Magistrate arrive. He shook my hand cordially and sitting at a desk, produced the shipping documents he had received in the morning. Again he seemed to be scrutinizing them and after bit of vacillation and muttering he stamped and placed his signature at the foot of each page, and handed them to the Company Manager who informed me that within three or four days we could be

265. The flag of the Falkland Islands and Dependencies

266. Edward Beveridge Binnie (1884–1956), South Georgia Magistrate 1914 to 1927

57: Tijuca *in Grytviken whaling station, mid-1920s (courtesy Jim McLaren)*

on our way to the South Orkneys.

The whale-catcher that would transport us on the second part of our journey was the latest of its kind. It had two engines, which gave it more speed and manoeuvrability, and had better facilities than usual for this type of vessel[267].

Tijuca moored at the main Grytviken pier[268] and the whale-catcher came alongside and secured. It then took two whole days to trans-ship all the cargo to the catcher. My colleagues made good use of that time to top up their tobacco supply and buy a few extra items of personal clothing from the company's slop chest, supplementing the several layers of protection that would be needed for that approaching polar winter.

In the early hours of 20 December, with all our gear stowed in the catcher, we sailed from South Georgia.

267. This was *Don Ernesto*, Captain Hans Olsen; built in Norway in 1910, 221 gross register tons, 37·8 m long; exceptional in having two steam engines driving one propeller each.
268. This is still known as the Tijuca brygge or Tijuca pier.

Once aboard I heard that the floating whale factory *Orwell*[269] was once again anchored in the South Orkneys, off Signy Island. One afternoon I was on the bridge talking to the Skipper and I asked him:

'When we sight the South Orkneys will you proceed straight to the Observatory or first go and visit *Orwell*?'

'I have to go to *Orwell* first', he replied.

'First to *Orwell* ··· and why is that?'

'Those are the orders I have', he replied.

'Well ···' I said with resignation, 'that means we will be spending Christmas in the company of your compatriots, aboard the floating factory.'

On the second day at sea we suffered bad weather and uncomfortable rolling associated with a severe gale from the south-east; but at nightfall on the third day, much to everyone's relief, we glimpsed the silhouette of the islands with which I was, by now, familiar.

At midday on the 24th we came alongside and secured to *Orwell* where, to my delight, I found I was greeted affectionately by the entire crew. Shortly afterwards I found myself on her deck surrounded by the Captain[270] and officers. Mr Rumbolds, the British Customs Officer, who was here once again supervising the taxation of this season's catch, shook hands with me vigorously.

I then saw the Skipper of the whale-catcher which had brought us here, handing over the documentation that the British Magistrate in South Georgia had signed and approved. Both men were walking towards the Captain's cabin that served as an office.

Meanwhile, the Captain of *Orwell* invited us to the Officers' Mess where, shortly afterwards, we did full justice to a sumptuous meal. The Captain sat at the head of a long table, surrounded by Officers and I struck up an interesting conversation with him in English.

'The *Lancing,*' he said 'a pelagic factory ship as big as this one and of Norwegian registry, is whaling at sea, and happens to be off Cape

269. This is the second *Orwell*, also owned by Tønsbergs Hvalfangeri of Norway which operated at the South Orkney Islands from 1925 to 1931 with two whale-catcher's; 7968 gross register tons. See § 78 for the earlier *Orwell*.

270. Captain Søren Berntsen (as in 1924)

Murdoch, not far from the base at Scotia Bay ….'

'Very interesting …' I interrupted.

'She is carrying two Argentines aboard.'

Perplexed I stopped eating. Two Argentines aboard *Lancing,* … at sea in a whale factory ship! .. Ah yes! Now I remembered.

'I am familiar with the *Lancing*[271] affair'. I told him, 'it was an episode that was never satisfactorily resolved.'

86: With the Whalers

I then began relating the following story:

'You will recall that in the late summer of 1925–26, the factory-ship *Lancing* in its eagerness to catch whales which, on account of the advanced state of the season, were well to the north on their annual migration, had reached the coast of Patagonia. Accompanied by her whale-catchers, *Lancing* entered an isolated bay on the coast of Santa Cruz province, in Argentine territory. She was moored there with the intention of completing her cargo of oil, most of which had already been obtained in the Antarctic.'

'*Lancing's* clandestine activities reached such a point that her presence was denounced by a line maintenance officer of the Argentine telegraph system who, to his astonishment, saw on that remote coast of Patagonia a foreign whaling ship taking whales without any permission or authorisation.'

'The news travelled rapidly and reached Buenos Aires where it found its way into all the metropolitan newspapers immediately. The authorities took the initiative, and dispatched telegraphic orders to the naval transport *Río Negro,* which was on its way north to Buenos Aires after a voyage to Tierra del Fuego. The ship was ordered to investigate the veracity of the denunciation and, if possible, to arrest *Lancing* and escort her to Buenos Aires.'

'The orders were obeyed with alacrity and a few days later *Lancing*

271. *Lancing* was built in Britain in 1898, converted by the Globus company of Norway to be the first pelagic whale factory ship which could haul whales aboard in 1925; torpedoed and sank in 1942; 7866 gross register tons. The vessel had been involved in unlicensed whaling off South Georgia in 1925. See: Palazzi, 2005.

was seen being escorted into the port of Buenos Aires by *Río Negro.*'

'Nobody could refrain from expressing an opinion on the affair, until, after two weeks of it total silence. No one said another word, but a laconic snippet appeared in the papers to the effect that "the matter had been satisfactorily settled through diplomatic channels …".'

'*Lancing,* once again was free, and in the current whaling season carries an Argentine naval officer aboard and a naturalist from the Museum of Natural History in Buenos Aires. They will conduct oceanographic research for as long as *Lancing* remains in southern waters.'

'I do not know the Argentine naval officer[272]', I mentioned, 'but, yes, I know the naturalist who is with him, he is Alberto Carcelles[273]. He made a research voyage to South Georgia aboard the vessel *Pampa* when I was making my first voyage to the South Orkneys in 1923.'

When I finished my tale, the Captain of *Orwell* said to me:

'I was not aware of the details of that incident. In the frequent radio communications we have had with *Lancing* I was surprised that she had two Argentines aboard, and could find no explanation for this.'

We continued talking animatedly, and after dinner while we were sitting round the table, I received a message which left me guessing.

The Radio Operator of *Orwell* came to find me telling me that a radio schedule had been requested from *Lancing* and that someone was waiting to talk to me.

I went straight away without even thinking what it could be about. I reached the radio room, and was very surprised to hear Carcelles voice on the radio:

'Hello … Moneta … hello … hello!'

I responded:

'Hello Carcelles ! … this is Moneta speaking. How are you? … I am at your service.'

'Hello Moneta … All well here … we have the Observatory in sight, we are fewer than ten nautical miles [18 km] south of Cape Murdoch.'

272. The Argentine officer was José Schwarz: Palazzi, 2005.

273. Alberto R. Carcelles (1897–1977), an ornithologist of the Museo Nacional de Ciencias Naturales in Buenos Aires, made several voyages to the South Orkney Islands, South Georgia, and elsewhere in the far south between 1923 and 1933.

'How much longer will you be in the area?' – I asked.

'I am not sure. It all depends on how many whales the Norwegians find around here, I am thinking of taking the opportunity of visiting when you are at the base. Do you have enough bunks or mattresses to accommodate two men for about a fortnight?'

'Hello ··· Hello ··· Carcelles. I look forward to your visit with the greatest of pleasure. We do not have any spare bunks or mattresses ··· but don't worry about food, we have abundant provisions. Please be sure to visit us, it would be a great pleasure to have your company and the only chance we have of meeting.'

'Thank you', replied Carcelles 'we will do what we can to come over, although it will depend if we can get transport to go ashore. I think we will manage it ····. See you later!'

'Merry Christmas, see you soon!'

The Radio Operator then turned off the dynamo and our call ended.

87: The Ice-Blocked Base

In the Officers' Mess of *Orwell* a noisy and impromptu Christmas Eve party began. From the Norwegian sailors, who were in the business of converting whale blubber into cash, I learnt some interesting details about the industry, which must operate wholly at sea.

Pelagic whaling is the only practicable industry at those latitudes. On the desolate, inhospitable land of the Antarctic there is absolutely nothing, no resource which might be commercially viable or suitable for direct exploitation. In contrast this is not the case for marine life of the Southern Ocean, especially whales. The production of whale oil from their blubber is subject to rigorous control and taxation.

This is the reason why factory ships, under a variety of flags although predominantly Norwegian, are obliged to call at the Falkland Islands or South Georgia on voyages bound for Antarctic waters. This allows a Customs Officer to board the vessel who will carefully monitor the amount of whale oil produced to assess the duty to be paid.

At the end of the whaling season, in March and April every year, the floating factories return to the Falkland Islands or South Georgia, where the Customs Officer disembarks and they pay the authorities two shillings[274] for every barrel of whale oil produced.

For any collector of revenue, the idea of levying a tax in these circumstances is very attractive. You have to bear in mind that each whale produces on average ninety barrels [15 m³] and the capacity of each factory ship is about thirty thousand barrels [500 m³], sometimes more.

At Deception Island alone, in the South Shetland Islands, more than a dozen floating factories gather annually, and they produce a combined total of more than seven hundred thousand barrels [1170 m³] of whale oil.

It is right to ask: 'Is the size of the whale stocks so large to permit exploitation on such a scale?' The answer of those who have been there and seen what goes on, is immediate: whale stocks at present are still very high, although there are occasions when the number of sightings in a particular area may be adversely affected by ice conditions, or the intensity of exploitation inflicted.

Occasionally factory ships have to abandon their anchorages near the coast to go in pursuit of whales far out to sea, or to remote areas where they congregate.

However they invariably find them; there has been no season that has been an entire failure. However, it could very well be that, with the passing of a few years and with this intense exploitation, they will disappear from the Southern Ocean, in the same way as they had been exterminated from northern seas many years ago.

⋆ ⋆ ⋆

On 27 December we left *Orwell* at Signy Island and continued to try to reach Orcadas on Laurie Island. From what the crew of the factory ship were saying there was a lot of pack ice around the eastern part of the archipelago; but as we had to reach either the north or south side of Laurie Island, I advised the Skipper to try to approach around the north side.

274. This is a British pre-decimal currency unit, equal to ten pence, or a tenth of a pound sterling. In 1926 whale oil sold for about £ (GBP) 5 a barrel, thus the tax levied by the Falkland Islands and Dependencies Government was approximately 2%.

58: Whaling factory Orwell *[II] anchored off Signy Island*

When we arrived at the position off the straits on either side of Powell Island, I could see that the sea was covered in pack ice, but the Skipper skilfully managed to push these floating hazards aside and succeeded in navigating to the vicinity of the northern end of Washington Strait. There we had to stop; it was impossible to go farther. Northerly winds, which are relatively uncommon at that time of year, had blown in enough ice to impede our attempts to get anywhere near Saddle Island.

The whaler turned about and headed southwards with the intention of reaching the base via Scotia Bay. We passed Mossman Peninsula and were off Ailsa Craig without encountering anything worse than a few bits of brash ice; however there were also large numbers of tabular icebergs which had calved from the ice shelves of Antarctica.

While we were under way, I was closely examining the southern horizon for any trace of *Lancing,* but saw no evidence that she was still in the area. It is possible she had gone farther south in search of better catching.

The whale-catcher made for the middle of Scotia Bay, and in the midst of that familiar panorama I could easily distinguish the base buildings which stood out clearly on the horizon although they were still far away.

Here I discovered more bad news; fast ice still covered the inner bay, and I could see a white line of fast ice stretching between Point Martin and Point Davis.

The ship, ably handled, was brought alongside the edge of the fast ice. That was the closest we could get to The Isthmus and base, and the Skipper asked me:

'Will that do?'

'Let's wait until Uruguay Cove clears of ice, as it will be an impossible task to disembark all our cargo here. We have big heavy boxes and crates, which would be dangerous to transport by sledge across the ice. It would be better to wait a day or two and then try to get in through Uruguay Cove to unload.'

'How far is it to the base from where we now are?'

'Six kilometres, more or less ….'

I looked at The Isthmus through my binoculars and said:

'Sound the foghorn, it looks as if they still haven't seen us ….'

The call had no effect, as we were too distant from the coast and the same north wind which had given us so much trouble with the ice, prevented us being heard from the shore. A few moments later I saw someone appear at the front door of the dwelling. First he was looking north, but then turned around, and was so surprised to see us beset in Scotia Bay, he leapt into the air then ran back inside.

88: The Third Time …

We continued waiting aboard the whale-catcher. It was three in the afternoon, the weather was glorious, and good enough for an adventure. I turned around and told my companions:

'I am going to land … anyone want to come with me?'

Every man put up a hand, so then I had to say:

'It would be better if some of us remained aboard. Jaramillo and Baldoni come with me. I'll be back tonight.'

We climbed down a rope ladder to reach the ice. The winter snow that covered it was very soft, and we sank to our knees in it.

We were wearing thick knee-length boots, but the walk was strenuous and difficult. Beneath the snow, between it and the firm sea ice, was a deep cold layer of slush. Very soon we were soaked from our knees down.

Halfway to base, I saw two men emerging from the dwelling and coming to greet us. I guessed, from their silhouettes, it was Berg and Witt, the Leader and Assistant Leader of the expedition we were to relieve.

We took ages approaching each other as we had to make frequent stops to get our breath back. Strapped to my shoulders was a thick bundle of letters and the latest newspapers we had brought.

On one of those rests, I thought Jaramillo was having trouble with the contents of a bundle he was carrying under his fur-lined parka, and I asked him:

'What are you carrying there?'

'Our companions ...' he replied.

'What?'

'The ones we were given in South Georgia.'

'The kittens?'[275] - I asked.

'Yes,' he said laughing 'I never imagined the cats would travel in this way across the ice. The reason why they are bothering me, is that their claws are getting caught in my clothes and I can feel them crawling all over my body.'

'It is only to you that things like this would happen. You could have left them for later', I said smiling.

We continued walking, and half an hour later we met Berg on the fast ice of Scotia Bay. We exchanged effusive embraces, and as we were all thoroughly uncomfortable after the soaking from briny slush, we went straight to the dwelling to find a warm fire, and a dry change of clothes.

The other base members were waiting for us inside, and with the warmth of the stove, we had a cordial reunion.

I was extremely tired, and my two companions were also feeling the effects of the strenuous trek. The hours went by quickly, and

275. A cat had previously been brought to the station, it appears in a photograph from the 1913 winter.

suddenly I realized it was already ten p.m.; I alerted Jaramillo and he asked:

'What! ten o'clock already?'

'Does it surprise you?' I asked in turn.

'Yes but the Sun is still up as though it were four o'clock in the afternoon in Buenos Aires!'

'Remember now we are at a very different latitude, and on 21 December, a week ago, it was the longest day of the year.'

'You are right. I had forgotten.'

'Right ··· let's go back to the whale-catcher', I said.

'Why don't you stay the night?' – Berg asked.

'I would, with pleasure ··· but we should get back. We have brought a large amount of valuable cargo, including all the equipment for the radio station.'

The presence of Jaramillo and Baldoni aboard the whale-catcher was not necessary, I decided they could remain ashore until the following day or however long they wished. Thus they could start work preparing the foundations for the new buildings we would have to erect.

They willingly accepted my proposition, and I went back alone towards the ship; but this time I put on ski to make the journey easier.

Large numbers of penguins could be seen crossing the fast ice on the bay, but I found myself looking at them mechanically, as if hardly noticing their presence. Everything looked so familiar, I felt as if I had lived there all my life.

'Back again in the Orkneys' ··· I said to myself while skiing across the ice; 'but this time with a different responsibility; very different from previous years. This year there will be a radio link and the station will function with the appropriate Argentine call-sign. I have been honoured with being the first Argentine to command an expedition ··· will it be a success? Will there be a radio contact before 31 March? Will we be able to keep one step ahead of the winter?'

With these thoughts I made my way back to the catcher which was still in the same place, secured at the edge of the fast ice.

I was very lucky as, during the night, or more precisely during the hours when it is normally dark at lower latitudes, a strong wind came

59: Jaramillo working at the desk in the parlour

from the south, presaging the clearance of ice from Uruguay Cove which later broke out and left the way clear for us to get to the shore.

At five in the morning, the Skipper woke me and said:

'We will sail directly. From the crow's nest I have been able to confirm that the ice has cleared from the bay on the other side of the isthmus. So we will go there, approaching through Washington Strait, and round Mackenzie Peninsula.'

Thus it was. On 29 December we started unloading the expedition's essential cargo on to the ice-foot at Uruguay Cove.

We had arrived two months earlier than the usual date for annual relief voyages. Two months during which we would have to work round the clock to win the race which, from that moment, would be against the weather.

89: Preliminary Work

The departing party returned north, and we remained alone. I had full authority to make changes in the organization of our tasks, and immediately gave thought to the equitable distribution of work to each of us.

My primary concern was to move all the cargo and equipment from the beach to a safe place. Food and the fuel first; then boxes containing scientific instruments and any machinery, and finally all the rest of the supplies with which we had been generously provided.

I limited the routine magnetic and meteorological work to the basic minimum, in this way I could ensure that everyone could devote as much time as possible to the construction work. This involved demolishing what used to be the pantry built of empty paraffin cases. In its place, on the south side of the dwelling, two new wooden rooms were built, with double walls filled with sawdust and clad on the outside with Ruberoid. These would be the radio room and generator room, as well as bunk-space for the Radio Operator, Baldoni.

Then came the most difficult part of all; the construction of the galvanized steel antenna pylon, which had to be thirty metres high.

It often happens that the calculations made in laboratories by specialized technical staff fail when they are put into practice; either through inadequate testing or from lack of experience. On other occasions it can happen that the failure of projects in the field is due to ignorance of the terrain and climatic conditions in which they will have to operate. Because of these inherent problems I found myself having to use all my ingenuity to erect only two sections of that heavy pylon framework. Getting these in place came close to causing some irreparable accidents, as we had to climb to the topmost level on the ice-covered lateral steel struts. The provision of a ladder of the right length had been overlooked[276].

Nor was the construction of a suitable foundation for this pylon taken into account by those who planned the project, which caused us enormous difficulties. However the whole project was saved by the enthusiasm and prodigious effort everyone put into the work, and it was thanks to that we would later be able to communicate with the

276. Dust jackets of the 1939, 1940, and 1944 editions indicate the form of the antenna pylon as originally planned. It was, however, augmented by one stage in 1930 (Pierrou, 1975).

outside world.

Thus eventually, in the race against the forthcoming winter, we managed to get everything finished. Steel struts secured with plates and bolts, rocks removed and deep excavations made in the frozen ground; new buildings constructed, generator mounted, the new rooms roofed, and finally the external walls clad. Later came the installation of a long antenna suspended between the extreme top of the pylon and a picket on the slopes of Mount Mossman. The result of this hard labour, eighteen hours of unremitting toil, was that, at the end of each day, we were left completely exhausted.

Many times, quite worn out, I would climb into my bunk without undressing, and my companions did the same. Then, at five in the morning I was up and about preparing breakfast for my companions and then we would all go out and work somewhere regardless of what the weather was doing.

We all had a common desire, to beat the winter and keep ahead of it. On that one critical factor would depend whether or not, during the long winter nights and the entire year, we would be able to receive news from the outside world.

90: Failure?

Two years in the Antarctic had been enough to inure me to that way of life in which anything new, or any change from routine, was entirely subordinated to the business of amicable coexistence with my fellow Argentines. A large part of each day was absorbed in the difficult work involved in the installation of the radio, as well as getting all the terrestrial magnetic instruments functioning properly again after an interruption to the record during the previous year.

In our rest periods, which were limited to a few moments when we were all seated round the table to dine, I recounted to my companions all I knew about the South Orkneys. That way they gradually began to accumulate knowledge through personal experience.

By the end of February we had successfully completed the most important outside works. While Baldoni continued with the installation of the radio apparatus inside, the rest of us spent time stowing the

60: Antenna pylon, Casa Moneta ir the background

winter coal, and resuming the full scientific programme which had been partly curtailed by the building work.

By now we were more organized, and soon could return to a more normal sedentary routine. This would help us regain the few kilograms we had lost working like slaves, but with so much zeal and effort, during the last three months.

The last fortnight of February was approaching, and one morning, while helping Baldoni install a transmitter, he said:

'I think soon it will be ready.'

'When?' I asked him.

'Possibly the week after next.'

The end of the month arrived, and the transmitter still did not work, despite the hard work and long hours Baldoni and I had dedicated to it, day in day out, without a moment of rest; we had both regarded it as a matter of personal pride.

Under the direction of that tireless lad, who was a true technician in every sense of the word, we took the transmitter apart time and time again, but always with the same depressing result; it didn't work.

The days went by while we made minor changes here and little adjustments there. Amidst our puzzlement, with the threat of failure looming over us, there was one thing we had achieved to date: now we had electric light in the dwelling. It was provided by the new Diesel dynamo set which had replaced the old one installed in 1925. Thanks to that there was even an electric light in the middle of the isthmus on the track to the Variation Hut. We had installed a mast with a light atop, which would shine like a beacon during the dark winter nights.

Despite the evident nervousness Baldoni had because of the transmitter failing to work, I had full confidence in him; I was an optimist, despite the evidence that lay before us. One afternoon, while we were working amongst the mass of different electric cables with our hands blistered from continuous use of screwdrivers, Baldoni finished adjusting the transmitter and said to me:

'We will start the generator, and if this does not work now ...'.

There was a tinge of dejection in his words, and while the Diesel engine started with a bang and began to make my head spin from its exhaust fumes, I saw my colleague connecting several levers and then, pressing the Morse key, he turned to tell me in a disappointed tone:

'There is no charge … nothing, nothing at all!'

Then he looked fixedly at the ammeter connected at the socket of the antenna cable.

He signalled me to stop the generator; then he said:

'Frankly … I don't think I understand anything at all! I don't understand how this could have lost so much of its charge … this thing is definitely not working ….'

For my part, I did not have much knowledge of radio-telegraphy, but allowed myself to hazard an opinion.

'You say there is always energy loss. Could this be from the electric lighting connection?'

'No, it's not from there. I have checked the contacts and even if I disconnect the electric wire, the loss of charge on the antenna persists.'

'Next where …' I asked?

'The anomaly must be somewhere else, and I cannot find it.'

'I want to ask you a question: does it transmit?'

'Yes, it transmits, it must do …' he replied, 'but not with the necessary power'.

'Let's suppose something else,' I said 'with the strength of transmission at present would it be possible to reach anywhere? Would it get as far as Ushuaia, for example?'

'I could not say, I would have to transmit and then "listen" myself.'

'Let's try something', I proposed. 'Tomorrow I will go with some of the lads to Point Martin or Ailsa Craig, or somewhere far afield but visible from base. I will take one of the receivers, and at an hour that we agree, you send some signals with the transmitter and I will endeavour to interpret them. I will reply with semaphore flags and let you know the strength of your transmission.'

'Good idea,' replied Baldoni 'that way I will get an idea of what is going on inside this apparatus.'

91: General Radio Call

On 21 March, accompanied by Fállico and Jaramillo, I boarded the

61 : Jaramillo and Fállico with a radio receiver at Point Martin

boat which had the outboard motor. Navigating south-south-east on Scotia Bay where the fast ice had broken up at the beginning of the year, we steered towards a large penguin-covered hill on Point Martin.

We landed and Fállico was able to visit the penguins for the first time while I prepared the radio receiver, lit my pipe and stood by waiting for the time we had agreed for the radio signals.

When the time came, we put the phones to our ears and moving the dial, searched for the wave length we were expecting.

Baldoni was punctual, and at the same instant I heard the characteristic dots and dashes of the Morse code that indicate letters and numbers, and then form words. A bit later I heard a few questions.

'Can you hear me? Is there a continuous signal?'

At once Fállico hoisted the signal flags on one of the oars of the boat, which Baldoni could see from base.

To each question, the correct answer was transmitted, so I was convinced, contrary to Baldoni's opinion, that the transmitter was working well.

We returned, and when we were together I said to him:

'Now we will transmit continuously, until we get a reply from someone.'

Baldoni's satisfaction with the field test result was evident; his good natured face was calmer, but as usual, he did not want to appear an optimist and said:

'Yes ··· no problem calling, but will anyone hear us anywhere?'

'Of that I have no doubt. I am convinced that all the Argentine coast stations are eager to hear us, and they are aware that about now we should be initiating communications.

'Will this machine work?' he asked ··· 'I personally am not convinced.'

He shook his head.

Then began the blind transmissions. It was enough to start the generator, at any time of day or night, for everybody to burst into Baldoni's little domain in a mob avidly asking:

'Did you contact anyone? Did they hear anything on the radio? Are they receiving us?'

I did my best to calm them saying:

'No ··· nothing. Still no word. Just a test; go back to sleep ···.'

Reluctantly they would leave that confined room, but not before reminding me that if there were any news at all would I let them know immediately; which was just as it should be.

During the hours that I was free from any other activities, I would pass the time with Baldoni in attempting new, but always unsuccessful, transmissions.

My companion used to spend many hours sitting in front of the receiver, with the earphones on his head and his fist within reach of the Morse key while he gently adjusted the controls with his other hand.

So as not to bother him, I would sit farther away, near the generator, and though I wanted to appear calm in the midst of all that uncertainty, my state of mind was anything but; I was constantly chewing the stem of my precious pipe in anxiety.

Suddenly, Baldoni would turn around and say:

'Start the generator please. We will make another attempt.'

So I would turn the fly wheel again and hear the rhythm of the generator engine firing. The filaments of the transmitter's valves would light and the Morse key sent another call to all stations.

CQ[277] *··· CQ ··· CQ ··· from LRT*[278] *··· LRT*

LRT ··· Orcadas ··· Orcadas ··· South Orkneys, general call ···

Stop!

Then silence, while Baldoni switched to the receiver and we quickly put on the earphones again to listen.

Nothing. Isolated signals from whaling vessels far out to sea. Remote stations exchanging radiograms. Unknown foreign call-signs ··· nothing for us.

Back to the beginning again with the transmissions, which had they been able to translate into words, would have had an anguished hint of bitterness.

No one answered those desperate calls. The receiver dial went from one extreme to the other within the six hundred metre range for international wave lengths. It was in this range that we could expect to receive anyone answering our transmission; but no one paid any attention to our calls.

Our companions now ceased coming into the room every time they heard the generator start.

'What is the point?' they said 'if there is nothing to listen to.'

Every day, after dinner everyone retired to bed, with only the man on the night shift staying up.

Baldoni would go to the radio and I would follow in his footsteps.

'Shall we call?' I would ask him.

'Yes, let's call!'

We did that every day throughout March which was now drawing to a close.

277. CQ is an international radio signal, a call for unknown stations or a general call to all stations in the vicinity.

278. LRT is the Argentine identification, or call-sign, for the Laurie Island station, Orcadas (Argentine stations were allocated call-signs in the 'LIA' to 'LRZ' range by the 1913 International Code of Signals).

92: Emotion

This was becoming a nightmare. Baldoni did not have much to say to me when we met at the radio. I found him confused and perplexed by the failure that was upon us.

For my part, there was nothing I could do except to support him with the laborious task of checking, a thousand times, the cables, each wire, each terminal, every connection of that intricate mass of wires, equipment, and engines.

On the night of 30 March, just twenty-four hours before the day of the deadline I had set for establishing contact with the mainland, at about eleven p.m., we made one more call to all the coastal radio stations of Argentina.

The key, held firmly in Baldoni's fist transmitted:

CQ ··· CQ ··· CQ ··· from LRT ··· LRT ··· LRT ··· Orcadas, Orcadas, Orcadas.

General call from Orcadas ··· ··· ···. Orcadas calling.

Please respond with a very long tuning call[279].

This is Orcadas calling.

'Stop! That's enough chief', Baldoni told me.

Again back to listening. I always put the earphones on, even if the receiver dial was in Baldoni's hands. His face showed the deepest concentration as he said to me:

'No, this is not the one ··· Puerto Belgrano communicating with a ship ··· nor this one, Cabo Vírgenes transmitting to San Julián ···.'

'Wait a moment!' – I exclaimed! 'what about this one?'

'No, not that one either, it is the Falkland Islands and South Georgia.'

We continued the patient search which had already been going on for more than half an hour; then again:

'Generator on?'

'Ready to go!'

After a few seconds, we heard the raucous sound of the generator

279. A long tuning call is to allow time for the oscillator (tuner) of the Orcadas transmitter to be synchronised with that of the receiving station, so both can work on the same frequency.

firing and causing the walls of the little room to vibrate.

While the generator was running Baldoni continued making anxious transmissions, I wandered over to a window, putting my forehead to the cold frosted pane, could see the bluish flame issuing from the exhaust pipe, which, helped by some extraneous sparks, reflected brightly from the snow.

'Stop!' – shouted Baldoni, in the midst of that deafening racket.

Back again to listening to signals that were not for us, and which to me seemed like voices coming from disembodied spirits.

'Here we are in the South Orkneys,' I thought, 'listening to sounds of the world suspended in the air, and the world doesn't take the slightest notice of our existence ···'.

Again I would put on the earphones and search the wavelengths.

Again I heard the Falkland Islands transmitting; then Montevideo replying ··· then a British ship making a general call similar to ours.

The Argentine stations seemed to have gone to ground, there was not a squeak from anywhere.

'Another night without contact', was Baldoni's comment.

'Let's listen a bit longer, it could possibly ···' I hinted.

I thought my suggestion would come to nothing. I realized that Baldoni was very nervous and at the end of his tether, as he adjusted the dial from one end to the other.

Suddenly:

'What is it?' I shouted.

'Just a moment - one moment!' – replied Baldoni.

'Don't move the dial!' – I shouted again 'Don't move the dial!'

Again, on the earphones, we could detect in Morse code:

LRT ··· LRT ···

These letters were being repeated in rhythmic succession.

'It's us!' – we shouted in unison, our fists clenched with nervousness.

'It's us! At last. At last!'

'Start up!' – shouted Baldoni, and then, stopping me with his hands corrected himself.

'No, wait a moment ··· who is calling us?'

On the earphones we could hear:

LRT … LRT … from LIK[280] *… LIK … LIK …*

'It is Ushuaia calling us!' – we exclaimed simultaneously. 'Now yes, start the generator!'[281]

In fewer than ten seconds the generator was running at full power, while Baldoni replied to the prolonged call from Ushuaia, I ran into the parlour and shouted with all my might:

'Get up lads! Up everyone! We have contact, communication with the world!'

I ran back to keep an eye on the generator.

Then, through the receiver, LIK Ushuaia replied in Morse:

'I receive you very well. I have been receiving your transmissions for a week, and have been answering.'

I looked at Baldoni in astonishment.

'They have been receiving us for a week', I said, and he nodded his head while continuing to transcribe the message:

'The personnel of the Ushuaia Radio Station hope that you are all well. We await your traffic. Send me any messages you have, and I will re-transmit them at once. Standing by ….'

My companions, some of whom were in their underwear, stood around us in silence, their faces clearly showed the emotion which seized them.

I took the pad of radiogram forms, and quickly, wrote the following; it was my duty:

H.E. The President of the Argentine Republic, Buenos Aires.

On the occasion of the inauguration of the most southerly radio station in the world, the members of the Argentine Expedition of the South Orkneys, motivated by the present opportunity, are deeply honoured to present their respects to your Excellency.

I signed with both my official stamp and manuscript signature. Then I wrote three more radiograms; greetings addressed to the Ministry of Agriculture, the Minister for the Navy, and the Director General of the Meteorological Service. It was the last on whom we

280. LIK is the call-sign for the station in Ushuaia.
281. The accumulator batteries, when charged, provided power when receiving signals but the extra power provided by the generator was necessary for transmitting.

most depended.

I did not want to exhaust Baldoni on that memorable first contact, and ordered the other radiograms with personal news deferred until the next day. That was enough to prove that I had kept my word and attained the set target date. Then I looked at the calendar and, to the evident emotion of my companions saw the date was 30 March.

There was only one day left to meet the deadline that had been agreed.

Baldoni arranged a schedule with the Ushuaia station for them to call at the same time on the following day. When he finished the day's work, I said to my companions:

'Now lads, let's have a few drinks to celebrate!'

We trooped out, and as soon as we reached the parlour, a dozen bottles appeared on the table.

'Before anything else,' I said, 'we will have a bottle of champagne.'

I uncorked the bottle, and once it was served, I raised my glass and made a brief speech:

'Lads! We now have communication with the continent and this success is due to every one of you, equally … Lads! We are not alone any more! … Everyone stand and – *¡ Viva la Patria!*'

93: Contact with the World

On the night following that memorable communication, Baldoni handed me five dispatches. The first one I opened was from the President of the Nation; it was addressed to me and said:

I gratefully acknowledge the kind greeting which you and your team transmitted to me on the occasion of the inauguration of the radio-telegraph station and express my fervent hope that this new means of rapid communication may contribute towards guaranteeing all possible success in the objectives of the Argentine Expedition in that distant place. I express my best wishes for the personal welfare of every one of you.

Alvear[282]

282. Máximo Marcelo Torcuato de Alvear Pacheco (1868–1942); Argentine President from 1922 to 1928

The remaining radiograms were official replies from the Ministers of Agriculture and of the Navy, and the Director of the Meteorological Service.

My companions, read the dispatches with satisfaction as they passed them from hand to hand, and watched me attentively when I began to open the fifth radiogram. It was short and to the point; it read:

Moneta, Expedition Leader, South Orkneys

Families of everyone well.

Plate. Director of Meteorology.

We had a moment of silence, as we looked at each other without venturing to speak a word.

'Of all the dispatches,' said Jaramillo, 'this is the one with the greatest boost to morale'.

Mechanically he repeated.

'Families of everyone well.'

Only someone down here, in the South Orkney Islands, can imagine the significance of those words. That chivalrous gesture left us wondering; the radio station, established with such sacrifice, troubles and bitterness had dispatched only the four signals I drafted on the day of the inauguration. I had not asked a single question about our families; I dispatched only factual news of the way in which every one of us had fulfilled his duty.

That trait, replying to something we had not actually asked in the first place, although the idea of asking was in every one's mind, received our admiration and approval, because it showed that our employers had interpreted our intimate thoughts accurately.

It took several days to get the system of receiving and transmitting signals properly organized. Every day, at nine a.m. Orcadas time, we transmitted the official dispatches with the meteorological data, and from time to time, so as not to overload Baldoni with work, a personal radiogram was added to the transmission if any of us was keen to have any news from his family.

News was received from the mainland in much the same way. With the arrival of winter the communications were occasionally disrupted by the rime ice that formed on the insulators, the aerial cable, the

62: Kahl in the radio room

antenna and contra-antenna wires[283] themselves.

The accretion of that mass of solid ice added greatly to the weight of the wires, and they broke frequently leaving us isolated from the rest of the world.

A single wire could occasionally have two, three and even up to four centimetres deposition of rime along its entire length, making it ten times heavier and causing the breaks.

The problem of deposition of this rime ice was exacerbated because it was formed by condensation from the moist sea air, which had a high proportion of salt. When it covered the insulators of the aerial system it allowed the current to discharge to earth, and this was the reason for weakness of the transmission signals.

Whenever the antenna and the contra-antenna, became rimed, there was a procedure necessary before a radio schedule: the complete de-icing of all the wires and insulators. For this we had to lower the long heavy wire, beat off the ice, then hoist it back up, with no more assistance than that offered by a simple block and tackle on which we all pulled together.

There were some extensive periods of bad weather during the long winter when we had to perform this arduous procedure every day. We all put much effort into this, as we knew that without it we could not communicate with Ushuaia. Perhaps, at that moment, there was a message from a mother, wife or girlfriend, ready at that moment, to be retransmitted to Orcadas.

Nor did it end there. Often, after the tedious work of clearing the ice off the antenna while snow fell copiously, another thin layer would already be forming on the parts we had barely finished clearing. Eventually, after the antenna had been cleaned and raised into position, and everything seemed assured to stay that way for at least the next half hour, we encountered another formidable difficulty.

Baldoni, who had already started the generator, would stop it and, in a disappointed tone, say:

283. The aerial emission of electromagnetic radiation from the antenna is balanced by a contra-antenna at ground level which compensates for variation in the dielectric properties of the substrate.

63: Communicating with the world

'I don't think there will be any communication today.'

'Why not?' I would ask.

'Currently there is a strong magnetic storm[284]. The static is terrible and makes it impossible to hear anything. Listen …' he said and would hand me the earphones.

I would place them on my head, but I had to take them off again. I could not bear the noise. It was appalling. A deafening background of electrical interference made it quite impossible to distinguish any signals in Morse code. I lifted the cover off the receiver and could see innumerable bluish sparks jumping in between the plates of the tuning condenser. At such times the electrostatic charges in the air must have been strong.

Many days went by like that; the whole morning spent working with gusto on the insulation and antenna, only to fail later by being unable to discern a single word of the message that we were straining to comprehend.

Afterwards would come a fleeting calm period when we made the

284. This is a result of excitation of the Earth's magnetosphere by bursts of strong electron emissions from solar flares. These phenomena also influence magnetic instruments of observatories.

most of it to tune to a news broadcast from Buenos Aires. In the circumstances this was the best we could do. Even though it was limited to listening to the distant voices of professional broadcasters whom we had never seen, it filled us with joy.

94: Orcadian Olympics

On every expedition the hobbies of the participants are a measure of individual temperament or nationality. During my first year in the South Orkneys the most popular game was bridge; we played endless rubbers simply to take our minds off the fact that there was still had a long time to go.

During my second winter, with the unforgettable Guerra Boneo, an expert player, we played chess exclusively. It is not an easy game, but he excelled at it.

During this expedition there was no preference for any particular game to fill the times when we were not working. None the less, it became apparent that cards accorded best with all our individual temperaments, especially poker. At the start of the winter poker games began for no better reason than to simply to pass the time. However it did not take long for us to realize that, with poker the game loses its appeal – unless, of course, some incentive could be found to keep on playing.

That was how, one evening after dinner, we had the following discussion:

'Poker is all very well' – said Casariego; 'but we need a motive to win ….'

'There is no money here,' Jaramillo added, 'and even if we had it, no one would risk a centavo on the favourite game of card sharps.'

'We could perhaps play for chips,' said Baldoni 'but we would still be no better off.'

'Yes' – Fállico broke in 'at the end of the year, when we leave the South Orkneys, we would go back owing money to our opportunist colleagues.'

'There must be no playing for money,' I said 'We are not down here to fleece each other, and anyway I would not permit it.'

64: Becker at work in the kitchen

The discussion continued for a long time, until Jaramillo slapped his forehead and said:

'That's it, I've got it! We'll play as for poker.'

'Poker, if there is no money? ··· and we will not play for money either', several voices agreed.

'It's not a matter of playing for money,' Jaramillo continued – 'we will play for something that is common property.'

'Common property?' I interrupted a bit surprised.

'Every one of us,' continued Jaramillo, 'since we bought it with our own money, the winner will have the right to take it.'

'What are you talking about?' I asked.

'It is an object ···.'

'Yes, but which?'

'It is over there', he said and he pointed towards a shelf.

'The soda siphon!'

'Yes that's it, the siphon, and the empty carbon dioxide bulbs that go with it.'

He was right; we had bought a siphon complete with its charger and four hundred bulbs in Buenos Aires, to make soda. We had used this only on Sundays to make drinks.

'Now that is not worth anything!' I exclaimed.

'You are wrong there,' said Jaramillo 'in market value it is not worth much, but remember with the bulbs, even the empty ones, there is a deposit on them when they are returned ….'

'However you would not get fifty pesos for the lot', I commented.

'That's not the point,' said Casariego, 'I think the idea is good.'

'Moreover,' continued Jaramillo 'we will have the incentive of competition to stop someone else winning the siphon.'

'True enough,' I added, 'in the South Orkneys it will be a matter of pride who gets that glass bottle.'

'Good …' said Jaramillo, getting up.

Then he asked;

'Accepted?'

'Unanimous', said Casariego.

'Well let's get on with it.'

There was one detail he had forgotten and I interrupted, saying:

'For this to be legal, I propose one condition.'

'What is that?' asked Baldoni.

'It's quite possible,' I replied, 'that the friendships we have formed here could work against the opposition. Let us suppose, either secretly or by open cheating, even at the expense of my own chance of winning, I were to cause Fállico to win the game.'

'True enough' – my comrades agreed.

'Well,' I continued 'in a case like this, Casariego will keep a detailed tally of what each man wins, or loses, on paper. The winnings will be calculated at a nominal rate according to whatever value we give to the chips, but individual losses will be noted as a net proportion at whatever rate is decided, one percent for example. From this we will form a common cash pool from which we will make a proportional compensation for each man's losses.'

'Also accepted', everyone agreed.

'This will be a guarantee of fair play', I continued. 'Furthermore, it is established in the rules that winnings will not diminish the daily

65: A birthday party during the 1927 Expedition

losses, they will be noted only to record our respective positions in the contest.'

'What will be done with the common fund that forms from the losses?'

'That fund,' I said, 'will be reserved to buy a lottery ticket when we return to Buenos Aires, whose potential winnings will be divided amongst every one of us.'

'Let's start to play tonight', someone comments.

The first round was under way, and suddenly Baldoni said to me:

'There are five of us, can Becker, the Cook, come and join us?'

'No problem', I replied and called:

'Becker!'

A few moments later, the Cook appeared and sat down with us.

'So that we do not get fed up with this game, I propose that we limit it to three times a week.'

My proposal was accepted, then straight away we began playing.

On the following morning I saw Jaramillo had pinned a note on the parlour wall in big letters saying:

Orcadian Olympics
League table:
Results of the last round:

66: A game of poker in 1927

followed by the positions each man occupied in the table as a result of his winnings in the previous night's play.

'Orkney Olympics' … even the expression used to describe that odd contest, indicated the prevailing good humour among my companions. If they are to reach the end of the year still bonded together in goodwill and friendship, they will need all the resourcefulness they can muster.

95: Phantom Light

It must be remembered that we were all Argentines on that expedition, the majority of us of Latin origin[285]. One afternoon, early in May, which was already fully winter, and just as darkness was beginning to envelope the snowy scene, we were gathered round the stove chatting about the day's events. Suddenly the parlour door swung open and Fállico appeared covered from head to foot in snow, and, judging by his face, clearly shaken.

From the door he shouted at us:

285. The concept of bad or evil light, *luz malo*, appears in several Argentine folklore stories.

'A light! ··· I saw a light!'

'Where?' I asked, leaping from my chair and followed by the others.

'Out there, in the direction of the cemetery ··· near the beach.'

'On the beach?' – I asked.

Fállico, out of breath from running, added:

'Yes ··· by the beach ··· a light!'

I took my fur-lined hat, leather coat and gloves then ran out followed by the others.

As I scrambled through the entrance hall of the dwelling, I heard Baldoni working the radio and shouted to Jaramillo, who was behind me:

'Tell Baldoni to come with us.'

Jaramillo caught up with me as soon as I was outside. While we ran in the direction of the beach to the place Fállico indicated, he asked me:

'What do you think? ··· What might we find? ··· A whale-catcher that has run aground and whose entire crew are coming here to ask for help?'

'I do not know,' I replied 'It might well be, there is a lot of ice farther north and even though Uruguay Cove has not yet frozen over ···.'

Casariego and Fállico reached us, and then I noticed Baldoni running in our direction.

It was already dark, and the falling snow obstructed the visibility. We reached the beach, but in the water there was nothing other than large floes of ice.

'You two go that way,' I said to Jaramillo 'Fállico and I will go round here.'

We continued in the direction of the precipitous seaward face of Mount Ramsay, and on the way I questioned Fállico:

'What type of light did you see?'

'I am really not sure. I would not like to say exactly ···.'

'Did it give you the impression of being some kind of signalling light or lamp?'

'Yes, somewhat similar. I saw something like two or three beams of light, very faint, then they vanished.'

'When you saw the light, where were you?'

'I was just entering the front door after returning from the Variometer Hut, and instinctively I turned around.'

'In what direction?' I asked.

'In the direction of the cemetery ··· looking exactly towards the spot where we are now.'

'It's certainly odd' I remarked ··· 'there is nothing here.'

'Could it have been an animal? A penguin?'

'I don't think so,' I replied 'by now they have all migrated north.'

'What about the cemetery ?' – Fállico asked me.

'Very difficult. You think that light could have been produced by an *ignis fatuus,* a will-o'the-wisp ?'

Fállico nodded affirmatively while we continued searching all parts of the beach.

'No,' I told him 'here there is no *ignis fatuus*. The bodies of those unfortunate fellows are encased by pebbles, which in turn are frozen together by ice, so that they have taken on the consistency of concrete. That would prevent any gas from escaping, if there was any. There is a metre of hard packed snow over the graves; but there is nothing to stop us taking a look over there ···.'

We went up the slope to reach the cemetery, and walked between the four crosses, as I carefully examined the snow around them.

There was nothing; just very solid ice.

I suspected Fállico may have been victim of an optical illusion, but he so firmly believed he had seen that tremulous light that I returned to the base more perplexed than ever.

We gathered beside the stove again, and Baldoni returned to work in the radio room.

It was five p.m.; winter and pitch dark. Outside the wind was whistling through the wires that guyed the chimneys, and snow pellets pounded the windowpanes. I was sitting next to the stove, changing into dry socks, when I saw Fállico going outside to make his hourly meteorological observations.

A few minutes later I heard him rush back shouting:

'The light! ··· again ··· the light!'

96: Aftermath

Casariego emerged from his bunk-room into the parlour and was followed by Baldoni from the radio room. They had both heard Fállico's shouting.

'I too saw it!' said Casariego 'When I heard Fállico I looked out through the frozen windows and saw, far off in the direction of the cemetery, a momentary brilliance which immediately faded.'

We dressed in two seconds, I went out grabbing the Winchester rifle and noticed Jaramillo armed himself with a Mauser.

We went out and started to run towards the beach; I was determined not to return until I had solved the mystery.

This time it was Jaramillo who came with me, while Casariego, Baldoni, and Fállico formed another group whom I sent to examine the western shores of Uruguay Cove.

We walked quickly and Jaramillo asked:

'In confidence, what do you really think about all this?'

'I do not know what to tell you,' I replied, 'I am baffled.'

'Do you think that a whale-catcher has struck a rock?'

'It is not completely out of the question,' I answered 'but at this time of the year no ships sail in these seas. I have heard of a few vessels that have been wrecked around the South Orkneys, but the crew always attempted to reach *Orwell;* as there was no radio here at that time, they would have had to remain throughout the winter without anyone knowing of their plight.'

'Yes, but now we do have a radio, and everyone knows about it.'

'A whale-catcher – at this time of year?' I reflected.

Nevertheless, anything was possible, and while we examined every piece of ice, every rock that emerged above the white surface, I thought about contingency plans which might result from an accident of the kind that Jaramillo was contemplating. He asked again:

'What if it was the shipwrecked crew from a vessel that had sunk?'

'We would have to receive them and provide food and shelter.'

'Well; but we would have to request by radio for a vessel to come and collect them, wouldn't we?'

'That would be impossible; remember that at this time of year no vessel could reach the South Orkney Islands, not even a powerful icebreaker. Nothing can get in before the arrival of the next spring.'

'So we would have to live with all these people here?'

'There would be no alternative', I replied.

'So we would have to share our provisions ?'

'As well ....'

'That would have severe consequences for every one of us. I understand the provisions are only enough for six men, but suppose an additional six arrived ....'

'Then there would be nothing for it but to start losing weight', I replied.

'All right, but what if they decided to take over the base by force? Remember, if it is to defend life, anything is justified and, in this case, life depends on the food store.'

Even though I did not want to show it, I was worried, or rather, perplexed.

Two men had seen this mysterious light that did not appear to come from anywhere they could identify.

I don't know exactly how long we wandered over the ice; only when I was exhausted by the effort of sinking in the snow at each step, did I decide to return to base without finding the slightest trace of life anywhere in the vicinity.

A thousand conjectures spun themselves about that mysterious light, which had failed to reappear. When we left the table after dinner, I gave instructions that anyone who caught sight of the light again, should immediately raise the alarm, and anyone who heard this should drop everything, to look out from the windows and doors to determine the exact location from where it came.

While the others went to bed I remained in the parlour with Fállico, who was on duty, while Baldoni went to the radio room. In a few minutes we heard the raucous sound of the generator as he turned the equipment on.

Baldoni had told me he would be trying to establish radio contact that night.

'Frankly,' I thought, 'all this has me baffled. What on Earth might this light be? What might it possibly be? Never in my time in the South Orkneys have I seen anything like it, neither had I ever heard mention of any apparitions of this nature.'

At night, I continued thinking, there are many animals that reflect light from their eyes when a torch is shone on them. Could it be that there is an unknown animal of this kind here that has that quality …?'

'No, that is just not possible.'

'It might be the dogs' … I supposed … 'but they have spent the day shut inside … all things considered I just don't know …'.

I got up, thinking that I would go to the radio room to keep Baldoni company, as I crossed the corridor by the front door I stopped and glanced at the dogs that were huddled together, keeping each other warm.

I made my way to the front door, as I wanted to see if the severe storm had done any damage. I half-opened it and …

I had a terrible shock and my blood went cold; there to my horror, I saw the light!

However, I quickly recovered and burst out laughing. I called Fállico who was at the table writing. I was aware that the generator was running and the radio operating.

Fállico came running and I said to him:

'There you have the light. Is that it what you saw?'

'Yes; exactly the same as the one I saw! Is it possible that I could have mistaken it for something else?'

'It could happen to anyone', I replied.

I saw he was about to fetch Baldoni, but I checked him saying:

'Leave it to me to deal with, I will explain the situation to him.'

I closed the door and went to the radio room. There I found Baldoni hard at work, but I interrupted him saying:

'Do you remember that light we installed half way along the track to the Variometer Hut which we can switch on from here at any time we need to go there?'

'Yes,' he replied, 'what's wrong with it?'

'You recall that it was impossible to deploy the cables aloft because of the gales, so we used armoured cable, lined with lead and a steel

67: Magnetic huts, store house, and post bearing the phantom light on The Isthmus

mesh which we laid along the ground.'

'That's right', he said.

'Could that cable lose an amount of power?' – I asked.

'I don't think so. It is well-insulated.'

'Nevertheless,' I intervened, 'do you want to see the "phantom light"?'

Seeing he was still mystified I continued:

'Leave everything running and go to the door. Do not switch on the light to the path. I will press the Morse key and then you will see the 'phantom light'.

Baldoni came back at once and said:

'Now all is clear. The direction of the radio antenna is parallel to that of the light cable along the path. The transmitter's signal produces a powerful induction current, so great, that it actually causes the filament in the electric light bulb, situated on the pole in the direction of the cemetery, to glow with a ghostly pale light that appears when the Morse key is closed. This is where the current loss occurs which I have been trying to locate for two months!'

'How is it possible that something so insignificant has caused such a stir among us all?'

'It is just a consequence of the strange life we lead here', I responded.

97: Words through the Air

One day I received a radiogram which read like this: 'Please let us know what radio station you receive best, so that we can organize a special programme for you'.

It was signed by my father. I showed it to my companions and asked Baldoni to monitor the news bulletins on all the stations operating at that time, to be able to give a definite answer to my father's question.

I also helped Baldoni make that difficult selection. We did not really have a favourite station, as there were very few that broadcast on a wavelength which could reach the South Orkneys. On these few we used to listen to the news, always assuming there were no magnetic storms occurring, or that the antenna had not been broken by the huge weight of ice accreted during the day.

We were not motivated by any desire to supply publicity for a particular radio station. I merely wanted to make sure that our families' voices reached us loud and clear.

In the end we opted for the station whose identification, in those days, was 'LOT Olivos', as this was the one we could hear the best[286].

I sent my father a radiogram to that effect, and from that day, we never missed the news programme that they broadcast from eight p.m. Generally Baldoni and I sat by the receiver, and then one night, on 20 May 1927, I noticed the atmosphere was free from static, even though the voices or music from the station were lost at times when the signal faded.

We had synchronised with the transmission frequency of LOT and waited anxiously, with Baldoni, for the radio news to begin.

At eight p.m. on the dot, Buenos Aires time, and at nine South Orkney time, I heard the announcer say:

'We will begin the programme with today's news broadcast.'

They gave the foreign news first and after a while: 'News of our Country'.

Without a single word to Baldoni, we followed the announcer's words closely, and even with the huge distance that separated us, I

286. LOT Olivos is a radio station in a north-western riverside suburb of Buenos Aires.

thought I could detect something unusual in the clear measured tones of that invisible man whose voice was reaching our ears.

He enunciated distinctly with unusually long pauses. He made a brief, concise, announcement, then continued to say:

'Now we have some news which will be of great interest …'.

He paused again, which seemed endless to me.

'Orcadas … Orcadas. Attention!'

His voice was slow and deliberate, grave, and seemed to come from a person putting all his effort into reaching us clearly.

He continued to call:

'Orcadas …'

'We inform our listeners throughout the country, that this radio station, a participant in the forthcoming independence anniversary celebrations, has prepared a special programme which it will dedicate to the heroic lads of the Argentine expedition in the South Orkney Islands.'

There was a moment of radio silence.

Outside the wind was howling.

'We hereby advise that all those who wish to take part in this programme can do so by addressing their applications to our editorial offices, or by contacting the Editor of the newspaper *La República*.'

'Desimone and my father have kept their word', I thought.

'Orcadas … Orcadas … we would be grateful if you could let us know by radiogram how you have received this transmission.'

No sooner said than done, I drafted a radiogram which Baldoni transmitted to the Ushuaia station, whose operator promptly replied to us:

'I also listened to the LOT news broadcast. I congratulate you. Here in Ushuaia we are all thinking of you.'

On the following day, 21 May, I received another radiogram from my father that said:

'On the 24th, at ten p.m., listen for a special transmission to all of you. Greetings to everyone.'

It was impossible to describe our feelings, there in the South Orkneys, when we received this confirmation.

At last I would be able to hear the voices of those I loved, draw close to them, and, in heart and mind, travel to Buenos Aires … at last!

I had lived in the South Orkneys for two long winters and, during all that time, never had the remotest idea of what was going on in the world outside. I would now be able to hear voices, other than those of my immediate companions; they would be speaking exclusively to me and only for me ….

From the South Orkneys we could not reply in the same way, we would be limited to just listening, but on the following day I could contact Ushuaia, or anywhere else on the Patagonian coast, giving my impressions or replying to my family's call.

Compared to the unfathomable silence of previous years, this was a huge improvement. In this way one of my greatest aspirations in returning to the South Orkneys was fulfilled.

98: A Broadcast

From early on 24 May, I organized the task of thoroughly checking all the installations of the radio equipment. We examined everything carefully: the antenna, the contra-antenna, the generating set, and the receivers. As a precautionary measure I decided that the antenna should not be raised until the last moment, in case a sudden gale left us without reception.

Baldoni fine-tuned all the instruments, while my companions and I worked on the outside cables.

Then came evening. Everyone's impatience was evident as the day progressed and as the scheduled time for the transmission approached, the nervous tension that surrounded us in that cramped space became more evident.

During the eight p.m. news bulletin, the announcer reminded listeners that, at ten, he would be addressing the expedition personnel in the South Orkneys.

To make the time pass we dined late, and as Becker was finishing clearing the plates I went to the kitchen and asked him:

'Could you prepare a late-night snack or supper, something for us to have in the early hours. No one has eaten anything this evening, they are all on edge and impossible to deal with ....'

As we were quite unable to put up with that prolonged wait, we went into Baldoni's radio room where he already had the receivers switched on.

I put on one of the pairs of earphones, and the others immediately followed suit. I then tuned the receiver to the 'LOT Olivos' station.

We were listening to a concert being broadcast, but whenever there was an interval, the announcer said:

'We remind listeners that tonight at ten p.m., we will be broadcasting to the South Orkneys. The families of expedition members are already arriving.'

The music played on, while we all fidgeted in our chairs.

Eventually the time arrived.

'Orcadas ... Orcadas ... Attention!'

'We are about to begin the transmission of the special broadcast which this station is devoting to those brave men stationed in the South Orkney Islands.'

Then the announcer, with a grave and unhurried tone, in which again I detected a tinge of emotion, added a few words of greeting and said that the next to speak would be the Director of the station. After giving a brief greeting the Director passed the microphone to the hands of those who would be addressing us:

The announcer continued:

'We will play the March of San Lorenzo, the martial strains of which, on this frequency, will reach far to the Antarctic and those brave lads stationed there.'

Immediately, through our earphones, we heard the beat of martial music.

Our impatience and nervous tension were evident. A profound silence ruled in that small room in which sat six men looking peculiar wearing earphones, with their heads in their hands, as they strained to catch the sounds from the distant transmitter.

Baldoni quickly turned to me saying:

'I will connect the loudspeaker, so we can listen more comfortably. There is no static tonight, and it is calm outside.'

68: Variometer Hut during the spring melt

A few seconds later we took off our earphones to listen directly to the words of the announcer coming from the cone of the loudspeaker, and heard him say:

'Now the father[287] of the Leader of the Expedition will say a few words.'

They all looked at me in silence, while I remained impassive, keeping my thoughts directed north. After a few seconds which seemed to last an eternity, I heard my father's voice saying:

'My dear José Manuel and esteemed young friends:'

'You are based in the most southerly inhabited region of the world. It is due to the happy initiative of the newspaper *La República,* as well as the benevolence of your Director, Captain Plate, and the kindness of the Board of Directors of LOT Olivos, who most willingly agreed to take part, that you are now, at this moment, listening to the persons who have you ever uppermost in their thoughts. Your mother, your sisters and your young brother are by my side, and I speak for them all.'

'You cannot imagine the emotion I felt when I received your first

287. Manuel Moneta

69: Base at night; Uruguay Cove and Mount Ramsay on the right

communication by radio, which, consistent with your prediction, reached me one day before the date you had stipulated on the day of your departure.'

'The satisfaction, both as a father and an Argentine, that I experienced at that moment, and that goes for the members of the families of you all, is not something which is easily explained in front of a microphone. The only thing that can convey sincere emotions is the happy and joyful expression of the speaker – and a microphone does not have that facility.'

'This happy result of the project to install the radio link is due to your perseverance and energy, which I well know, and to that of the rest of your comrades. No doubt you have had some good times doing it, though more likely it will have caused you the difficulties and annoyances inherent in the success of any venture. No doubt it will be another merit on the record of your career, which I hope your superiors will take into account in the future. This expedition you are leading has long been your fervent dream, composed as it is entirely of Argentines who have supported you so well. You have another achievement in the already long and distinguished history of the

Meteorological Service of our country. The advent of radio will put an end to the romantic concept of a solitary immaculate wilderness of ice and snow. It is equally true that the radio waves can draw our hearts and minds together with the affectionate, if laconic, messages that you send over the vast distance.'

I was all concentration but glanced at my companions as they silently listened, to my father's moving words. My father continued to comment on some more general items of news which he thought might be of interest to us, before finishing with these words:

'Well, dear son. I do not have to remind you and your compatriots of your great responsibility to your leaders, to your families, and to your country. This winter party, so uniquely and essentially Argentine, is something for which you have struggled hard. Considering that the majority of your companions were introduced by me; I feel, to some extent, that I am the expedition's godfather. May you, and the individual members of this winter party, never fail for a moment to have the highest concept of what it means to do your duty.'

'As for you my son, your brief must be that of an Argentine officer who manifests his education, preparation, and effective leadership in every sense of these words. When you eventually relinquish those solitary wastes, those who follow will find the way clear. They will advance the benefits already attained by the drive and determination of Argentines like yourselves, benefits which one day, will redound to the exploitation and enjoyment of one of the greatest natural resources of our country.'

'At the age of twenty years you were the youngest amongst foreigners in that inhospitable land. You repeated the experience a second and third time, becoming the youngest leader of any expedition. This was your aspiration. I exhort you to continue to act in such a manner that I shall be proud to be the father of a man of action who, single-handed on his own initiative, has constantly striven to dominate his future. I warmly embrace you, as I also do your companions in the adventure.'

Prolonged applause resounded through the loudspeaker, and when it faded, the announcer said:

'In a few moments we will be hearing the clock strike midnight, this time instead of the traditional chimes, the station's orchestra will usher in a new 25 May!'

'Attention please!'

We looked at each other in silence, without even attempting to say anything; we did not want to miss a single sound that emerged from the loudspeaker. Outside a hurricane-force wind was howling in the darkness, whistling as it passed between the cables of the antenna.

Every so often a whirlwind shook the frail building to its foundations, and snow pellets splattered violently on the window panes.

The stove in the corner, glowed red from the burning coals, the air was thick with spirals of smoke emanating from the pipes and cigarettes which we smoked continuously.

Jaramillo brought a bottle of whisky, saying to each man as he filled and passed the glasses:

'Here is a tot to keep you going ....'

Suddenly, from the loudspeaker, there burst on our ears the solemn strains of the National Anthem of our distant country reaching us from a distance of four thousand kilometres to the north, and we rose to stand.

We, there in the south, in a world of ice, stood to attention listening.

I saw Casariego's eyes misting over with tears, and when I glanced towards Fállico and Baldoni, they were deeply moved. Jaramillo's lips moved as if he was about to say something, and I realized he wanted to intone the verses of the Anthem, but first he had to overcome the turbulence in his soul before his warm tenor voice could be heard.

We were all standing stock still; without moving, we followed Jaramillo's lead. We must have looked a strange choir, chanting that anthem together, with the music being transmitted for us over such a vast distance.

When the applause transmitted through the loudspeaker fell silent, I heard Jaramillo say to Casariego, as he patted his back and passed the bottle of whisky:

'Dry your eyes and have a drink. Don't worry; we are all in the same state. It's manly to be able to cry!'

The loudspeaker then went on to announce:

'The lady, mother of the Assistant Leader, will now say a few words:'

There was another silence and then we heard a male voice in the background, as if trying to veil his words:

'This way, madam ··· don't become emotional ··· just speak slowly ···.'

I heard a woman's voice, faltering and with words choked by emotion, trying to say:

'My dearest son ··· Miguel Ángel ··· I want ··· I want to tell you ···.'

I felt a lump in my throat and sat with my elbows on the table supporting my head so I could cover my face with my hands.

'··· All well here ··· this is your mother!'

Then another pause.

'Miguel Ángel ···.'

Behind me was my companion, her son!

Jaramillo placed his head on my shoulder and I heard him sob: I did not want to turn around and look at him, nor indeed would I have been capable.

'I am fine ··· I am your mother speaking ··· my son!'

I heard the voice on the loudspeaker crying.

The announcer said:

'Mrs Jaramillo, mother of the Assistant Leader, is unable to continue, she is too overcome by emotion. If possible, she will try again later.'

Then followed a short interlude, with a song from a well-known group, and the announcer returned:

'Miss Casariego will now say a few words to her father.'

I heard a child's voice saying:

'Papi ··· can you hear me? Mama is here; after me if she feels strong enough, she will speak to you. We are all well. Do not worry about us, we have been strong and brave enough to bear this separation. Papi ··· dearest papi! How are you – is there lots of snow? Papi, still eight months to go!'

As far as these broadcasts were concerned, I knew the limits. They profoundly affected our emotions; we would have been quite capable of killing anyone who dared to interrupt them.

After that they spoke to Baldoni and then to Fállico.

In between all these speeches, interrupted several times by the uncontainable sobbing of mothers and sisters, they transmitted a selection of music from Buenos Aires, performed by orchestras who had generously volunteered their services.

A moment later the announcer continued:

'Now *Frau* Becker will speak to the Cook of the Expedition.'

A woman's voice, with a strong German accent, followed:

'*Mein Konrad!* I received your radiogram and here I am ··· I'm fine. The hens are well and the dog ··· he is fine too. Don't drink too much beer ··· well ··· there isn't any down there. Better you drink something else. Take good care of yourself, you know I don't like your drinking habits ··· Just behave, and listen, you too are an Argentine now ··· Greetings ··· *Auf Wiedersehen!*'

I looked at Becker, while he sipped his whisky, and he said to me with a twinge of irony:

'She is always like that. She doesn't even let me have a glass of water – and she swigs anything that's going as long as it is alcoholic!'

While raising his glass to the loudspeaker he toasted:

'Prosit, mein Gretchen!'

He drank it in one gulp. Then, looking incredulous and thoughtful, he added:

'It was her ··· that was her voice!'

The broadcast continued, and our eyes, fixed on the loudspeakers, seemed to want to penetrate it to bring our inner beings closer to the voices reaching us with such warm and affectionate words.

By the time that memorable first broadcast ended it was four in the morning, South Orkney time[288].

We felt quite drained by the emotional anguish we had experienced, yet it was not without pleasure that we shed those manly tears whenever we had a chance to hear the warm emotional voices of our loved ones.

288. The late timing of the programme is because radio communications are usually far better at night, when solar interference is minimal.

70: Celebrating Argentine national day, 25 May 1927; left to right: Baldoni, Casariego, Moneta, Becker, Jaramillo, and Fállico; (the portrait, in front of the Argentine flag, is of General José de San Martín)

99: Good Humour

The radio broadcasts generally, especially those of 25 May and 9 July, had the virtue of breaking the monotony and loneliness of life to which members of previous expeditions had been obliged to submit.

How different did all that seem now! I doubt if my colleagues could possibly be aware of the vast contrast there was between that new system of rapid communication with the inhabited world, and the absolute solitude of earlier expeditions. In former times we had to wait for the relief ship to arrive, only then did the already out-of-date news of what had been happening in Argentina reach us.

Today any problem that arose could be resolved with the help of the radio-telegraph. Personal communication by voice had to be restricted to reduce the load on the equipment and keep it serviceable. However there was nothing to stop us sending or receiving Morse code radiograms from family and friends. Sometimes, in this way, they were able to consult us on matters of a personal nature.

The isolation, the mysterious charm of those icy solitary islands forgotten in a remote region of the world, had probably gone forever. There is no longer any place for poetic contemplation of the unknown, for imagining daily life beyond the South Orkneys. Like a chain which now bound and united us to the distant world, radio communications gave us enormous self-confidence. It was good for our morale to know that we were inexorably linked to those who could always be in contact by radio.

This was the main reason why good humour and a calm atmosphere prevailed throughout that winter. Just occasionally, tempers frayed and men disagreed over passing points of contention, but these faded as quickly as they had begun. Any interesting news item which Baldoni brought us, had the beneficial effect of diverting our thoughts to distant places where our imagination was distracted. This resulted, for a few moments at least, in our forgetting about the isolation of where we were living. Nevertheless this isolation was something which we all bore with stoicism.

Thus it was with equanimity that I greeted the arrival of winter; a winter no different from all the others. As I was the only veteran, my companions used to quiz me about the date of return of the wildlife which had migrated north before the critical period of maximum cold.

'In September,' I would tell them, 'within a month, we will see petrels again; shags will follow, and after them seals and penguins.'

'Will we be able to visit their colonies?' someone asked.

'Yes; you will find it a well-worthwhile trip to the places where they gather for the breeding season.'

In anticipation of the trips I would have to make with my ciné camera, I devoted time to training the dogs to pull the sledge. At first I had to struggle against their undisciplined nature, but, with patience I took them for training one at a time and, by the end of the winter, they had become so accustomed to being attached to a sledge that whenever they saw one of us going outside carrying a pair of ski on our shoulders they would come over to us.

I continued with these methodical training sessions while we awaited the arrival of better weather to begin field excursions.

The most common topic of conversation among my companions was the low temperatures that were registered during that winter. It was almost as if I could sense a spirit of rivalry between them to be the first to record the lowest temperature of the winter.

Every time anyone went outside to make the observations, the others pounced on him as soon as he was back:

'How low?'

'Thirty-eight point six degrees below zero.'

'Is it still going down?'

'It looks like it ….'

One day, when a severe wind was blowing from south-west, whipping up the surface snow and casting it against the dwelling, we were gathered in the parlour, when Jaramillo came in from recording the meteorological observations.

While he took his windproofs off, he said.

'Could you prepare another thermometer ….'

'What for?' I asked.

'Forty point two degrees below zero[289]', was the reply.

As we remained none the wiser, he added:

'The thermometer we have outside is defeated ….'

'Done for? Damn … you have broken the wretched thing!'

'No. The thermometer is defeated. The mercury has disappeared into the bulb ….'

It was necessary to change those in the psychrometer and install thermometers calibrated to lower temperatures[290].

A few days later we were gathered round the table after dinner in conversation; Fállico opened the gramophone and played some records. I knew by heart every single note; it was as if they had been recorded in my head many years ago.

289. This record does not appear in the published data which reports the minimum temperature registered at Orcadas Station to 1951 is near midnight on 3 August 1904 (-40·0°C in: Acuña, 2015). See the reference in the footnote of § 29 and: Davis, 1905. Subsequently, during the period from 1961 to 1990, a minimum of -44·0°C was observed, the record low.

290. At temperatures below the freezing point of mercury (-38·8°C) thermometers with a different liquid, usually alcohol, become necessary to measure minima.

Casariego, who was patiently rolling cigarettes for smoking during the next day, listened attentively to the tangos played by well-known Argentine performers, and suddenly said to Fállico:

'Hey, Luís.'

'What is it?' he replied.

'Seeing you are determined to have music, you might at least put on a decent record.'

'Well what do you want then?'

'It would be better to listen to a woman's voice ···.'

'*El gordo* is right,' said Jaramillo 'it is quite a while since we listened to a female voice.'

100: And Another Year Passed ···

Time was flying. September arrived; then October, during which my scientific work was doubled by the annual calibration of the geomagnetic instruments. Then came November, with the annual ritual of gathering penguin eggs.

During the first days of December I had an idea which I was keen to broach to my companions at the earliest opportunity. Sure enough, during lunch a few days later, I said:

'What do you think of the idea about buying tickets in the National Lottery which has a prize of a million pesos.'

'Brilliant idea - but how?' they asked.

'By radio', I said.

The proposition was accepted at once, and I telegraphed my bank asking them to acquire three complete tickets for the Christmas draw. One of the three tickets was to be entered jointly in all our names; the second in Casariego 's name, and the third in mine.

As a security check I insisted that the bank, after it received the ticket stubs, should telegraph me the serial numbers, indicating to whom each belonged. The tickets were to be delivered to our families.

The night of the lottery draw arrived, and in a mood of tense expectation we listened to the news. Great was the general disillusionment when we heard that the winning number was quite different from any of the three of ours.

Nevertheless the attempt had proved a point. Amidst our general disappointment, there was a reason for satisfaction; it was yet another proof of the potential benefits of having established radio communication.

In January I received the first definite news about our relief. This summer they were again unable to secure a ship strong enough to make the direct voyage from Buenos Aires to the South Orkneys. Again there was no other way than using a whale-catcher from South Georgia to transfer us there, and then board another ship to continue to Buenos Aires, which I bet it would be no other than the veteran *Tijuca*.

Regarding relief, well, at least I knew what to expect. No longer that dreadful apprehensiveness I had suffered on previous years waiting day after day for a relief ship to suddenly come into view. I had no reason to be anxious, now, I knew perfectly well it would arrive in its own good time.

One afternoon, Baldoni said to me:

'I have just received a message from the Ushuaia radio station saying the lads there already know who will be the next Radio Operator.'

'Anyone in particular?' – I asked.

'No; but he is a Warrant-Officer of the Argentine Navy, currently in charge of the radio station at Cabo Vírgenes[291]. His name is Fortunato Escobar.'

'Do you know him?' – I asked.

'I had the opportunity to call him once, but it was only a brief radio contact.'

'Very interesting,' I replied 'every day seems to bring more news about our forthcoming relief.'

Actually that is exactly what happened: a few days later the station LOT Olivos announced the men of the next winter party would be introducing themselves to us that evening and sending their best wishes.

291. Cabo Vírgenes is a radio station and lighthouse at the extreme south-east of continental South America, on the northern entrance to Magellan Strait from the Atlantic Ocean.

At the appointed hour we heard the voices of Bruhns and Chakí, the Leader and Assistant Leader of the team who would relieve us. They spoke to us in person, and for the other members of their team who were not present, and said they would soon be embarking for the South Orkneys.

Early in February there was an announcement that the expedition would depart on the 15th of that month for South Georgia. When Baldoni spread this news, animated conversations began about the duration of that voyage. Despite the peace of mind that news brought us, the apprehensiveness amongst us was only too evident. No one could stop himself thinking of that date, which in our private thoughts was approaching all too slowly.

At last, on the 15th, we heard from the loudspeaker:

'Orcadas ··· Orcadas ··· Orcadas.'

Shortly afterwards, came a single message, which said:

'Orcadas ··· today at four in the afternoon, the relief ship left Buenos Aires'.

That was what we all needed to know. Everything was in readiness to receive them; the base had been painted just as in every previous year, maintenance work around base completed, and the outgoing cargo ready.

Fifteen days later *Tijuca* reached South Georgia, and although we were not in direct communication with the powerful radio station at King Edward Point, we were kept up to date with the progress of their voyage through news broadcast by stations in Buenos Aires and Montevideo.

Another day came the announcement:

'The whale-catcher[292] has departed from Grytviken.'

I said to my companions:

'Within three or four days it will be here. It all depends if they are lucky enough not to encounter any ice during the voyage and on how long they spend with *Orwell*.'

I asked Baldoni:

292. *Don Ernesto* on this occasion

71: Miguel Jaramillo, Assistant Leader in 1927

'Has it been possible to communicate with *Orwell* yet?'

'On none of the attempts I have made so far was it possible. *Orwell* works constantly on short wave, directly with Norway; it appears they do not listen, or do not want to listen, on the 1500 metre band which is the most suitable one for our transmitter.'

'Nevertheless' – I added 'it is important to remain alert in case there is a call from *Orwell*.'

They are bound to call when they know that the whale-catcher from South Georgia is on its way with mail for them.

'In that respect they are better off than us', Jaramillo commented. 'They have been in the area catching whales for less than two months and yet they still receive news and mail from their country. In contrast, we have been wandering about on the ice for the last thirteen months with not a scrap of any news apart from what the radio provides.'

'It's a bit weird,' added Baldoni, 'that they have not been round to visit us.'

'Perhaps the whale catching has proved very productive and they have just been too busy' I replied.

101: On the Way Back

The whale-catcher sent on the relief did not have a radio, so there was no way of telling exactly what day and at what time she would arrive in Uruguay Cove. At dusk on the third day we scanned the horizon closely; it was partially clear of ice, but, in that scene of desolation, there was no indication of anything out of the ordinary. When it became fully dark, no light could be seen that might reveal the presence of the long awaited ship.

The next morning, after the usual radio transmissions at eight a.m., Baldoni came into my room unexpectedly:

'Chief, please come straight away, as *Orwell* is calling us.'

'Surely,' I thought, 'the whale-catcher must be there'.

I ran to the radio room and arrived just in time to hear:

'Meteorological Station ··· Orcadas ··· Argentine Expedition ··· *Orwell* calling ··· *Orwell* calling ···.'

When the call had finished, indicating that the operator had switched off his transmitter and was listening for us to call, I started our generator again and went back to Baldoni, who with the Morse key in his hand, was calling the whaling factory ship.

Before Baldoni had finished transmitting, we heard the voice from *Orwell* again saying:

'Relief expedition here, we will be arriving at the Observatory tomorrow afternoon. Greetings to all.'

When I heard the greeting I had no difficulty in recognizing Bruhns's voice, and as that was as much as I needed to know and *Orwell* had ceased transmitting, I went to the parlour to give the latest news to my companions.

The next day the whale-catcher arrived. Needless to mention the indescribable joy that came over us when we saw the catcher appear round Point Lola.

I embarked on the motor-boat with Baldoni and went to the middle of the bay to welcome them; this evoked similar scenes to those I had experienced in previous years.

Of the newcomers, I knew only Bruhns and Chakí. The others, who gave the impression of being good lads, were gazing in amazement at the panorama that surrounded them for the first time.

I observed Baldoni's replacement carefully. I knew his age was forty-two, but he looked older ····.

When he was introduced to me, he said:

'This voyage has made me suffer a lot. I feel a bit sick.'

I did not pay much attention to his words, but a few months later, while I was in Buenos Aires, I had good reason to recall them. It was after I had learnt of the total radio silence from the South Orkneys.

Then, while Escobar exchanged impressions with Baldoni, I heard him say:

'I have brought a new short wave transmitter. I will try to communicate directly with Buenos Aires to avoid the usual relaying of transmissions from the Patagonian coast stations.'

An hour later they proceeded to unload all the cargo of the new winter party, and in the early hours of the next dawn, we departed for *Orwell,* where we would take aboard their mail bags destined for

Europe, via South Georgia.

We spent another day aboard *Orwell,* where I was cordially received by the Captain and Officers. There was also time to have another chat with Rumbolds, the British Customs Officer, and we shook hands effusively.

At last we reached South Georgia, but things took longer there than I thought they would because we had to wait for an entire month before *Tijuca* was fit enough to sail for Buenos Aires.

While we staved off our boredom one afternoon by taking a walk on that picturesque island, Jaramillo said to me:

'Do you realize that our expedition is the one to have spent more time in the South Orkneys than any other?'

'That's certainly true,' I replied 'we arrived in December 1926; spent the whole of 1927 there, plus a portion of the present year. Fifteen long months at Orcadas, to be precise.'

We continued chatting about things and suddenly he said:

'You, who have been in the South Orkneys for three winters, may be the right man to answer the following question:'

'I am listening', I replied.

'Would you think of returning again?'

'Why not?' – I contended.

However, in truth, at this particular moment I had no definite idea whether I wanted to return to that solitary island, or even whether it would be possible for me to do so.

I had achieved my aims: in the twenty-three years during which annual expeditions had been dispatched to the South Orkney Islands I was the first Argentine to lead one. I was also the first leader to recruit an expedition exclusively of Argentines. For them I had the satisfaction of breaking the absolute isolation of those remote Antarctic islands: I established a successful modern radio communications system that linked us to the mainland. In previous years others had failed in undertaking such a project[293].

However, to return to the South Orkneys! …. When all is said and

293. See also the footnote in the prologue for earlier details of Antarctic radio transmissions.

done …. Who can foresee the surprises or the bitter shocks – and even a few good things – which come with life in the course of time.

102: Northbound

The wait in Grytviken seemed interminable[294]. Two weeks had gone by since our arrival and *Tijuca* showed no signs of being at all ready for the voyage to Buenos Aires. Jaramillo, who had urgent personal matters to attend to, was in a hurry to return home, and had no hesitation in embarking aboard a ship that was about to depart directly for Montevideo. We were not able to follow him, his hasty decision left me with insufficient time to ask for permission from Buenos Aires for us all to embark together.

One morning the manager of the Compañía Argentina de Pesca informed me of the reason for the delay:

'We are waiting,' he said, 'for the arrival of a vessel direct from Norway bringing two new engineers who will repair the defects of the auxiliary engine of *Tijuca.'*

'What about the engineers who are already here?' – I asked.

'Their contracts are finished and they will be returning to Europe; I was unwilling to renew the contracts as they wanted more pay.'

There was nothing to do except to fortify ourselves with patience and await the arrival of that ship. Then, we received a radio message saying she would be arriving the following day. I was immediately told that, as soon as the ship arrived and the new engineers transferred to *Tijuca* we would sail.

The orders were given to have everything ready and wait aboard for the moment of departure.

I was delighted to be returning to my homeland after so many long months of absence, but the idea of making another voyage on the ancient *Tijuca* did not altogether please me, and from what I could

294. Coincidentally, on 22 February 1928, Arnold Weinholt Hodson, Governor of the Falkland Islands and Dependencies, made an official visit to South Georgia where he reported meeting some of the personnel from the South Orkneys at Grytviken: Hodson, 1929. Moneta's films included scenes of the unveiling of Shackleton's memorial and other events during the governor's visit.

72: Return voyage aboard Tijuca *(note the angle of the horizon)*

gather, this opinion was general.

Despite the obvious elation the thought of our imminent return gave me, I was somewhat apprehensive as a result of a thought going through my mind. That evening I had the following conversation with Baldoni about it:

'You realize,' I said 'that we have two new engineers aboard who are taking over from the previous two on this sailing vessel'.

'I have heard something to this effect …', he replied.

'Well, I doubt whether these new engineers, however expert they may be, can familiarise themselves with the ship's auxiliary engine in such a short time between their arrival and our departure.'

'That's right. They have barely any time at all ….'

'This is what makes me uneasy. For several days now I have been thinking that there is something not quite right going on. Let's suppose that the engine has a problem, or that they have done something that prevents it from working properly ….'

When Baldoni heard this, he looked at me in surprise and said:

'Do you think the two engineers who are leaving might have spiked the engine deliberately?'

73: The 1927 Expedition aboard Tijuca *in South Georgia*

'I reckon that the two engineers who are leaving are capable of something. They are far from happy.'

'I don't think so', Baldoni said.

'However,' I added, 'I have the feeling that there is something about the atmosphere aboard this ship that is not quite right.'

I did not have long to wait to prove my intuition was not deceiving me. As soon as *Tijuca* was under way, leaving the pier from where the people of Grytviken were waving farewell, I happened to glance at the funnel and saw the exhaust from the engine, and became very anxious when I saw it belching thick black smoke.

Baldoni was standing beside me, and I drew his attention to it, and said:

'Look at that! Far too much smoke from an internal combustion engine. It is not burning the oil properly. There is something not quite right about it ….'

We moved farther away from the pier as *Tijuca* was manoeuvring out of the bay. We passed King Edward Point and turned hard to port, keeping close to the shore. Once past Hope Point it was no longer possible to see the whaling station, although it was only a couple of kilometres away.

My companions went below to the mess, though I preferred to stay on the bridge aft with the Master.

At that instant I had a shock. I could hear the engine start to stutter and instinctively looked towards the shore to see how distant it was from the vessel.

'Barely a hundred metres', I said under my breath.

Then, at that moment, the engine stopped. A few anxious moments went by, and I regarded the slight figure of Captain Thorsen[295], who, with all the calm in the world, took his whistle to his lips.

Several of the crew responded to his summons and I heard some orders being given but I was unable to understand what was said as they were in Norwegian.

My companions ran back to the bridge and pestered me with questions:

'What's happening? What's the matter?'

'Nothing ··· keep calm! The engine has broken down.'

'Well what is wrong with the sails? Why do they not raise them?' asked Casariego.

'Impossible to set them in an emergency like this,' I replied, 'we have a cross-wind which will undoubtedly cast us on the coast.'

I gave the order:

'Every man stay here. No one is to move from my side!'

I turned to Baldoni and said:

'Go to the radio and help the operator. He may be trying to call Grytviken. We are only three kilometres away and they may be able to send a catcher to tow us before we ground. The shore is getting closer!'

Baldoni ran off in a flash and I went over to the Master, and asked him:

'Trouble with the engine?'

'Yes.'

'Can it be fixed quickly?'

'I don't know', was his laconic reply.

295. Captain Gotfred Thorsen, as in 1926

I peered through the grating to look into the engine room where I saw the two new engineers, running from one side to the other, opening and closing valves, moving levers, talking hurriedly to each other.

Ten minutes passed during which, *Tijuca* slowly drew nearer to the shore, drifting at the mercy of the wind, which was catching her amidships.

Baldoni reappeared; and dispiritedly reported:

'They are not answering our calls. We are so close, yet they are unaware of our perilous situation.'

I had already made up my mind, went over to my companions and said:

'Keep calm men! If we ground, which we are sure to do any moment from now, take off your boots and heavy clothing and immediately make your way to the starboard lifeboat. There is not much swell so this operation will be quite straightforward.'

103: Risky Navigation

I assumed charge from personal responsibility and concern for the safety of my companions. I was none too pleased with the lack of consideration shown us. The situation had been anticipated, everybody knew what was likely to happen, but I had no idea that the fault would appear so soon after leaving port.

At sea, under sail alone, things could be remedied; but here, close to the coast with rocks around, it was far more dangerous. Losing my own personal equipment was of lesser importance, my chief concern was to get the box with the observatory data we had recorded during the year, out from the hold and on to the deck.

Then, just as the situation was at its most critical, I suddenly heard the engine spluttering into action again, but under my feet I felt the intermittent vibration and again thought:

'This will not last long; too much smoke and it is too black!'

The revolutions of the propeller were sufficient to turn *Tijuca's* bow seawards and we gradually made way out of Cumberland Bay.

We drew away from the coast, but my mind did not rest easy until nightfall, when the Master announced:

'We are thirty nautical miles [54 km] to the east of Cumberland Bay, and if the engine stops again, we will be able to avoid the outer islands under sail alone.'

'I would like to have your opinion', I said.

'I am listening.'

'In the event that the engine stops functioning again, how long will it take us to reach Buenos Aires?'

'It depends on the winds that we encounter on the voyage. If they are favourable twenty days will be enough, but we could be in for delays of a month … or two, even three ….'

Shrugging his shoulders, he added with a smile:

'I am not in any hurry ….'

'Yes, that is quite clear …' I replied, 'but what about us!'

The Master was a good man. Small and thin; with a florid complexion, but with traces in his countenance of a man with great strength of character.

During meals I sat at his table and we exchanged a few words.

The following morning, after lunch, I invited him to have a glass of brandy, but he declined, saying:

'There is nothing I should like better, but not at the moment thanks. There are icebergs in sight and I have to be alert. Many icebergs have drifted from the south this year and we happen to be in a region where the largest number of sightings has been reported by the whale-catcher's.'

I went on deck, from where I could indeed see that a boundless number of them surrounded us, enormous icebergs scattered all around the horizon.

At daybreak on the third day I was surprised to be awakened by the severe rolls *Tijuca* was making. I dressed and went on deck, where I had to take great care so as not to be swept overboard by the waves washing over it.

I reached the bridge and found the Master, who announced:

'Very bad weather. Wind from the south-west, many icebergs, fog ….'

That was the worst of all: the fog which limited visibility to about one kilometre. Fog was treacherous; it could suddenly obliterate any view ahead and, without warning, conceal any enormous island of ice lying directly in our course.

I went below to the mess to eat my frugal breakfast, and then I made myself as comfortable as practicable wedged with a cushion; and held a book with the intention of reading.

Shortly before midday I was startled to hear the engine suddenly slowing and then stopping.

I put on my jacket, leather hat and gloves, and in two bounds was on the bridge. Captain Thorsen was there.

We looked at each other, and without exchanging a word, understanding exactly what each other was thinking. He pointed to starboard and said:

'A huge iceberg.'

'It seems to be of considerable length,' I replied 'it is about two thousand metres away from us, I reckon?'

'It's one nautical mile [1·85 km]. I have measured it', he replied.

'Have we been sailing parallel to it for quite a long time?'

'About four hours.'

'Four hours,' I said, 'at an estimated speed of seven knots [12·5 km h^{-1}], that gives it a length of twenty-eight nautical miles [50 km] ....'

'Yes,' affirmed the Master, 'and we still do not know how much of it is left. The fog prevents making any estimate.'

Meanwhile, below work continued feverishly on the defective engine. I went below and could see the engineers, who were covered in soot, had taken the heads off the cylinders which they were in the process of cleaning.

Baldoni came to find me and said:

'We are going to be here for a while.'

'How long?' – I asked.

'I reckon it will be a whole day at least.'

I returned to the bridge and did not move from there.

Under minimum sail, because the strength of the wind did not permit setting more, the vessel was making slow progress on a course parallel to the continuous wall of ice which lay a nautical mile

[1·85 km] off to starboard.

Hours passed and still that persistent wall of ice lay off our side. Most dangerous of all was the cross wind which caught us amidships blowing us closer to the ice all the time.

Every few moments the Master blew his whistle while he remained positioned one step from the helm, which was safely in the hands of a sturdy sailor.

The crew, demonstrating impeccable balance, went aloft to the top-sails, while the wind whistled around them and through the rigging. Enormous waves swept over the entire deck. On that wet slippery surface, inclined at a sharp angle, it was almost impossible to maintain one's balance.

Darkness fell and our fate remained precarious.

I was summoned for dinner, and went to the mess for the usual sweet soup, the customary fare of Norwegian sailors. I then returned to the bridge.

The storm raged with unabated fury, and when my eyes became adapted to the dark of that unforgettable night, I could discern that we were fewer than five hundred metres from the wall of that vast island of ice, I could clearly distinguish the waves breaking against its flanks.

Ahead, off the starboard bow a dull whitish blotch of ice could be seen disappearing and reappearing in the fog. I watched the Master, using an instrument to measure the vessel's drift and calculating the time it would take to pass a particular section of the iceberg, and I asked him:

'Will we manage to clear it?'

His reply was brief and to the point as he shrugged and answered:

'I do not know, this is very bad.'

With the engine stopped, I thought to myself:

'If at least the engine was working, it would help us get away from this imminent danger.'

I remained on the bridge until dawn, living anguished moments as I noticed how, slowly we were closing on that island of ice, whose inaccessible vertical walls, were over one hundred metres high. They would not provide us with the slightest refuge in the event that we had to abandon *Tijuca;* assuming we did not drown in the process.

The die was cast, everything depended on the engine and on that wall. We had been running almost alongside for a good twenty-four hours now. Would it end and leave enough open water for *Tijuca's* drift to carry us clear to safety?

At three a.m. the engine started again, and a passing smile of relief appeared in Thorsen's face. However he remained serious, gazing fixedly at that vast ice wall from which we could not separate ourselves. I asked him:

'We will be able to get clear of that iceberg now, don't you think?'

'Yes, I think so. The engine will definitely be of some help, but the wind remains very strong and if it does not slacken, we will inevitably smash against the iceberg.'

More whistles and more orders. It was impossible to go astern; such a manoeuvre would only have brought us even closer to that gigantic wall of ice.

I decided to await our fate on the bridge.

Wind, huge waves, ice, snow, intense cold, darkness! The hellishness of the ocean all united in a single night!

My companions were below, in the mess, lying in their bunks. Aloft, in the rigging, the agile movements of several shadows could be seen.

The Master gave another order. Three men on deck dodging the waves that flooded over the deck, ran and climbed onto the bowsprit. At each wave its tip disappeared under the seas, only to re-emerge violently at the same time as a sheet of spray was scattered by the wind.

I looked at Thorsen, this order seemed like an act of folly. The Master, understanding my thoughts, said:

'We have to lower the jib sails ….'

From the bridge I followed closely the movements of those fearless men who were risking their lives. I could not distinguish their faces because of the distance that separated us, and because the whole scene was immersed in the blackness of night. I could well-appreciate the desperate efforts they were making to comply with the order, while every few seconds they found themselves submerged bowsprit and all, beneath the waves.

Nearly simultaneously, two triangular sails dropped, but at the same instant the foc'sle of the barque was struck hard by an almighty wave.

The bowsprit disappeared under the waves and rose up again a few seconds later, but on the way up, I could see only two men holding onto the life lines. The third man was no longer there ….

Over there … I signalled to Thorsen, while I ran to starboard to release a lifebelt.

In one bound we reached the rail from which I looked out intensely. I could discern a dark form floating in the spume that ran slowly alongside the hull.

I cast a lifebelt just as another huge wave put more distance between us and the shape which, in all that spray and spume became lost from sight. The darkness was very intense, and although it allowed us to distinguish the huge white mass of ice, it prevented us from distinguishing any details on the surface of the sea.

Anxiously I looked at Thorsen, who evidently in deep distress, took me by the arm saying:

'There is nothing we can do. Impossible to turn or heave to …. Nothing, absolutely nothing … poor lad!'

The helmsman, clinging to the wheel, remained at his post with his eyes fixed on the compass. From time to time he glanced at the Master as if anticipating a further command.

Thorsen drew me over to the middle of the bridge and then gave a long shrill blast on his whistle. Several men arrived at the double, and amidst them I could see the two who had been on the bowsprit. They were drenched from head to toe.

The Captain addressed a few words in their native language, they then doffed their hats and uttered a few words in unison, which I was unable to understand.

It seemed to me that they were praying, I looked at Thorsen; who had tears in his eyes; tears swept by the hurricane from a visage tanned and weather-beaten by the winds of all the oceans of the world.

He gave another order, the sailors began to disperse from the bridge. When they had departed, he shouted something and remained motionless while continuing to gaze at the iceberg.

I went over to the helmsman and asked him:

'What did he say?'

'Everyone is to go to their post and nobody is to look aft.'

Several hours went by and a new dawn arrived. The storm continued unabated and the iceberg had not disappeared from our immediate vicinity. We had neither drawn any closer to it nor moved farther away, the assistance from the engine was limited to merely keeping us on course.

The fog lifted at five a.m. when, with joy, I could see that, about four nautical miles [7 km] ahead, that wall of ice came to an end.

'At last!' I said to the Master 'Now we can all rest!'

'You may certainly be able to do so, but I can't.'

'Why?' – I asked.

'We are not yet out of danger.'

'What do you mean?'

'You cannot be sure that what you are looking at is really the end of the iceberg. It is quite possible that it is just the entrance to an embayment in the island of ice, and if we end up drifting in there with this crosswind, it will be extremely difficult to get out again, we would have that hurricane right on the bow.'

That old sea dog was right.

I was very tired but remained where I stood, although there was nothing I could do to help in the manoeuvring of the vessel. I was resolved to stay where I could see straight ahead to await the arrival of whatever fate would apportion to us.

Only by midday did I manage to retire and get some sleep. The fog had lifted, and a wide horizon appeared before us.

Several icebergs were revealed off the bow, but they could easily be avoided. The mass of the giant iceberg was still visible astern. Its length, calculated from the time taken to navigate past it, could be estimated as approaching a hundred kilometres. Undoubtedly it was one of the largest icebergs ever encountered at that latitude.

Because of its characteristic tabular shape it could be said to have originated from a massive calving of an Antarctic ice shelf, and had already travelled a huge distance under the influence of Southern Ocean currents.

This encounter was by no means exceptional: on reaching Buenos Aires, I was told that in the same summer of 1928 several ships on the regular run from Europe had encountered icebergs off the coast of Brasil and off the Río de la Plata. The latter sighting was confirmed by holiday makers, spending the summer in Mar del Plata, who were amazed to watch, from their beaches, several imposing fragments of Antarctic ice drifting slowly northwards.

104: Some Notes of Interest

One afternoon, after I had been back in Buenos Aires for a few days, I went in to the Meteorological Service where I found two of my former colleagues from the South Orkneys, Casariego and Fállico, waiting for me in the hall outside the Director's office.

We exchanged greetings then they asked:

'We wanted to meet you to find out if there has been any answer to our job applications.'

'I have heard nothing about it officially', I told them.

'I imagine,' said Casariego, 'that we will have no difficulty finding stable employment here. Although everyone knows that when we enlisted to go to the South Orkneys, we were not members of the permanent staff here, but I think we gave a sufficiently creditable performance that we might now be considered ….'

'Have you had a word with the Director?' – I asked.

'Yes we did, but I don't think anything came of it.'

'Well,' I added 'I will try to have a meeting with him, but I cannot guarantee that I will have any more luck than you. Come and see me in a couple of days.'

Some time later I had the opportunity to speak to the Director, with whom I had the following dialogue:

'I would be grateful for your opinion,' I said 'regarding the future of two of my companions who were on my expedition to the South Orkneys.'

'I am already acquainted with the matter; Fállico and Casariego have spoken to me about it.'

'Yes indeed,' I replied 'I feel they have both acquitted themselves well enough to be given work here that will provide them with a modest income. As you are well aware that a sojourn for a full year in the South Orkneys neither constitutes an automatic advancement, nor is it sufficient entitlement to claim an increase in our very modest pay. However these two men resigned from remunerative jobs to join the expedition, whose undeniable success is manifestly clear within this establishment.'

'I understand your reasons,' the Director said, 'but I am under no obligation towards these two men.'

'I am sorry, Director. I ask you to please excuse me for having troubled you …'; and I walked out of his office.

I have a good memory, and that is exactly what happened.

It was certainly true that the personnel sent to the South Orkneys were not permanent staff of the Meteorological Service. They were considered under contract solely for the duration of the expedition. It was no less true that those brave and enthusiastic lads deserved to be treated with more consideration; and on return to Buenos Aires, be offered a modest employment as a reward for their motivation and hard work in a job which they could not have accomplished without major personal sacrifice.

I never went to the South Orkneys in the hope of improving my salary when I returned. I neither expected, nor received, a pay rise after I returned from any expedition. On this occasion, a few days after the conversation with the Director, I had the satisfaction of hearing that both my former companions had found work in different offices in that important government department.

* * *

During the first few months of 1928 I was involved with preparations for the showing of the ciné film I had taken in the South Orkneys.

As with so many things, that reel of celluloid which I exposed with the expenditure of so much labour and sacrifice, is a story in itself. It influenced my decision to go back once again to the South Orkneys on my fourth expedition; I must mention it here to set the record straight on a few points. This is the story.

When I returned from my first expedition in 1924 I got together one night with several friends at a central location in Buenos Aires.

Someone invited me to watch the showing of a North American ciné film, which was very successful; the plot dealt with the adventures of an Eskimo on the coast of Labrador[296].

'We are in the middle of summer', I told my friends. 'Although I always find it interesting to watch a successful film, on this occasion I prefer to decline the invitation for the following reasons: first, because it is much too hot to be shut in a cinema; second, because I am too fed up with ice, snow and polar adventures to waste a night seeing them played by others.'

'You are right', said my friends.

It seemed, however, that I was doomed by fate to see that film; on one of my frequent visits to Córdoba, while walking through the streets of that provincial capital without any idea of where I was going to pass the time, I walked into a cinema without even bothering to look at what was showing. Great was my surprise, when I saw the title of the film being shown was the one of ice, snow and polar adventure that I had not wanted to watch in Buenos Aires.

I resigned myself to sitting through the film and when the snowy white scenery was at its best and most interesting, I overheard a few opinions from the spectators sitting nearest to me:

'This is amazing', said someone.

'Fascinating panoramas, most interesting …' others added.

I heard several more comments like this all around me.

I found myself thinking:

'If this is all that marvellous what I have seen in the South Orkneys is much more interesting, grander, and more beautiful!'

296. The film was *Nanook of the North* by Robert Joseph Flaherty which appeared in 1922. Coincidentally Flaherty's first attempt at filming Eskimo was destroyed by fire, as was Moneta's (§ 83), and he returned to the Arctic to produce another.

Then I thought: 'If I return to the South Orkneys, I will try to make a ciné film; I am already certain it will be a success.'

In October 1924 I was notified that I would be included in the expedition for 1925. When I returned to Buenos Aires, to prepare to join it, I managed to find a business house which dealt in cinematography. The directors were interested in my proposition, and supplied me with camera equipment and more than two thousand metres of negative film.

I had to learn the trade and become a 'cameraman'. By the time we left port heading for the South Orkneys, I had a reasonable idea of what ciné photography was about, and how to bring to the screen some of the more outstanding features of our unusual South Orkney way of life.

105: Business ··· is Business

At the time I had plenty of youthful enthusiasm, but no professional experience. I had already surmised the potential financial benefits to be gained from making the ciné film. Quite unintentionally I had become associated with an Argentine ciné company[297] that had its own name and trade mark, and felt that if this company were to write a simple business letter proposing the arrangement, and I were to write a reply giving my acceptance, this would correspond to a contract and would be sufficient guarantee of a reward for my labours. That was how, armed with this weak guarantee of a pseudo-contract, I went to the South Orkneys to make my film

When there, I was able to use any free moments from my scientific observations and general work of the base. Gradually I developed the film, metre by metre, depicting everyday scenes at Orcadas, including some of the local wildlife.

I returned home early in 1926 with all the reels of film entirely used, and handed it over at once to the film company so they could carefully make a positive print.

Great was the surprise of those who saw the preliminary version,

297. See the footnote in § 79

even though still in negative form. There were scenes of which nobody could have dreamed. Everyone who came to these private viewings agreed my film would enjoy great success before the cinema-going public.

It was never shown. One day, a raging fire destroyed the company's premises, and that was how my first attempt to bring scenes of Antarctic life to the general public came to an end.

The newspapers devoted long columns to the disastrous fire including extensive descriptions of the loss of the film about the South Orkney Islands.

After that I returned to the geophysical observatory at Pilar, in the province of Córdoba, where I stayed until my presence was required in Buenos Aires, to make preparations for the expedition of 1927, of which, for the first time, I would be the Leader.

I was keen to take advantage of this new opportunity to make another film. This time I didn't have to waste time finding a ciné company willing to supply me with the necessary equipment. The same Argentine firm, with which I had dealt previously, and had kept in touch, offered to make available two ciné cameras of the latest model and all the negative film that I would need.

I accepted their offer gratefully, but guided by the advice of friends whom I could trust, I proposed to the company that we make a more comprehensive formal agreement over the distribution of the film and apportionment of any profit. I succeeded in getting the directors to sign a contract that I had drafted in my own hand.

With this agreement in my pocket, and with the ciné equipment in my luggage, I returned to the South Orkneys with the 1927 expedition.

Few people know, or can even imagine, the sacrifices I had to make to obtain a few metres of exposed film in the depths of the raw winter, when the extremely low temperatures caused the film to snap as if it were glass, or when the driving mechanism of the camera completely froze.

The difficulties were all resolved, and in June 1928 the première of the film *Entre los Hielos de las Islas Orcadas* [*Amid the Ice of the South*

74: Poster advertising the 1931 film Entre los Hielos de las Islas Orcadas.
(The image, adapted from an old illustration, was not intended for publication. A copy is held by the Museo Marítimo, Ushuaia.)

Orkneys][298] was screened in a fashionable cinema crowded with spectators, who were astonished by the scenes of seal and penguin colonies as well as difficult ascents of icy peaks, Antarctic blizzards, and everyday life at Orcadas. When it came to our celebration of 25 May the entire audience burst into prolonged applause.

Not all was glory however. Shortly after those screenings, where all tickets were sold out before the film began, I had a cruel surprise.

Despite the fact that the box office receipts were lush and the profits from the film were increasing, especially on the company's account, it was no less true that the outgoings, the operating costs, increased vastly and out of all proportion. The result was that the film never, at any time, showed a credit balance.

In such a murky business as this there was little financial control. Until then I had no experience of such shady methods, but I was learning rapidly, the hard way.

However, the film was a growing success and it was shown throughout Argentina. I also found that copies had been sold overseas; but nowhere was there any sign of any profits.

Tired of struggling to try to establish some kind of control over these activities, I showed my contract to an acquaintance of mine, a well-known lawyer in the Argentine Court of Justice, who after considering several options, called me and said:

'There is nothing that can be done. Although it is correctly drawn up, this paper has no legal value because it does not carry the seal of a notary at the time it was made. It bears a date, 26 November 1926, and we are now coming to the end of 1928. There would be a large cumulative fine to pay, and in a case of this kind, I do not think any useful financial benefit can be obtained from dealing with people of this kind who are totally unscrupulous. Nevertheless, I will see what can be done ….'

However, the reality was that he was not able to do anything. The

298. The 1928 version of the film, titled *Islas Orcadas, Antártida,* lasted 55 m 36 s; a 1931 revision, *Entre los Hielos de las Islas Orcadas,* was 18 m 52 s. In the 1940s the stock held by Cinematografía Valle was sold to a manufacturer of celluloid combs (Capdevila, 1996). A copy of the 1928 film was found in Punta Arenas in 2014, one of the revision is in the Archivo Nacional, Buenos Aires.

only thing that he managed to secure was a thousand pesos which I shared among my companions, in compensation for their help and collaboration in making the film, in which I was the victim of gross exploitation.

106: Hypothesis

It was 1928 and I was already thinking of returning to the South Orkneys for the fourth time. In October, the authorities of the Naval Military Hospital received a radiogram from Orcadas which made my heart sink. It was laconic and read:

I am ill. I have high fever, and severe headaches.

Escobar, Radio Operator.

I was immediately consulted on the chances of finding a suitable treatment in the medical kit on base. A reply to the radiogram was sent, requesting more details in order to diagnose the illness.

There was no immediate reply, but a few days later we received another radiogram from Escobar which read:

There has been an interruption in transmission. I was too ill to keep the schedule. Only today have I been able to get out of bed momentarily.

I asked him again to send more details of the ailment from which he was suffering so that we could help alleviate it, using what was available on base.

On 26 October, we received another radiogram from Escobar saying:

I am in a very bad way. I have severe pains in my kidneys. I am urinating blood.

There was no room for doubt, and the medical officers who were meeting in the Naval Hospital affirmed:

This is very grave, nephritis ….

Nevertheless, on the following day, 27 October, a message with clinical observations was received. From that day on radio communications from Orcadas ceased completely. There was no reply from Orcadas Station to the insistent calls from radio stations in Patagonia.

The South Orkney Islands maintained a disturbing silence.

Several journalists, on the lookout for news, came and asked my opinion about the possible reasons for the radio silence. Few of them knew about the radiograms sent by Escobar, and so as not to alarm his family I declared:

'Radio communications from Orcadas may be interrupted by many things. Some of them I experienced in my previous expedition.'

'What were they?' asked the journalists.

'The antenna may have fallen as a result of the weight of ice accumulated on the wires, or the cable that holds it up may have come away from its mooring. In that case, the repair will be long and slow, it will require working outside when the weather allows it, and ascending the hill to reach the anchoring point of the main guy line of the antenna.'

'Are there any other reasons why radio communications might be temporarily interrupted?'

'As long as there is nothing wrong with generator. Anything is relatively easy to repair out there; but there is another possible reason for the interruption which could be even more serious.'

'For example?'

'Failure of the transmitter's valves: in that case, there will be no more communication until the relief expedition arrives with spares.'

'What about the short wave radio that Escobar took?' – they asked.

'I understand it has not been put in operation.'

'So now they are operating with the old long wave transmitter installed by you on the previous year?'

'Yes, that's right', I replied.

I added 'there could be another reason for the silence. One cannot discount the possibility of a fire on base. You must remember, the dwelling is wooden, and with the help of the wind and the fuel for the generator, it could burn down in minutes. This would be a very serious situation for the winter team, even if they escaped unscathed from the disaster, they would have to spend four months living in the little huts used to house the magnetic instruments.'

I continued 'there is no point in thinking that this is what may have happened. They always take the greatest precautions to avoid such disasters.'

75: Winter panorama: background Mossman Hills; left Point Diebel and frozen Scotia Bay; right Uruguay Cove, also frozen

'Is that all?' – they asked me.

'That is everything I can tell you based on my previous experience. After that one may think of many other hypotheses which would be too distressing to contemplate.'

That was what the journalists were led to understand. They took their leave, promising not to report anything else in their newspapers.

107: My Fourth Expedition

The radio silence from Orcadas was a contributing factor to the 1929 expedition being enlisted hastily. It was vital to discover the reason for the break in communications, and to arrive there with enough time to repair any damage that may have occurred with the equipment and installation. This was the official version of the situation that was circulated, and it made no mention of the alternative possibility, that Escobar was dead.

On the morning of 21 January 1929, we assembled aboard the naval transport ship *1° de Mayo,* which would make her second voyage to the South Orkneys. In command was Lieutenant Francisco J. Clarizza, and among the other officers, whom he had personally selected, were several who had already voyaged on the Southern Ocean aboard the corvette *Uruguay*.

The outbuildings of the Naval Arsenal on North Dock were completely surrounded by crowds who had come to bid farewell to the ship.

All the cargo for my fourth expedition had been loaded in good time. Thus in the final moments there remained nothing else for me to do but shake hands with family and friends who had come to say goodbye.

From time to time the inevitable journalist would manage to approach me, and after obtaining a few graphic details to fill out the news of the day, he began to interview me:

'I have been informed,' he said, 'that this is to be your fourth winter in the South Orkney Islands ....'

'That's right', I replied.

'What are the names of your companions?' – he asked.

'Félix Celestino y Monti, Jorge Páez Montero, Yakob Orlovski ....'

'Radio Operator ?'

'To fill this vital position I am bringing an excellent young fellow, very well trained and a good colleague ....'

'What is his name?'

'Carlos Kahl', I replied and then added, 'an Argentine, like Monti and Páez Montero he is barely twenty years old ....'

'Cook?'

'That will be Rómulo Devoto; a good Argentine. That is him over there surrounded by girls and boys ... they are his children.'

'Another question, sir. Do you have anything to say about your future companions?'

'There is always something worthwhile to say about the men who, in the prime of life, abandon the pleasures of the world to enlist in an Antarctic expedition.'

Those who happened to be around me seemed inclined to listen so I continued on the subject:

'I refer to Páez Montero, and Félix Monti. They are both young lads full of life. Two years ago they finished their studies in the School of Agriculture at Casilda. To obtain their degrees they had to have a year of practical experience on a farm, working in their speciality. They did not hesitate to go to Misiones, to the El Dorado colony, where they worked a whole year as labourers, enduring scorching sun, mosquitoes and all the other vermin of that region, just to cultivate *yerba maté*[299] for the rest of the country.'

'The future for these two men looked dismal indeed. Unfavoured by fortune and with no one to help them, there was small chance they could prosper. They were good friends and inseparable companions, perhaps one of them thought that in time, they could themselves become landowners instead of working for others.'

'However they had no money to realize their golden dream. Nevertheless, they were not unaware of the fact that the modest pay they would receive as members of an expedition to the South Orkneys, would be enough to start a business and they did not hesitate to take the necessary steps to contact my friend Jaramillo, who introduced them to me. The result of that meeting was that I included them in the list of candidates for the expedition.'

'Their education, the fact that they were "rounded men", their natural courtesy were the qualities that impressed me, and I selected them from all the other candidates. I saw too, as part of my duty as leader, to try to help these two fellows who were enthusiastic and ready to do anything, and did not shrink from the risks in the Antarctic; I wanted to help them obtain the wherewithal to earn a decent wage.'

'They are a fine example to many of our countrymen', said the reporter, who with all those standing around us listening with interest, then added:

'Good luck to them!'

He continued asking questions:

'I have seen some beautiful dogs on the bow. Are they part of the expedition too?'

'Indeed they are,' I replied 'I thought taking them to the South Orkneys would be useful, they are needed to haul sledges and, this year, I have some other projects for them. Apart from that they are there for the good company they provide.'

'They seem like pedigree dogs. From what country did you bring them?'

299. *Maté* is variety of strong green tea traditionally sipped through a tube from a gourd. It is made from leaves of *Ilex paraguyensis* and is a familiar social beverage in southern South America.

I burst out laughing, and to the bewilderment of the bystanders, said:

'These animals also have a story behind them. It was too expensive to purchase thoroughbred huskies to haul; their cost is far beyond the funding for the expedition. So I had the idea that I might be able to find something in the Buenos Aires municipal dog pound! I went there one afternoon with my Orcadas companions, and among the motley pack of hounds, I selected three dogs and a bitch. They are strong pups. They all have good long coats of hair, so it will not be difficult for them to acclimatize to the very low temperatures of the South Orkneys.'

'How interesting', said several of them at the same time, and then I heard them saying:

'Dogs born in Argentina, native *porteños* as well, and they are off to render their service to the expedition in the South Orkney Islands!'

The hour of departure arrived. I joined the swarms of people edging around the deck of *1° de Mayo* under a blistering Sun. In the midst of this I went to say farewell to my family and friends, who one by one left the ship to go ashore and stand on the wharf.

Only a few officers and crew remained on deck. They dexterously started to cast off, and with the assistance of tugs, headed towards the channel out of North Dock.

I remained on deck waving my handkerchief in acknowledgement of the greetings I received from those who tarried on the wharf.

We were quite far from shore when I decided to go below to the mess of *1° de Mayo*. Just as I was about to step through a hatch to descend, I felt a hand on my shoulder holding me back.

I turned around and faced a gentleman dressed in an immaculate 'palm beach' suit. He smiled at me, took off his Panama hat with one hand, and stretched his other hand to greet me:

'Although we have never met, I know a lot about you. I am Dr Víctor Rodríguez[300] ....'

300. Víctor Rodríguez is arguably the earliest tourist to visit the South Orkney Islands, conceivably a relative of Carlos V. Rodríguez who wintered in 1928 and who would have returned aboard *1° de Mayo* with him. A more substantial tourist group arrived on 12 February 1933

76: Chichilo *and* Kataplum

I replied to his greeting saying:

'I have also heard about you. I know you are making this voyage to South Orkneys to study Antarctic flora and fauna.'

He nodded in acknowledgement, and I continued.

'There is very little to see down there. I suppose you will return with this ship?'

'I am certainly not thinking of staying for winter ….'

'In which case you will not have much time to complete your work.'

'No matter. What I really want to do is this voyage; when my occupations leave me the time, I do what I can to travel the world. So far I have never made it to the South Orkneys.'

'That may be so …' I replied laughing 'but remember, doctor, the South Orkneys are so remote they seem to be on another planet ….'

'That does not matter to me either. Some day I will have to make an excursion to that other world that nobody knows.'

I was interested in such a genial personality who did not hesitate to visit the most inaccessible parts of the Earth. I took him by the arm and said:

'Let's go to the mess.'

Thus together we made our way down the narrow companionway.

---

aboard *Pampa* which included the first women to reach the islands: Palazzi, 2005.

108: Southern Ocean

Altogether I have sailed many thousands of nautical miles on sea voyages. I have traversed some of the world's oceans, and although I do not have the uncomplicated outlook of a professional seaman, I have learnt about some of the disasters that can occasionally occur at sea, on the unforgiving salty waves.

I will always affirm that the most terrifying part of Antarctic expeditions has been precisely the voyage itself. Many of these I made aboard ships that had no right to be at sea, and many in hulls which were designed to stay afloat on seas very different from those of the far south.

The *1° de Mayo* was a sea-going ship in every sense of the term. She pitched and rolled atrociously, even in the most tranquil waters of the Río de la Plata, but she had the reputation as a sound sea vessel, which, for me, was reassuring.

Moreover, I had sailed aboard her on my second expedition to the South Orkneys. Although the hull was already quite old, I had complete confidence in her stability, even if some frightful lurching occasionally tipped her to an angle which, on other ships, might be considered dangerous.

The voyage as far as Puerto Cook, on Isla de los Estados, was quite uneventful. There we found the fuel tanker *Ministro Ezcurra*[301], and were able to fill our tanks and give us autonomy to reach the South Orkneys and return to Ushuaia.

After bunkering we departed from Puerto Cook with the hope that if the good weather held, which at that latitude is extremely problematical, we should arrive in seven to eight days at the South Orkney Islands.

The inevitable had to happen. As soon as we lost sight astern of Isla de los Estados, a strong south-east gale arose that caught us on the bow. *1° de Mayo* was being to be put to the test.

By the second day the wind had increased in violence and the heavy

301. *Ministro Ezcurra* was a bunker vessel built in Britain in 1914; in service until 1966; 80·5 m long.

sea made navigation difficult. The main deck was almost awash, while below reigned total disorder. Every object not lashed down careered across the decks with each roll.

The ship continued to make progress through that raging sea until, on the third day, it became so violent that the Commanding Officer ordered a reduction of speed as a precautionary measure.

The storm, far from abating, actually increased, and at times the huge seas towered over us. Occasionally the propeller came clear of the water, and the shuddering caused by these sallies induced a horrendous feeling of the hull cracking under our feet.

We found ourselves in mid ocean, halfway between the South Orkneys and Tierra del Fuego; and to add to our risks, apparitions of icebergs began to loom out of the murk.

At dusk on the fourth day, revolutions of the propeller were further reduced to the minimum to keep the ship under way. It was impossible to remain standing upright, and when I went on deck to try to reach the bridge to ascertain our position, or rather the estimated position, I was surprised to see that a section of the port rail had been torn from its mountings by the enormous waves sweeping over the main deck.

From the bridge, the spectacle was terrifying. Never on any of my voyages on the whale-catcher's, not even *Tijuca*, had I experienced a storm of such proportions.

Huge waves of enormous height, vertical walls of water raced towards the foc'sle and, for a split second, towered over it. In the darkness of night I could just discern the crest, by the white curtain of spray blown from it by the hurricane force wind.

At each wave I instinctively closed my eyes, and wedged myself against the bulkhead of the bridge; it seemed quite impossible that the vessel would be able to raise herself over the crests of the waves and surmount them. Wave after wave, wall after wall threatened to dash us, vessel and all, and cast us into the abyss of the ocean.

The Commanding Officer remained on the bridge for two days, surrounded by the other officers. The peril of their precarious situation was apparent on their faces.

'Keep her head into it' was the order and it was followed faithfully.

I heard the Commanding Officer ordering:

'Engine room – five revolutions less!'

Then the repeat order came through the speaking tube 'five revolutions less'.

Suddenly, another wave, more violent than the rest swept the deck and reached the bridge. The wind tore at anything and everything, screeching through the wire cables that stayed the two masts.

The Commanding Officer asked:

'How many revolutions?'

'Twenty, Sir.'

'Five less!'

'Five less' replied the voice from the engine room.

Suddenly another officer, who was beside the helmsman declared:

'She's not responding!'

'Increase five revolutions!' – ordered the Captain.

The situation was looking very bad. I could see the anguish on the officers' faces, and I felt it on mine.

I went below to my cabin and took something from my suitcase which I put in the pocket of my leather jacket. I returned to the bridge determined to remain and see out the storm.

The waves succeeded each other uninterrupted; suddenly I saw two mountains of water approaching very close together and very high. I looked at their crests and this time, defying all the elements of the sea, I awaited them without closing my eyes.

I felt a terrific shock against the vessel's hull, and while I clung to the bulkhead with one hand, I put my other one in my pocket and felt for my automatic pistol.

'If the end has to be now,' I told myself, 'at least I won't suffer'.

Suddenly, we all looked at each other in terror. Beneath our feet, on deck, we felt a powerful wrenching of something breaking loose. I looked down and immediately realized what it was.

'The boat' I yelled, 'the boat! One of its shackles has broken loose …'.

An immediate order and five sailors leaped on deck to secure it.

Those men were risking their lives, but it had to be done, as the blows against the superstructure were threatening to destroy it. In the middle of that furious tempest, drenched by the icy Antarctic water, I saw them working valiantly, while my emotion intensified.

It was those courageous Argentine seamen who, in the service of their country, compared favourably with the best of any foreign sailors, braving the severity of the oceans of the world.

For me that night was full of anguish and terror. Only when it became sufficiently light and the storm abated, did I go below to sleep following the example of the Commanding Officer, who during three difficult days had remained constantly on the bridge.

109: In the Gloom

On 3 February 1929, *1° de Mayo* was very close to the South Orkneys. It was calculated that at some time on the following day we would be anchoring in Uruguay Cove. The storm that had lashed us on the previous days had diminished in intensity, permitting the ship to continue progress but with all the attendant difficulties and discomforts resulting from navigating through severe seas in fog, with icebergs suddenly looming out of the gloom wherever we looked.

Eventually, on the morning of the next day, the islands were sighted, enveloped by their characteristic fog banks, and with an overcast sky obscuring their summits.

On one of my walks through the ship, my presence was requested by the ship's Radio Operator who said to me:

'Something very odd is happening. Several days ago, when the storm was at full force, I could hear various stations with Norwegian call-signs that seemed to be working in areas close to the South Orkneys. Now everyone has gone silent, and, although I am making desperate efforts to obtain communication with anyone to request information on ice conditions around the islands, I have been unable to get a reply to my persistent calls.'

'That silence does not surprise me,' I replied 'you have to take into account the fact that the factory ships and whale-catcher's that ply these waters are monitored by the British government and it is they who exercise authority in these waters.'

'Do you think,' he asked me, 'that the British government would try to stop those vessels communicating with us?'

'I don't think their supervision extends that far. Perhaps the whale-catchers themselves do not want to reply to us so as not to get into trouble with those who are monitoring them.'

'Something of that sort must be happening,' said the Radio Operator 'every station has gone quiet, as if the sea had swallowed them.'

'This means,' I added, 'that they know perfectly well that an Argentine ship flying the Naval Ensign is in the area. Continue calling, perhaps someone will reply.'

However there was no reply to our persistent calls.

I went on deck. It was quite cold, though there was not much pack ice close to the islands; but large icebergs were drifting, at the mercy of the icy south-east winds.

We increased to full speed all morning and for part of the afternoon, hoping to reach the coast before dusk. By the time we passed the north-east side of Saddle Island it had already begun to get dark. From there it was still two hours to Uruguay Cove. I proposed to stay on the bridge, in case our worthy Commanding Officer required any practical advice or information.

I remained there a long time surrounded by officers, watching the approach to *Manchón Austral,* then Mount Ramsay and finally Point Lola.

Someone asked, pointing in the direction where the base should appear:

'What could have happened there?'

I shrugged my shoulders indicating that I did not know, and said:

'Something serious must have happened. I have a bad feeling.'

'Fire? Destruction of the radio station?'

'No. I rather think we will find another cross.'

'Is the little cemetery visible from the middle of the bay?'

'With daylight it is all visible, but now it is getting dark, it is a bit difficult to discern.'

The ship was level with Point Lola. A few minutes later, we all directed our binoculars to the coast.

I could distinguish the silhouettes of the buildings in the middle of The Isthmus: the metal frame of the antenna pylon, and farther across on the right, above the beach, the site of the cemetery; but darkness had set in and it was impossible to distinguish how many crosses there were.

I indicated to an officer standing beside me, pointing to the shore, 'there is the cemetery, in that direction. Look closely and tell me how many crosses you can count'.

He took the binoculars and looked carefully, then, after a while, said:

'I cannot see very well, there is very little light .... There appear to be three ... no, four ... five ... no; four crosses ....'

They all did the same, but no one could be confident enough to come up with a number on which all agreed.

On the other hand, it seemed that we had not been sighted from the base. Only one light could be seen at one of the windows of the dwelling; everything about the base looked motionless, shut, almost lifeless.

The *1° de Mayo* was now stationary in the middle of the bay and the anchor was dropped.

The Officer in Command said to me 'Let's go ashore, then we will know what has happened.'

In a few minutes we were in one of the boats with the sailors rowing vigorously toward the beach.

'There,' I pointed 'that is the best place to land.'

'That is where the cemetery is', someone commented.

'That doesn't mean it is not the best beach to land the boat.'

I could see the line of the beach, the irregular outline of the magnetic huts, and the mounds of the uniform row of graves. Suddenly the crosses stood out, and that view was so engraved on my memory, I realized, as soon as I saw them, that there was one more.

'Five crosses', I said to the Commanding Officer who was standing beside me.

Wanting to be an optimist until the last, I added:

'That does not necessarily mean anything. Possibly they have added a fifth cross in memory of Bache-Wiig, who disappeared in the winter of the 1915 expedition.'

At last the boat touched the smooth sand of the beach. I leaped out and, without waiting for anyone to follow, went up the rise and toward the crosses. My steps took me in a straight line towards the fifth cross, that new cross was not there when I last left the South Orkneys.

At last I reached it and stopped. I took my torch from one of my jacket pockets, and shone the beam on the centre of the black painted cross, and could read the inscription carved into it:

Fortunato A. Escobar
30 - x - 1885
27 - x - 1928

I took my heavy leather cap off, and that was the first salute I made to the brave men of the South Orkney Islands.

110: In my Home for the Fourth Time ···

When we reached the dwelling, we were greeted by all the expedition members with delight. I had a long conversation with Bruhns, the Leader, whom I was to relieve, and he gave me all the details of Escobar's death.

'He fell gravely ill,' he told me, 'yet he retained enough firmness of character not to reveal any hint of the true seriousness of his state of health. I myself had no knowledge of how serious it had become. The radiograms you said he sent to Buenos Aires asking for a remedy, were completely unknown to me'.

'All I can say,' he added later, 'was that Escobar had a complete private medical kit in which he had a range of medicines which he made use of when he needed something. This has been confirmed by my expedition colleagues.'

Bruhns handed me the certificate in which Escobar's death had been elaborated. It recorded:

Declaration
Referring to the Death of the Radio-Telegraphist, Warrant Officer Fortunato A. Escobar on 27 October 1928 in the National Observatory of the South Orkney Islands.

On 27 October, at 19:00 hours, we found the Radio Operator Escobar lying on the floor of his bunk-room which was the radio station. His body was turned on his back, in a natural position with his arms lying alongside his body. His eyes were closed and his facial features contracted. He appeared to have vomited, but probably after losing consciousness or dying, because the vomit had gone down both sides of his face. The front window of the room was open and the heater was found turned off. Immediately we lifted the body of Escobar and put him on his bunk, applying vigorous massage with ammonia in the area of his heart. Meanwhile the window was closed as best it could be, and the heater was switched on. We were soon convinced this was in vain, that Escobar was dead: no heart beat could be detected, no breath escaped from his mouth. It was impossible to open his eyelids which had become stiff. Only in his armpits was there any vestige of warmth. Shortly afterwards we verified that the whole of the back of his body was reddened. Bearing in mind that Escobar did not go to the dining room at 12:30, nor did he answer the cook's calls (which sometimes happened), and that he had the engine running to charge the accumulators until 11:00, it may be supposed that death occurred roughly at 12:00 hours local time (11:00 hours official time). I too called Escobar twice after 13:00 hours, from the entrance to the radio station, to ask him a question or two about the antenna, but I received no reply. I did not attach any importance to that, I thought that he might be sleeping, or did not feel like answering. I then absented myself to make the horizontal magnetic observations and continue with my work. During the whole day very bad weather prevailed, a strong gale with snow and with sudden spells of unusually high gustiness. The mean temperature was seven degrees below zero. Considering that there was nothing more that we could do for him, all Escobar's personal effects were packed, sealed and signed. From midnight until daybreak a coffin was made. At approximately 09:00, his body was placed in the coffin which we transported to the cemetery on a sledge, where he was left. The whole of the 29th was employed excavating a grave, for which ten containers of fuel were burnt to thaw the shale. At 18:30 hours we lowered the coffin and buried him in his grave.

Laurie Island. South Orkney Islands, 30 October 1928.

It ended with the signature of Ernst Bruhns and all his companions. Such is the document referring to the loss of another member of the expeditions.

⋆ ⋆ ⋆

It must have been a very sad episode in the history of that winter; its memory was still fresh in all the minds of all the departing men. The blow to morale, which still affected them, even extended to my new companions, the young lads who contemplated the new cross in silence. They had not heard any previous mention of the fifth cross.

However, when the departing winter men and the relief party came to say their farewells, a mood of optimism kept spirits high. Only when we remained alone did I devote my energy to ensuring that the recollections of what happened to the unfortunate Escobar did not affect the minds of the new winter party.

In truth, this was not an easy task, as I too found myself pensive and distressed. I was saddened and my morale had been sapped by the circumstances that brought me back for a fourth adventure in the South Orkneys. I found it a disquieting experience to have to look out on one more cross in the icy polar waste, and that increased my unease.

Being well-schooled in all life's contingencies and accustomed to surviving its rude blows, I tried to think objectively and ignore my own moral sufferings. I had to remind myself that, once again, I was leader of the expedition. I had the responsibility to look after the welfare and safety of my young colleagues, as well as faithfully accomplish all my obligations to that particular winter party.

III: First Things First

The radio station had not been functioning for four months. Outwardly everything appeared to be in order. When Escobar's companions realized that it would be not be possible for them to master the complexities of the radio equipment, they had left it and continued with their usual activities just as in the days before the radio had existed in the South Orkneys.

Carlos Kahl, the young Radio Operator, gave proof of his efficient training in getting the apparatus in working order in only a few days. At last, one morning he announced to me that everything was ready and on that day he would try to make contact with Ushuaia. He was successful and from mid-February we were in continuous communication with the stations in Patagonia.

For my part, there was little or nothing left to learn about the South Orkneys. I knew it all; everything was thoroughly familiar to me. There was not a single corner in the dwelling in which on some occasion I had not driven a nail, and no part of the isthmus or the nearby islets where my feet had not stepped nor my ski furrowed.

It was possibly I who suffered most from boredom during that winter, but at least I understood this for what it was. I imposed on myself an intense schedule of work, which as in previous years, kept me busy most of the time.

My new companions demonstrated that they easily adapted to the Antarctic environment. After the first few months, only a few picturesque places remained for them to discover which I used to describe frequently in conversation. Perhaps it was the experiences of that particular the year of isolation that caused me to strive to find ways to make it less difficult, both for myself and the others.

Winter arrived, and with it the extreme cold, and thus the contest to beat the record of the lowest temperature. Then came the ice fishing, with a competition to catch the greatest weight of fish within a given time. The prize for the winner was a carton of cigarettes which I donated.

Then came the training of the dogs to haul sledges. We had only three of those animals left, as one had disappeared during the first months of our stay. Monti and Páez Montero, commenting on this subject said:

'That poor hound must have suffered when he became lost.'

'Perhaps,' replied Páez Montero, 'it would be interesting to see the faces of the naturalists who arrive one day and, while exploring the islands, find a skeleton or scattered bones of a strange animal ....'

'Of the dog that vanished?' – interrupted Monti.

'That same one. There is nothing extraordinary about it. You yourself would have a bit of a surprise, if during a distant field trip, you were exploring a place in the hills and found on the rocks the bones from of an animal that supposedly couldn't have reached that place.'

I interrupted the conversation and said:

'To find large bones of a quadruped is not impossible in the South Orkneys ....'

'How?' – they asked 'Are there quadruped mammals here?'

'They do not exist, but I have it on good authority that many dogs from previous expeditions have gone missing, having become disoriented with the lack of tracks on hard-frozen snow. On one of my previous expeditions we brought four lambs with the intention of fattening them during the summer months[302]. We had four bales of hay for them. Well; we were never able to eat those animals; one night they absconded, and a subsequent snowfall obliterated all their hoof prints which might have indicated the direction they had taken. In spite of a systematic search of Laurie Island, I have never found a single bone. So just take good care not to bring me back a sheep bone thinking you have found an exotic creature.'

'If I may ask you a question,' said Monti, 'how is it possible for a dog to completely disappear from here?'

'There are various hypotheses,' I replied 'you have seen the lengths we have to go to prevent these heartless dogs chasing poor defenceless penguins. For the dogs the meat of these little creatures, and especially the blood, is a delicacy. It is quite feasible that, while a dog was chasing one, it leapt on to an ice floe afloat near the shore. This ice slowly begins

302. Presumably the 1925 expedition.

77: Summer 1929; left to right, Moneta, Monti, and Kahl

to drift with the wind and moves away from the coast. If the dog could reason it would plunge into the icy water, and swim safely back ashore in a few moments, but it prefers to remain where it is. Unknowingly it is getting closer to death, as the floe never drifts back to land. On the other occasion,' I continued, 'I have seen dogs become disoriented on the wide and featureless surface of glaciers, and especially on a field of fast ice. On one occasion, during my third expedition, a dog that had been born here was seen wandering off across the sea ice of Uruguay Cove. When I noticed him rounding Point Lola and turning in an easterly direction, there was no doubt in my mind that the animal was disoriented. I put on my ski and went after him trying to catch up. Only when I reached Jessie Bay did I notice the animal had stopped to rest and turned his head. As soon as he saw me, he came racing towards me. Had I not realized he was completely disoriented and if I had not reached him skiing across twelve kilometres of frozen sea, he would have been utterly lost and never returned to the base. You will have seen yourselves,' I added, 'how the dogs inevitably follow behind you and walk in your ski tracks.'

We continued talking on the subject of dogs, they were the only friends we had down there, when Páez Montero asked me:

'Does the cold not affect them?'

'Not too much. They must suffer, naturally, but they acclimatise quite well. There is one interesting thing that I recommend you to watch for; it is something that only happens once to a dog new to polar regions. We have three new dogs, and hence there are thus three potential chances of its occurrence.'

'What is that all about?' – they asked intrigued.

'When you go outside when it is very cold in the depths of winter with a hand which is moist, and touch an exposed metal object, you will have the unpleasant sensation of your fingers freezing instantaneously to the thing you touch. The moisture on your hands freezes on touch, which causes the effect. Now suppose that one of those dogs decides to lick a metal runner of a sledge .... When its wet tongue freezes to the runner, or any other metal object licked, you should hear how the animal yelps .... I have seen some break free immediately, and run away howling desperately. It is something only done once ...'.

112: Adoptive Father

Of all the men who accompanied me to the South Orkneys Rómulo Devoto, the Cook, was one of the best-natured and deserves a chapter to himself. As a member of the expedition, he has a story to tell. At the end of 1926, while I found myself in Buenos Aires, preparing for my third expedition, I had to select someone who would accompany me as Cook for the winter. It was a crucial appointment; on which depended the morale and well-being of the six men during their South Orkney year. I had known cases in which the expeditions suffered cruel deprivations because the man supposed to have been capable of putting the contents of the food store to good use, proved unable to do so.

That is the reason why, on that occasion, when making the appointment of a Cook, I chose Konrad Becker. His previous experience showed him as the most suitable candidate to join expedition. He had already acquired experience aboard several ships under a variety of flags. I knew that our kitchen in the South Orkneys was not so dissimilar from the routine of a galley aboard a ship.

However one of Becker's rivals for the job was an older man, dark skinned, with a good-natured face, he was going grey and his perspicacious look revealed a wealth of experience.

'Are you a professional cook?' I asked.

He haltingly replied:

'Yes, I … think I am.'

Argentine?

'From Buenos Aires province.'

Where have you worked?

'Various places, presently I am in La Boca[303] ….'

Do you know anything about preserved food?

'A … little', he muttered in reply.

Correcting himself, he said to me.

'Yes. I know what you are worried about. That I don't know how to make a *puchero* down there.'

I had to laugh, and I realized this fellow would not be of any use if he aspired to make a *puchero* in the South Orkneys, and I remained with Becker.

The years went by, and in 1929, while preparing for the fourth expedition, I found myself again in the presence of Devoto. I admired the perseverance he showed in wanting to go to the South Orkneys and I questioned him more closely.

He was well-informed about the privations to be expected on an expedition in the South Orkneys, and even had a tolerable idea of what he might be in for. All in all, and bearing in mind previous examples of inexperienced cooks and their disastrous effects on the base, I was not keen to take him on until he said:

'Sir, I am going to be of service to you … have complete confidence in me, I am *criollo*. Have no qualms about helping me ….'

'At the moment, it is not the point whether you are Argentine or

303. La Boca (literally 'the mouth') is an area of Buenos Aires at the marine entrance to the old docks; it has most of the facilities associated with any major port.

78: Members of the 1929 Expedition; left to right: Orlovski, Monti, Kahl, Páez Montero, Devoto, Moneta

not. It is about the security and well-being of all the others.'

The man was inspiring me with confidence; and I asked him directly, 'how many children do you have?'

'Twelve', he replied smiling.

'Well, in that case, you can come to the South Orkneys with me.'

That was how Rómulo Devoto came to accompany me on my last expedition.

He was the adoptive father of all the lads on the expedition, addressed them in the familiar way, and generally treated them like his own sons.

If any of them suffered from a stomach complaint, you would see Devoto heading for the medicine chest to get the castor oil, and then you heard a heated argument because the patient, who was not allowed to get up, refused to swallow a third spoonful. The next morning he could be seen preparing a special soup for the patient, and taking it to him in bed, he would wake him with the words:

'You take this; if not ··· I won't speak to you again!'

The lads all thought highly of him and none of us ever had to thread

79: Devoto and Páez Montero enjoying maté

a needle to sew a button on to our clothes. Devoto did it without being asked; he even repaired the soles of our boots. That good-natured man was a true father to all my companions.

His weakness was *maté*. In the afternoons he would sit and rest in his room, accompanied by Páez Montero. Together they chatted endlessly about life in a small town in the interior of our distant homeland, while they gradually emptied several large kettles of hot water.

'This man,' Monti would say to me, 'he not only is a father to us, he is also a mother, as he even darns our socks ….'

113: Quack Remedies

The days went by. Winter … only the parties introduced some variety into the monotony of that life. Everything had a sameness about it; only the faces of my companions were different. Actually it was only Monti's face which changed: he grew a beard.

On 25 May we heard our first special broadcast on the radio, repeated on 9 July.

After that, always the same. The observations, the changes of shift, the weekly bath, a special meal on Sundays, one or two outdoor excursions … everything the same.

To introduce a bit more variety into the routine, it occurred to me that once a week, apart from Sundays and holidays marked by the calendar, we would celebrate something out of the ordinary in our own lives, for example: the birthday of a loved one or friend, or someone who had become famous in any field, anyone at all.

This was what, indirectly, was pernicious to Monti, as we celebrated those days with a hefty main course. Devoto had already accustomed us to large meals.

After these innocent feasts I noticed that Monti went on a diet; and would not eat for a couple of days. One day, when we were sitting at the table, I mentioned it to him:

'I think you are eating too much, or you eat in a very irregular way.'

'I have also noticed that', added Páez Montero, and turning to Monti, he said: 'you should take it easier, if you do not want the same thing to happen again ....'

'What thing', I asked?

'No ... it's nothing', was the evasive reply.

However, on account of his youth and physical strength, Monti paid no attention to my observations or to those of his colleague, who knew him better than I did.

I remember one night, after a hearty meal, Monti rose from the table unwell. He went to his bunk-room and I thought I heard him getting into bed.

Moments later I heard the sound of groaning from his room; I was alarmed and went to inquire what the matter was. On opening the door, a distressing scene presented itself: Monti in bed, seized by strong convulsions.

'What's the matter? What is happening?' I asked.

As there was no reply, I asked again, just as Páez Montero came into the room:

'Not your stomach again?'

Monti nodded, while Páez Montero, showing evident concern, tried to do something.

Down the forehead of our patient rolled large drops of sweat, and with twitching hands he seized his bedclothes and bit them rabidly.

'What has gone wrong with him?' I asked myself. 'What am I to do? What should I give him?'

The convulsions increased[304]. I summoned my courage, as I realized everything depended on how I dealt with it. Rolling my sleeves up to the elbow I went to his side. I ran my fingers over his abdomen and it occurred to me to feel the region around his appendix, which could be very serious.

'Does that hurt?' I asked.

'No ··· not much.'

'And now?'

'No ··· No ···.'

Then another convulsion, stronger than the rest, completely took hold of his body; but these were no ordinary groans, Monti was screaming desperately.

'What is this?' – I asked myself 'My God, what is it?'

I looked at my companions, around the bunk, and realized that each one of them was waiting on my words and what I would prescribe.

I looked at them interrogatively and then asked:

'What do you reckon?'

No one replied, those brave lads understood even less about medicine than I did. With the possibility of appendicitis, which is always the spectre which hangs over any expedition, I ordered:

'Ice - get a bag of ice!'

Instantly a bag of ice was placed on Monti's stomach. I cannot verify whether this did any good or not. I feared it could be appendicitis, and this was the only thing I could think of.

After fifteen minutes of worry and indecision, the fits still continued. I took Páez Montero by the arm to the parlour, and said to him:

'We will give him a few drops of laudanum.'

'A good idea,' he replied 'at least that will calm him.'

However I had a doubt:

As there was no one who could advise me, I went in search of the

304. These symptoms, including recurrence, gastric spasms, sudden abatement, and subsequent sleep indicate this was a form of autonomic epileptic seizure. Monti spent five more winters on Laurie Island.

three medical books which were on the library shelf; one written in Norwegian, one in English, and the third one in Spanish.

I put the first aside, I did not understand a word of that language. I chose the one in English. I looked for the chapter describing laudanum.

Meanwhile, from Monti's room came groans. I could also hear the blows he was giving himself as he struck his head against the boards over the bunk; he was in the grip of despair, or so it seemed.

Then I read:

'Laudanum may be administered to the patient in ten to thirty drops per dose.'

I took a glass with a little water and added twenty-five drops of laudanum to it: I returned at once to the bunk-room where, with great difficulty, I managed to give Monti the liquid between his tightly contracted lips and clenched teeth.

We waited a few seconds, but ··· nothing. More convulsions, and then he vomited, getting rid of the laudanum.

In desperation I urged my companions to take care that Monti did not injure himself by banging his head against the wooden bunk, and I went to my bunk-room.

I placed the medical books on my desk, and sat in front of them, while Monti continued having his attacks in the room next door. I am not sure how long I read, but I can confirm that I learnt a few things that I never expected could possibly be of use one day. It took a considerable effort to concentrate on my reading while hearing the convulsions coming from the next room, my comrade was near his extreme limits. I will never forget those moments of intense crisis that had me deeply worried.

Turning page after page, I rapidly scanned the chapters until I reached the end of the book. They were all written in the same discouraging tone; just when I thought I had the solution, or at least made the diagnosis which would indicate what medication I had to use, I found the following phrase:

'If the patient deteriorates, call the nearest doctor.'

This seemed to be sarcasm; I threw the book away in disgust and referred to the one in Spanish, which was more or less equally useless.

'What do we do? My God, what to do?' I muttered to myself. These were not minor symptoms.

That is how, by studying the situation, I reached the conclusion that my companion was suffering from severe gastric-enteritis, a result of an excessive quantity of food in his stomach. Although I was perplexed by the convulsions he was suffering, I still came to the conclusion that the best medicine was laudanum …. 'Laudanum?' I asked myself … but when his stomach does not tolerate anything!

All in all, I had to give it another try. First I read the chapter on laudanum again and at the end I found a phrase that said:

'When laudanum cannot be tolerated by the patient, administer morphine.'

Morphine? That had to be the remedy; the first thing to do was to alleviate the acute pains he was suffering.

Confident, and ready for anything, I went back to the bunk-room and said to Monti:

'Keep calm my friend, this is brief and you will soon be over it.'

He looked at me imploringly and said.

'Please; something to calm the pain, I can't take any more.'

'Calm. Be brave. I cannot give you laudanum, your stomach will not tolerate anything.'

'What then?' he asked.

'What about an injection of morphine ?'

'Morphine?' he asked in dismay.

'It's the only thing; it will remedy the pain.'

'But ….'

'Don't worry, you will not become addicted …', I said, wanting to laugh.

'No … not morphine, absolutely not!'

'I repeat, there is nothing to fear … it will not harm you; on the contrary, it will calm the pain and that will help you rest.'

I left the room followed by Páez Montero and went straight to the medicine chest, and took out a package that said:

'Morphine hydrochloride, 0·1 gram.'

I noticed there was a date on the top of the container:

'December 1922'

I stopped in the act.

'One moment,' I said to Páez Montero, 'prepare the syringe and sterilize it, I have to consult the medical book again.'

Great was my astonishment to read in the chapter on morphine:

'If the morphine phials are very old, or are open, or not sealed properly, avoid using them, as it is likely they will have changed into apomorphine.'

Those phials had been lying there for at least seven years!

I remained baffled, not knowing what to do, while Monti, in the next room was waiting for the one thing that would calm him.

While I was in that state of indecision, Kahl arrived and said:

'What shall we do? I don't think he is getting any better; on the contrary he seems to be getting worse.'

I went back to the medical chest and rummaged again. I found a smaller package, which looked newer. I quickly read:

'Cocaine and adrenaline phials, for local anaesthetic.'

'This will not harm him', I thought.

Then I had an afterthought:

'I might just manage to convince him. It is worth an attempt ....'

I broke open one of the phials, prepared the syringe that had been boiling in a dish of sterilized water and went into Monti's room.

They all looked at me, my confidence and amazing aplomb must have surprised them. Páez Montero assisted and I injected my sick comrade's thigh muscle.

I emptied the syringe and withdrew the needle. After a few seconds, I noticed the patient was calmer, and in fewer than five minutes, he was sound asleep.

I could not be sure how much was due to autosuggestion, but the deception had a happy ending, as, the next day, Monti was apparently completely cured.

However this was not the end of the story. The next morning revealed the part that Devoto had played in all this.

I overheard what he said to Monti, as he lay resting in his bunk:

'Look, you little devil's lad, if you gorge yourself on spaghetti and

80: Monti in his bunk

meat sauce once more, you are going to oblige me not to make them again ….'

I heard Monti burst out laughing and replying:

'What do you expect old man! You cook so well even the devil would be tempted!'

Monti was right, for the first time in the history of the South Orkneys, there was an authentic *porteño* cook. Sometimes Devoto even made *fainá*[305].

114: Thus Life Goes On ···

I contrived to be constantly busy with work of some kind or other. My only preoccupation was how to make those boundless days go by without my even noticing them. In my four years in the South Orkneys, I always enjoyed spending most of the time outside, making excursions, living in the open air, contemplating the wonder of white wilderness, geologising amid the frozen rocks, walking on the frozen sea or along the shore, and on the limitless glaciers ….

That was my secret therapy, how to survive that singular life of imprisonment amid the polar ice.

I obstinately clung to the idea that this was the only way to live, tiring the body so as to have the luxury of a good night's sleep afterwards. I tried to inculcate this idea in my young companions.

Páez Montero, Kahl and Monti always used to accompany me on my long excursions over the sea ice. Occasionally we would be forced to retreat, muttering about the grim weather or the sudden storms that overtook us when we were some distance from the shelter and warmth of base, but they quickly learnt that this was the only way to achieve that peaceful and restoring sleep. That was how those lads, who had never previously seen snow on the mountains, or ice on the sea, were continuously adding to their knowledge. This, in turn, served to enrich their personal experiences of the Antarctic.

For the rest, everything developed predictably, and although, for my part, I had little more to learn from it all, I continued my studies

305. *Fainá* is a flat-bread made with chick-pea flour.

to help pass the time. I duly gave lectures to my companions, with the intention of helping them reach the right decision whatever contingency might arise.

* * *

With my enthusiasm for field trips I never hesitated in efforts to reach the glaciated summit of Mount Ramsay. Everyone deemed it was inaccessible, its summit of hardened snow was surrounded by a high vertical wall of smooth ice. When I conceived that undertaking, I began by making reconnaissance forays lasting just half a day with the object of getting to know the approaches to the ice cap to attempt to climb it.

One afternoon, engaged in this task, I was slowly walking on the extensive glacier that covers the slopes of Mount Ramsay, when, suddenly, the surface beneath my feet gave way; I thrust out my arms instinctively, and remained suspended over the white abyss. I realized I was straddling a crevasse. I dragged myself away from the dangerous zone, and started walking in the other direction, when, once again, I found myself suspended over another crevasse

Latterly I made many attempts to reach the summit of Mount Ramsay, and after a conspicuous lack of success, I eventually managed to reach the ice cap, not before I had risked my life by crossing a narrow knife-edge of ice with a vertical drop on each side[306]. I understood that Sigurd Schjær, during his expedition in 1919 was the first to reach the ice cap of Mount Ramsay, whose summit presents very difficult climbing conditions, even for the most experienced climber.

* * *

306. Mount Ramsay has a noticeably flat summit with a small ice cap. It is surrounded by steep glaciated slopes at the top of which lie bergschrunds, where the ice has separated from the rock, making access precarious. At 537 m elevation it is the highest peak on Laurie Island (see the footnote in § 20).

After dinner, in the evenings, we had prolonged social gatherings, in which the most interesting performances were given by Devoto and Kahl playing the traditional game of *truco*[307].

For music the old gramophone continued to be the preferred musical instrument, as musical radio broadcasts played through the loudspeaker were distorted by the variable atmospheric electrical interference.

The record collection had grown, but there were few new ones. On the contrary I was condemned to listen to tango. I had spent four years listening to tango, but to my companions it seemed the latest and most beautiful music of its type. They replayed it time after time, changing needle after needle and incessantly reploughing the tracks in the whingeing records.

115: Farewell! ··· South Orkney Islands!

With the passing of time and of life, that year too drew to a close. Then followed the anxious wait for the relief by the naval transport ship *1° de Mayo*.

The body of the Radio Operator Escobar, which during almost two years had remained buried in the South Orkneys, was exhumed and taken to Buenos Aires, having been claimed by his family who lived in the city of Santa Fé.

While the ship's crew dealt with that commission, we made our last-minute preparations to embark on the return voyage, back to civilization. The scenes thereby occasioned between the relief party and the departing team, were similar to those I had witnessed in previous years. This time, however, the unloading of provisions and coal was not as hectic as on previous reliefs. While *1° de Mayo* was there a hydrographic survey was to be conducted of the coast and offlying islets, even some of those I had hoped to visit during the winters but had been unable to do so. Therefore, on this occasion, the ship remained for longer than usual[308].

307. *Truco* is a complex trick-taking card game of Spanish origin popular in Argentina and surrounding countries.

308. The results of this survey appeared in a series of charts published by the Argentine Navy

81 : Food and other provisions landed for the 1930 Expedition

Thus I had time to spare to bid farewell in my own way to that Antarctic icescape, which I would probably never see again; that basic little wooden house which had been my home for four years and nine months; the desolate rocks which in summer appeared through the snow; that infinite sea horizon which became lost in the confines of the south; the penguins who cheerfully came to visit, even venturing inside our house; the seals, gulls and petrels; and even the hand-carved crosses fixed between the rocks of the cemetery.

I contemplated all this in silence. Many of those around me could not understand the anguish of my spirit at that moment, as, in truth, I was very sorry to be leaving it all behind - so much a part of me, so intimately mine. It was like that, because I was the only one who could know that in each rock, in each piece of ice, over the glaciers and eternal snowfields, everywhere had my footprints. Even the polar atmosphere contained my innermost thoughts and musings whenever circumstances in the dwelling drove me to seek that infinite openness of the Antarctic, where I could meditate in silence.

It has always been said that part of the enchantment and attraction

---

in 1930.

82: Five of the six men of the 1930 Expedition alone on the beach

of snow is its immaculate whiteness. For my part, I have to say this is true.

You do not have to be a poet, or have a sensitive mind, to feel attracted by the infinite white. Anyone who has spent one winter in the South Orkneys, and returns for a second, will certainly have experienced this. Even though several years had elapsed from my third expedition, I could not explain exactly what it was that took me back for the fourth. I have already said that it was not the attraction of the modest pay; nor was it the very human desire to do something which was not possible for the majority of mankind; nor was there any sentimental reason why I might have been drawn to this exile, seeking oblivion from my sorrows.

I always went South of my own volition; it was the adventure that pleased me. I derived no personal advantage from it, except living it. I had the chance to reflect deeply, perhaps more so than in any other part of the Earth.

That is why, in the midst of the general happiness of that last departure, amongst the bustle and exuberation my companions felt on returning to life and civilization, I felt myself sad to be leaving that forgotten part of the world, one which I doubt if I will ever see again.

However it brought immense satisfaction. I was the first Argentine to be in command of an expedition, which I did twice. By this modest work in the service of my country, I have contributed to strengthening our sovereignty in the Antarctic regions, which, by their geographical proximity, are a heritage of our nation.

Looking over the rail of the ship that carried us away from the ice, in which I had lived a significant part of my life, I saw for the last time, there on the beach, six men forming a single group. They were under the protection of the sky-blue and white flag, flying horizontally in the icy blasts coming from the immensity of the Antarctic.

83: Casa Moneta in 2007, now the museum of Orcadas Station (R. K. Headland)

84: Orcadas cemetery in 2007 (courtesy David Fletcher)

Map 2: The South Orkney Islands

Based on: Directorate of Overseas Survey 1963, *British Antarctic Territory, South Orkney Islands West Sheet*, 1 : 000 000. London DOS 510 (series D 601), and *East Sheet (Provisional),* (1988). Reconciled with Argentine Servicio de Hidrografía Naval, chart H-611, *Islas Orcadas del Sur,* corrected to 1993. A simplified map of the South Orkney Islands appeared in all editions of the book.

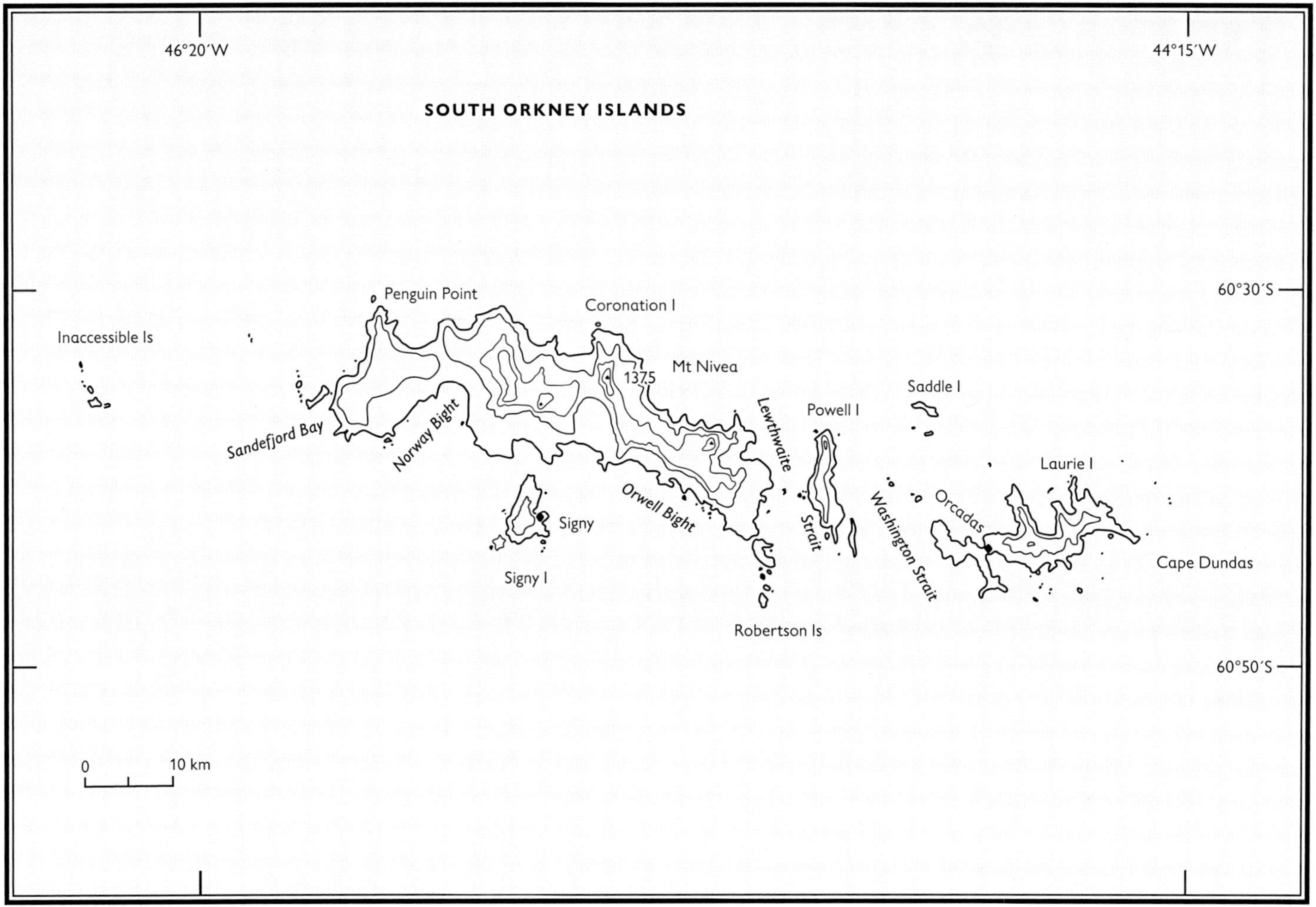
46°20′W
44°15′W
SOUTH ORKNEY ISLANDS
60°30′S
60°50′S
Penguin Point
Coronation I
Inaccessible Is
Mt Nivea
137.5
Saddle I
Powell I
Lewthwaite
Strait
Sandefjord Bay
Norway Bight
Laurie I
Washington Strait
Orcadas
Orwell Bight
Signy
Signy I
Cape Dundas
Robertson Is
0
10 km

Map 3: Laurie Island

Based on: British Antarctic Survey 1988, *British Antarctic Territory, South Orkney Islands East Sheet* (Provisional), 1: 000 000. Cambridge: B.A.S. 100 (Part of SP 23-24/1.2). Reconciled with Argentine Servicio de Hidrografía Naval, chart H-651, *Bahía Scotia,* published 2012. A simplified map of Laurie Island appeared in all editions of the book.

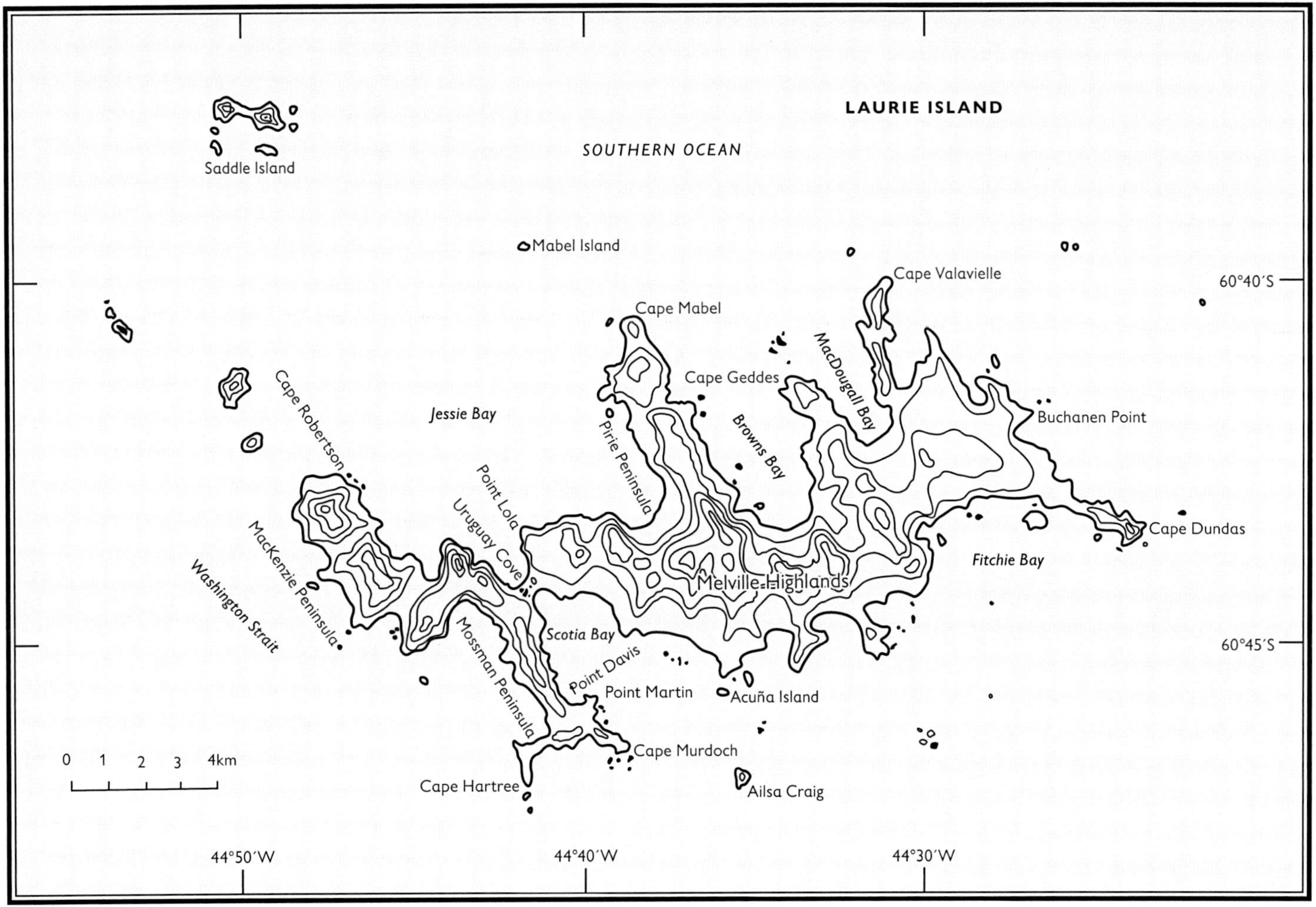
LAURIE ISLAND
SOUTHERN OCEAN
Saddle Island
Mabel Island
Cape Valavielle
60°40′S
Cape Mabel
MacDougall Bay
Cape Geddes
Buchanen Point
Jessie Bay
Cape Robertson
Pirie Peninsula
Browns Bay
Point Lola
Uruguay Cove
Cape Dundas
Fitchie Bay
Melville Highlands
MacKenzie Peninsula
Washington Strait
Scotia Bay
Mossman Peninsula
60°45′S
Point Davis
Point Martin
Acuña Island
Cape Murdoch
0 1 2 3 4km
Cape Hartree
Ailsa Craig
44°50′W
44°40′W
44°30′W

Map 4: The vicinity of Orcadas Station

Based on the same sources as Map 3.

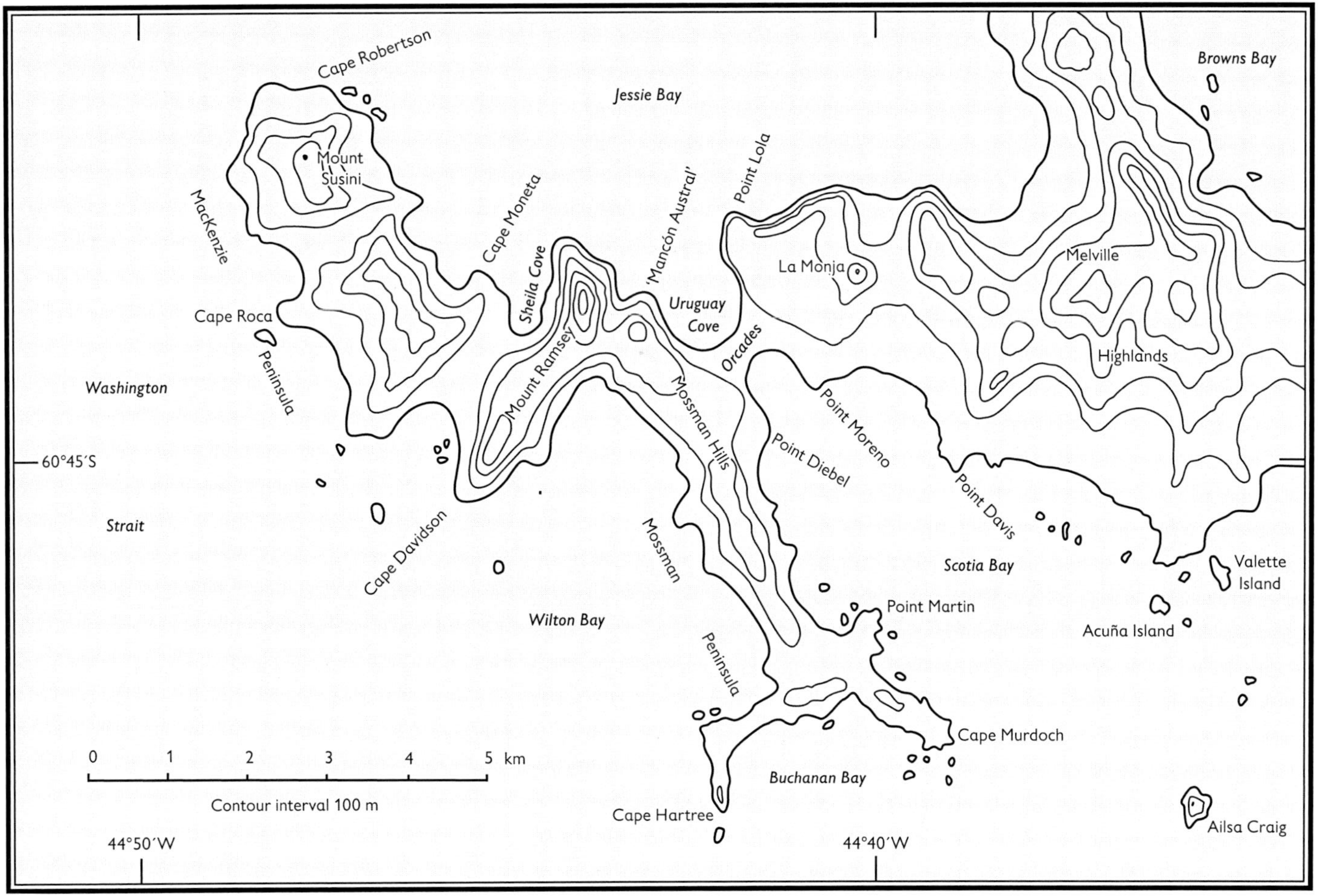
Cape Robertson
Jessie Bay
Browns Bay
Mount
Susini
MacKenzie
Cape Moneta
Point Lola
'Mancón Austral'
Sheila Cove
La Monja
Melville
Uruguay
Cove
Cape Roca
Peninsula
Orcades
Highlands
Mount Ramsey
Washington
Mossman Hills
Point Moreno
Point Diebel
60°45′S
Point Davis
Strait
Mossman
Cape Davidson
Scotia Bay
Valette
Island
Wilton Bay
Point Martin
Acuña Island
Peninsula
Cape Murdoch
0
1
2
3
4
5 km
Buchanan Bay
Contour interval 100 m
Cape Hartree
Ailsa Craig
44°50′W
44°40′W

Plan 1: Orcadas Station about 1927

This plan is based on one published in all editions of the book. These included a Decauville railway for man-hauled cargo trolleys running about 170 m from outside Casa Moneta to the shore of Uruguay Cove, near the cemetery (ending about 25 m from the magnetic huts). There is no reference to this railway anywhere in any edition of the book, even in the station description in § 4, and it does not appear in Moneta's film. Thus, because its existence during Moneta's period is apparently an anachronism, it has been expunged from the plan in this translation. The 1963 edition alone dates the plan as 1927. The railway was presumably constructed after 1930 but before 1939. It appears on several early photographs of the station dated between 1940 and 1950 which are exhibited at the Museo Marítimo in Ushuaia (some remains of the rails and track were apparent in 2007). As well as for cargo it was also used for carrying ice to supply fresh water.

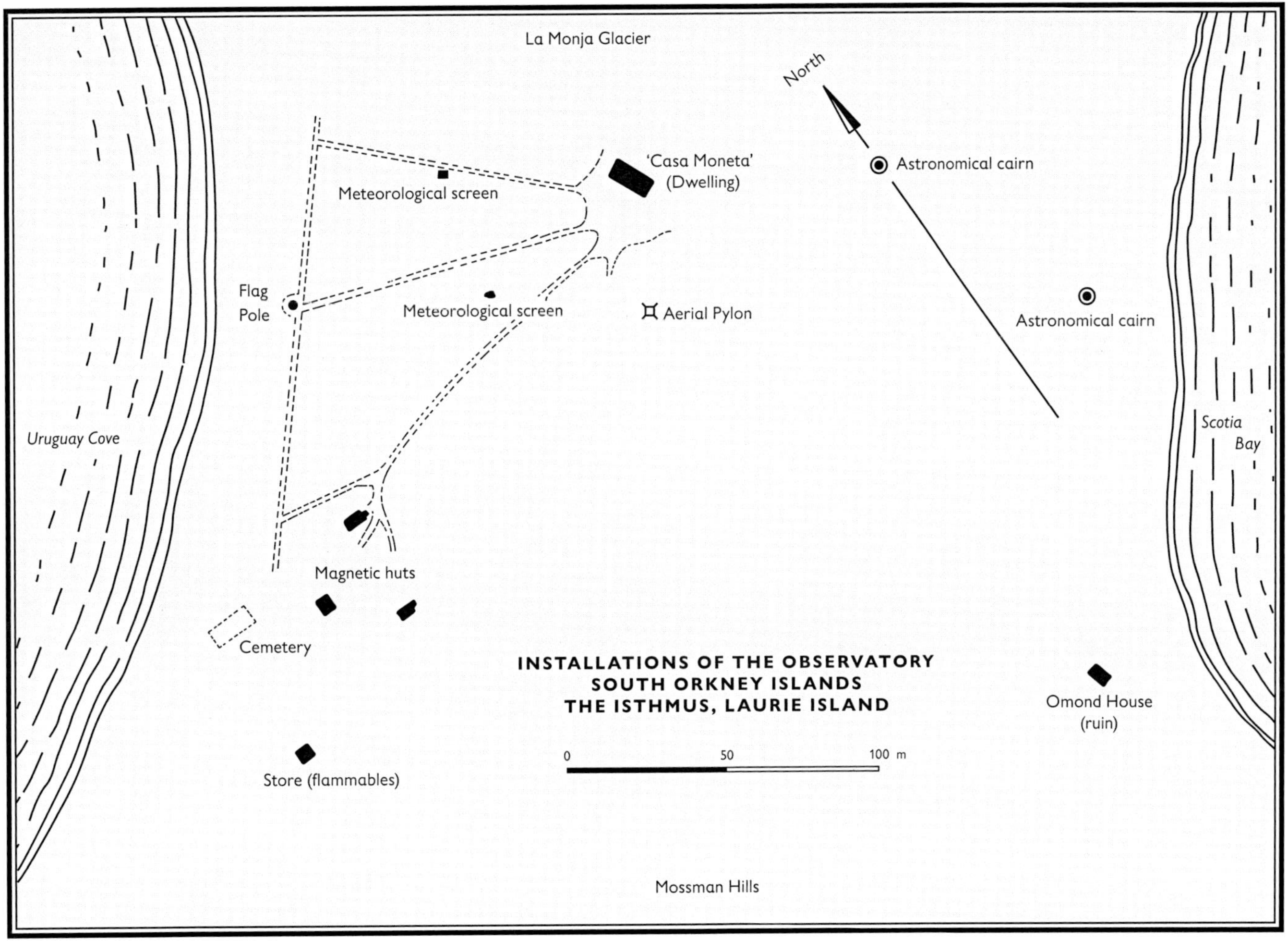
INSTALLATIONS OF THE OBSERVATORY
SOUTH ORKNEY ISLANDS
THE ISTHMUS, LAURIE ISLAND
La Monja Glacier
North
'Casa Moneta' (Dwelling)
Astronomical cairn
Meteorological screen
Flag Pole
Meteorological screen
Aerial Pylon
Astronomical cairn
Uruguay Cove
Scotia Bay
Magnetic huts
Cemetery
Omond House (ruin)
Store (flammables)
0
50
100 m
Mossman Hills

Plan 2: The basic structure of the dwelling [Casa Moneta]

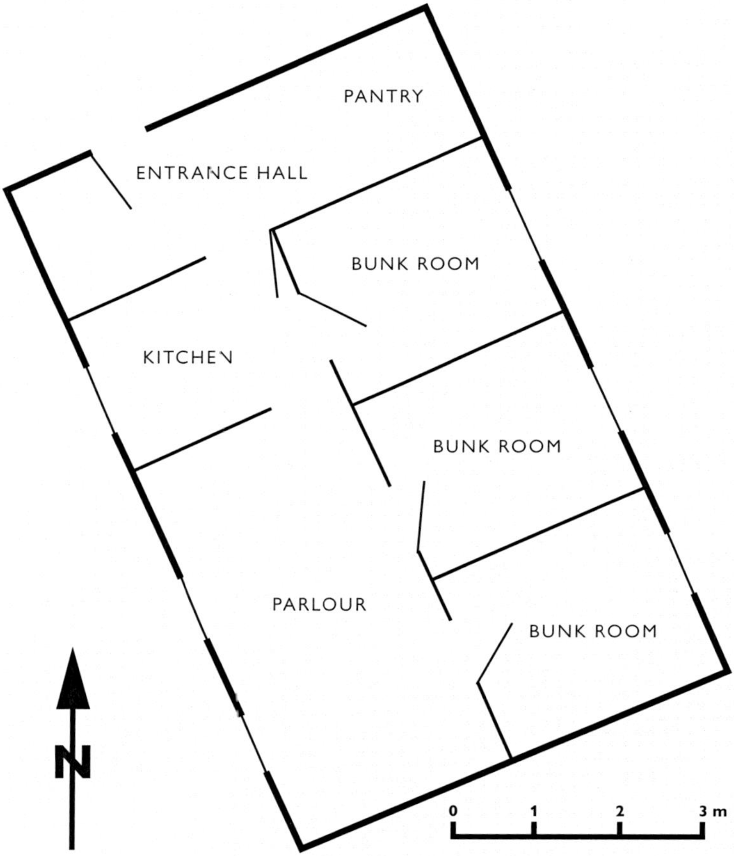

This shows the original basic form of the building, described in § 4, which has had various additions and extensions to the north-west (food store) and south-east (radio room) during many years. Adapted from the plan in: Pereyra, 2007.

Appendix 1: Toponyms of the South Orkney Islands and South Georgia[309]

These are the names that appear in this book, the majority of which are on Laurie Island. The position given of the larger features is approximately that of the mid-point. The principal source is: Hattersley-Smith, G. F. 1991. Positions have been reconciled with the map published in 1988: *British Antarctic Territory, South Orkney Islands East Sheet* (Provisional), 1: 000 000. Cambridge: B.A.S. 100 (Part of SP 23-24/1.2). Tenths of minutes have been interpolated from that map. '*Manchón Austral*' is from the chart extract in: Grunwaldt, E. G. M. 1959, reference was also made to: Morandi, M. C. 1993.

Acuña Island: 60°45·8´S, 44°36·3´W. An islet outside Scotia Bay just off Rae Point; named after Hugo Acuña who wintered in 1904 (originally named Delta Island by William Bruce).

Ailsa Craig: 60°46·9´S, 44°35·8´W. A prominent rock, 169 m elevation, to the south of Acuña Island, see the note in § 20.

Atriceps Island: 60°47·6´S, 45°09´W. A small island off south-east Coronation Island; named after the Blue-eyed Shag colony there (*Phalacrocorax atriceps*).

Coronation Island: 60°38´S, 45°35´W. The largest of the South Orkney Islands; named by George Powell after the Coronation of the British King, George IV, in 1821.

Cumberland Bay: 54°13´S, 36°28´W. A major, bifurcate, fjord-like bay in South Georgia, site of King Edward Cove and Grytviken whaling station; named by Captain James Cook in 1775.

Davis, Point: 60°45·0´S, 44°39·0´W. The eastern entrance to Scotia Bay; named by William Bruce after Walter Gould Davis, Director of the Argentine Meteorological Service in 1904, see § 2.

309. Inserted for the English translation.

Diebel, Point: 60°44·7´S, 44°41·5´W. The western coast of Scotia Bay; named after Otto Diebel (+1905 at the station, see § 61).

Factory Cove: 60°43´S, 45°36´W. An anchorage on the east coast of Signy Island used by a shore whaling station and floating factory site by Tønsbergs Hvalfangeri of Norway from 1920 to 1931.

Grytviken: 54°17´S, 36°30´W. Site of the whaling station of the Compañía Argentina de Pesca at King Edward Cove, South Georgia. A Norwegian term translated as cauldron bay from sealers' try-pots found there in 1902.

Hartree, Cape: 60°47·0´S, 44°42·0´W. The southernmost part of Laurie Island; the south-western extremity of Mossman Peninsula, see note in § 45; named in 1821 by George Powell and Nathanial Palmer (renamed Cape McVitie by William Bruce).

Inaccessible Islands: 60°35´S, 46°44´W. A small isolated group of rocks about 36 km west of Coronation Island; a descriptive name from George Powell, 1821.

Isthmus, The: 60°44·4´S, 44°42·0´W. A 300 m wide neck of low lying land, occasionally overwashed in severe storms, between Uruguay Cove in the north and Scotia Bay on the south where Omond House was established in 1903 and is now the site of Orcadas Station. Named Istmo Iberguren on Argentine charts. It is capitalised in the text when used as a proper noun.

Jessie Bay: 60°43´S, 44°42´W. A large bay on the north-west of Laurie Island; named after the wife of William Bruce, Leader of the Scottish National Antarctic Expedition, 1902–04.

King Edward Cove: 54°17´S, 36°29´W. A small sheltered anchorage in South Georgia with Grytviken and King Edward Point on its shores.

King Edward Point: 54°17´S, 36°30´W. The administrative centre of South Georgia, about 1 km from Grytviken whaling station. Base of the Falkland Islands Dependencies (now South Georgia and South Sandwich Islands) magistrate, a meteorological and research station.

La Monja Glacier: 60°44´S, 44°42´W. A glacier descending from La Monja Peak; named after the peak.

La Monja Peak: 60°43·7´S, 44°40·3´W. A snowy peak, 270 m elevation, north-east of Orcadas Station; a descriptive name for its resemblance to a nun wearing a cowl or domino.

Laurie Island: 60°44´S, 44°37´W. The easternmost of South Orkney Islands, site of Orcadas Station; named after Robert Laurie, a British publisher of nautical charts at the time of Powell and Palmer.

Lewthwaite Strait: 60°41´S, 45°07´W. Lies between Coronation Island and Powell Island; named after George Powell's London agent.

Lola, Mount: 60°43·5´S, 44°41·6´W. A snowy peak, 170 m elevation, above Point Lola; named after María Dolores [Lola] Bosch who became the wife of Luciano Valette.

Lola, Point: 60°43·4´S, 44°42·0´W. The eastern point of Uruguay Cove, named after the mount (originally named Russ Point by William Bruce after a dog of his expedition).

Mabel, Cape: 60°40·6´S, 44°38·5´W. The northern point of Pirie Peninsula; named after the wife of James Hunter Harvey Pirie, a member of the Scottish National Antarctic Expedition.

Mackenzie Peninsula: 60°44´S, 44°47´W. North-western part of Laurie Island, between Jessie Bay and Washington Strait; named after the maiden name of the wife of William Bruce.

Manchón Austral: 60°44·0´S, 44°43·2´W. A 40 m high prominent triangular rocky bluff surrounded by glaciers, at the north-west of Uruguay Cove, named in 1905 when *Uruguay* visited (in English the *Southern Mark*, see the note in § 16).

Martin, Point: 60°45·7´S, 44°40·0´W. On the south-west coast of Scotia Bay; named after William Martin, a sailor aboard *Scotia* who assisted at the observatory.

Melville Highlands: 60°44´S, 44°36´W. Central region of Laurie Island reaching elevations over 500 m; derived from a name applied to the islands by James Weddell in 1822.

Moneta, Cape: 60°43·9´S, 44°45·4´W. The western entrance to Sheila Cove in Jessie Bay, on the east coast of Mackenzie Peninsula.

Mossman, Mount: 60°44·2´S, 44°43·4´W. The highest peak of Mossman Hills, 290 m elevation, west of Point Diebel, named after Robert Mossman who wintered on Laurie Island in 1903 and 1904.

Mossman Hills: 60°45´S, 44°42´W. Range of foot hills, rising to Mount Mossman, along the northern part of Mossman Peninsula near The Isthmus.

Mossman Peninsula: 60°45´S, 44°42´W. Separates Wilton Bay from Scotia Bay, bifurcating at its southern end to Cape Murdoch and Cape Hartree.

Murdoch, Cape: 60°46·6´S, 44°39·0´W. The south-east extremity of Mossman Peninsula; named after William Gordon Burn Murdoch, an Antarctic colleague of William Bruce (named Cape Burn Murdoch by Bruce).

Nivea, Mount: 60°35´S, 68°30´W. A peak, 1265 m elevation, on Coronation Island; the highest peak of the South Orkney Islands, first climbed in September 1956.

Orcadas Station: 60°44·3´S, 44°41·8´W. The Argentine Meteorological Observatory on The Isthmus between Scotia Bay and Uruguay Cove; a Spanish translation of Orkneys.

Pirie Peninsula: 60°42´S, 44°38´W. The eastern limit of Jessie Bay; after James Hunter Harvey Pirie, a member of the Scottish National Antarctic Expedition.

Powell Island: 60°41´S, 45°03´W. A large island between Coronation Island and Laurie Island; named after George Powell, a British sealer and one of the discoverers of the South Orkney Islands.

Rae, Point: 60°45·5´S, 44°36·4´W. The nearest part of Laurie Island to Acuña Island; named by William Bruce after John Rae, a Scottish Arctic explorer.

Ramsay, Mount: 60°43·9´S, 44°44·1´W. A prominent peak, 537 m elevation, at south-western corner of Jessie Bay, the highest peak on Laurie Island; named after Allan Ramsay, Engineer of *Scotia* (+1903, see § 61).

Robertson, Cape: 60°42·7′S, 44°47·5′W. North-west limit of Jessie Bay on Mackenzie Peninsula; named by William Bruce after Thomas Robertson, a Scottish whaling captain, master of *Active* in 1892–93 and *Scotia* in 1902–04.

Saddle Island: 60°37′S, 44°50′W. An islet on the approach to Jessie Bay, about 25 km north of Orcadas Station; a descriptive name applied by James Weddell in 1823.

Sandefjord Peaks: 60°37′S, 45°50′W. High peaks, 635 m elevation, on Coronation Island; named after one of the principal whaling towns in Norway.

Scotia Bay: 60°45′S, 44°40′W. A large bay south of The Isthmus where Orcadas Station lies; named after the ship *Scotia* of the Scottish National Antarctic Expedition (1902–04) which anchored there in 1903 and remained, frozen in by pack ice, for the winter.

Shag Rock: 60°40·5′S, 44°39·4′W. An unofficial name of an outcrop some 250 m off Pirie Peninsula near Cape Mable. [Not to be confused with Shag Rocks some 270 km west of South Georgia in the Southern Ocean.]

Sheila Cove: 60°43·6′S, 44°44·7′W. A small cove south-west of Jessie Bay, named after the daughter of William Bruce.

Signy Island: 60°43′S, 45°38′W. A small island south of Coronation Island, used as an anchorage by whalers who had a small shore station there occupied in summers from 1920 to 1930; named after the daughter of Petter Sølle, a Norwegian whaler. Site of a British winter scientific station open continuously from 1947 to 1995 and in subsequent summers.

South Cape: 60°47·9′S, 45°09′W. The southernmost point of the South Orkney Islands on Atriceps Island at the south-eastern extremity of Coronation Island; a descriptive name.

South Georgia: 53°30′ to 55°S, 35°30′ to 38°40′W. A large mountainous peri-Antarctic Island with Grytviken, several other whaling stations, and the administrative centre at King Edward Point along its north-eastern coast; named by Captain James Cook in 1775 after the British King George III.

South Orkney Islands: 60°30′ to 60°48′S, 44°15′ to 46°45′W. One of the peri-Antarctic Islands in the Southern Ocean; named by George Powell in 1821 after the Orkney Islands north of Scotland.

Uruguay Cove: 60°44·0′S, 44°42·5′W. A cove north of The Isthmus where Orcadas Station lies; named after the Argentine ship *Uruguay* which first anchored there in 1905.

Washington Strait: 60°42′S, 44°55′W. Lies between Laurie Island and Powell Island; named by Nathanial Palmer, a United States sealer and one of the discoverers of the South Orkney Islands, after George Washington, first President of the United States.

Appendix 2
Concise Biography of José Manuel Moneta[310]

José Manuel Moneta was born in Buenos Aires on 9 July 1900. His father was Manuel Moneta and mother Ana Elena, *neé* Vivot. In 1912 he lived in Tierra del Fuego for a period where, when 19 years old, he saw *Uruguay* on her return from a relief voyage to the South Orkney Islands. He was appointed, in 1922, by the Ministry of Agriculture as a technical officer in the Argentine Meteorological Service (Servicio Meteorólogico Nacional) and remained employed by that organization until 1946 when it was incorporated in the Directorate of Meteorology, Geophysics, and Hydrography (Dirección de Meteorologia, Geofísica e Hidrología). His four voyages to the South Orkney Islands were made in 1923 (aboard *Rosita* to and, in 1924, *Karl* from Orcadas), 1925 (*1° de Mayo* to and, in 1926, *Foca* from), 1927 (*Don Ernesto* in both directions), and 1929 (*1° de Mayo* in both directions). Of these four winter parties he was a Meteorological Assistant on the first, Assistant Leader on the second, and Leader of the next two. He described the total of four years and nine months spent in Antarctic surroundings as profitable and beneficial; not only personally, but also for the idea that began to develop 'to discover and establish Antarctica for the Argentines'. Effectively that is what happened. After returning the eminent Buenos Aires newspaper, *La Prensa,* gave him the opportunity to write more than a hundred illustrated articles about the diverse aspects of Antarctic regions. In Buenos Aires he was a founding member of the San Isidro Club in 1935.

On 23 March 1946 he was appointed to the newly established post of Secretary General of the Comisión Nacional del Antártico of the Ministerio de Relaciones Exteriores y Culto, a post he held until 1949. His knowledge of the whaling industry also resulted in him becoming

310. Based on information from his family, with the notes on the dust jacket of the 12th edition and: Del Río, 1984.

the principal Argentine delegate to the meeting which established the International Whaling Commission in Washington during November and December 1946. He was one of the signatories of the convention. From September to December 1947 he was the Argentine substitute delegate to the United Nations in New York.

Subsequently, until 1954, he held a series of Ambassadorial posts representing Argentina in Czechoslovakia (from February 1949), Hungary (from February 1950), Turkey (from November 1951), and the Philippines (from August 1953). He remained with the Ministerio de Realciones Exteriores y Culto until April 1956.

In January and February 1956 he made another visit to the Antarctic aboard *Bahía Aguirre* as a guest of the Instituto Antártico Argentino. The ship visited Argentine stations in the South Shetland Islands and on the Antarctic Peninsula but, unfortunately, not that on the South Orkney Islands. In early 1973, however, he made a return visit to Orcadas and was greeted with honours fifty years after he was first there. Shortly after his return José Manuel Moneta died in Buenos Aires, on 30 March 1973.

Ediciones Peuser, founded in 1867 by Jacobo Peuser in Buenos Aires and closed in 1964, published the twelve editions of his book, *Cuatro Años en las Orcadas del Sur* the first of which appeared on 15 December 1939. For this José Manuel Moneta was awarded a prize by the Comisión Nacional de Cultura. The commission also gave him a grant following the second edition, published in 1941, to complete his studies in the United States. The book proved very popular with twelve editions published in Argentina during the next twenty-four years.

As well as the dwelling becoming known as 'Casa Moneta' José Manuel Moneta is commemorated by the toponym Cape Moneta[311] which was applied by the Comisión de Coordinación Geográfica Antártida Argentina in 1978. Subsequently it was accepted by British Antarctic Territory and other national authorities. This cape is visible from any ship arriving in Uruguay Cove to reach Orcadas station.

311. Morandi, 1993

Bibliography

Acuña, Hugo A. 1982. *Diario del Estafeta Hugo Acuña: Pionero de la Soberania en la Antártida.* Bahía Blanca: Universidad Nacional del Sur. Republished in 2015 with additional material. Ushuaia: Ojosvista.

Andersson, Johan Gunnar 1914. *Harald Wiström.* Stockholm: *Ymer,* 24 (2); 187–188.

Argentina; Dirección General del Servicio Meteorological Nacional 1951. *Datos Climatologicos y Geomagneticos, Islas Orcadas del Sur, … Periodo 1903–1950.* Buenos Aires: Ministerio de Asuntos Téchnicos.

Bagshawe, Richard W. 1947. *Postal History in the Falkland Islands Dependencies.* Cambridge: *Polar Record,* 5 (33–34); 45–59.

Basbous, José 1971. *Primer contacto radiotelefónico desde la Antártida.* Buenos Aires: *Antártida,* 1; 18–21.

Blyth, J. Douglas M. [Editor] 1954. *Antarctic Claims – Diplomatic Exchanges between Great Britain, Argentina and Chile in 1952 and 1953.* Cambridge: *Polar Record,* 7 (48); 212–226.

Brown, Robert N. R.; Mossman, Robert C.; and Pirie, James H. H. 1906. *The Voyage of the 'Scotia': Being the Record of a Voyage of Exploration in Antarctic Seas.* Edinburgh and London: William Blackwood and Sons.

Bruce, William S. [Leader] 1907. *Life in the Antarctic.* London and Glasgow: Gowans and Gray Ltd.

Bruce, William S. [Speak, Peter (Editor)] 1992. *The Log of the* Scotia. Edinburgh: University Press.

Bush, William M. 1982. *Antarctica and International Law.* London: Oceana Publications, 1; 551.

Capdevila, Ricardo 1993. *Vientenueve Hombres del Norte y un Argentino.* Buenos Aires: Dirección Nacional del Antártico, Publicacion 24.

Capdevila, Ricardo 1996. *Accion Argentina en la Antártida entre los Años 1916 y 1930.* Buenos Aires: Dirección Nacional del Antártico.

Davis, Walter G. 1905. *Observaciones de las Orcadas en el año 1904.* Buenos Aires: *Anales de la Oficina Meteorólogica Argentina,* vol XVI.

Del Río, Juan 1984. *Los años de Moneta en las Orcadas.* Buenos Aires: *Antártida,* 13; 2–9.

Grunwaldt, Enrique G. M. [Editor] 1959. *Derrotero Argentino Aeronaval – Antártida y Archipelagos Subantarcticos Argentinos,* Buenos Aires: Servicio de Hidrografía Naval.

Hart, Ian B. 2006. *Whaling in the Falkland Islands Dependencies 1904–1931.* Newton St Margarets: Pequena.

Hattersley-Smith, Geoffrey F. 1991. *The history of place-names in the British Antarctic Territory*, Cambridge: *British Antarctic Survey Scientific Report,* 113.

Headland, Robert K. 1984. *The Island of South Georgia.* Cambridge University Press.

Headland, Robert K. 2009. *A Chronology of Antarctic Exploration.* London: Bernard Quaritch.

Hodson, Arnold W. 1929. *Note on a Visit to the Dependencies of the Falkland Islands.* London: *The Geographical Journal*, 73 (1); 61–66.

Krause, Reinhart 2007. *Edmund Müllers Bericht über seine Überwinterungen auf Laurie Island, South Orkney Islands, Antarktis, in den Jahren 1926 und 1928 – ein Streiflicht.* Bremerhaven: *Polarforschung*, 77 (2–3); 89–94.

Marr, James W. S. 1935. *The South Orkney Islands.* Cambridge: *Discovery Reports,* 10; 283–382.

Mavaroff, José Otto 1954. *Por los Mares Antárticos.* Buenos Aires: Peuser.

Mey, Carlos 2017. www.histarmar.ar/Antartida/Orcadas/Orcadas-Cap10-dotacion.htm, read 20-II-2017.

Morandi, Maria C. 1993. *Nomenclador Antártico Argentino.* Buenos Aires: Servicio de Hidrografía Naval.

Mossman, Robert C. 1907. *The South Orkneys in 1907.* Edinburgh: *Scottish Geographical Magazine*, 24 (7); 348–355.

Murdoch, William Gordon Burn 1894. *From Edinburgh to the Antarctic.* London: Longmans.

Palazzi, Rubén O. 2005. *La Argentina del Extremo Sur, 1810–2004.* Buenos Aires: Dunken.

Pereyra, Pablo 2007. *La Primera Casa Antártica, el Museo Moneta*. pp 61–69; of Vairo, C. *et seq* 2007.

Pierrou, Enrique J. 1975. *90 Años de labor de la Armada Argentina en la Antártida*. Buenos Aires: Armada Argentina.

Powell, George 1822. *Notes on South Shetlands etc., printed to accompany the chart of these newly discovered lands, ···*. London: R. H. Laurie.

Rosenman, Helen [Editor] 1987. *Two Voyages to the South Seas by Captain ··· Jules S.-C. Dumont d'Urville ···*. Melbourne University Press.

Speak, Peter 2002. *William Speirs Bruce*. Edinburgh: National Museums of Scotland.

Stewart, John 2011. *Antarctica: An Encyclopedia (2nd edition)*. Jefferson: McFarland and Company.

Swinney, Geoffrey N. 2002. *William Speirs Bruce, the Ben Nevis Observatory, and Antarctic Meteorology*. Glasgow: *Scottish Geographical Journal*. 118 (4); 263 –282.

Tapia, Atols 1971. *Memorias de Augusto Tapia*. Buenos Aires: *Argentina Austral,* 14 (19); 29–38

Vairo, Carlos; Capdevila, Ricardo; Aldazábel, Verónica; and Pereyra, Pedro 2007. *Antártida: Patrimonio Cultural de la Argentina*. Ushuaia: Museo Maritimo.

Valette, Luciano H. 1906. *Viaje á las Islas Orcadas*. Buenos Aires: *Anales del Ministerio de Agricultura*, 4 (2); 3–70 and (3); 293–318.

Webster, William Henry Bayley 1834. *A Narrative of a Voyage to the Southern Atlantic Ocean ···*. London: Richard Bentley.

Weddell, James 1827. *A Voyage Towards the South Pole Performed in the Years 1822–1824*. London: Longman, Rees, Orme, Brown, and Green.

Woehler, Erik J. [Compiler] 1993. *The Distribution and Abundance of Antarctic and Subantarctic Penguins*. Cambridge: Scientific Committee on Antarctic Research.

Index

The index includes toponyms, most subjects, ship names, and personal names other than those who appear only in the personnel list in § 5. It includes references in the footnotes and illustrations, the latter with italic numbers. Some entries are of principal occurrences only for subjects where much is covered *in passim*.

1° de Mayo, 40, 273, 274, 282, 372, 375, 377, 380, 382, 401, 421
Abrahamsen, Captain Anders M., 282
Absolute Hut, 93, *129*
Acari, 170
Active, 419
Acuña Island, 188, 189, 193, 194, 232, 233, 415
Acuña, Hugo Alberto, 37, 38, *39*, 44, 45, 54, 415
Adélie Penguin, 108, *178*, 197, 199, *231*
Ailsa Craig, 96, 130, 300, 309, 415
albatross, 85
alcohol, 28, 60, 78, 90, 91, 103, 111, 134, 163, 169, 208, 209, 210, 234, 235, 246, 248, 266, 339, 341, 357
algae, 243
All Fools' Day, 248
All Souls' Day, 225
altercation, 142, 215, 278, 343
Alvear, President Marcelo, 316
Álvez, Ricardo Walter, 53
Andersson, Ernst W., 46, 148
Antarctic, 32, 63
Antarctic Peninsula, 54, 422
Antarctic stations, 27
Antarctic Tern, 115
Antarctic Treaty, 10, 24, 35
Antártida Argentina, 29
antenna, 275, 306, 319, 331, 334, 371, 384
Antona, Ángel, 44, 47
appendicitis, 164, 224, 394
Archivo Nacional, Buenos Aires, 369
Arctocephalus gazella, 30
Argentina, President, 36, 315, 316
Argentine Air Force, 49
Argentine Antarctic, 26
Argentine Antarctic bases, 27
Argentine Court of Justice, 369
Argentine Embassy, London, 22
Argentine government, 20, 36, 38, 72
Argentine Meteorological Service, 20, 36, *39*, 53, 55, 65, 315, 338, 363, 415, 421

Argentine National Anthem, 136, 173, 339
Argentine National Day, 134, 173, *342*, 369
Argentine national historic site, 42
Argentine Navy, 20, 21, 27, 33, 261, 273, 285, 346, 401
Argentine sovereignty, 37, 38, 175
aspirin, 161, 163
astronomical cairn, 147
astronomical time, 128
Atriceps Island, 178, 415
Austral, 42
award and decoration, 28
Bache-Wiig, Hartvig, 46, 52, 53, 94, 148, 149, 150, 152, 154, 223, 224, 382
Back, Kenn, 21
Bahía Aguirre, 27, 422
Balæna, 32
Balaenoptera musculus, 67
Baldoni, Emilio Alberto, 44, 47, *72*, *135*, 284, 301, 303, 305, 306, 308, 309, 310, 312, 313, 314, 315, 316, 317, 319, 321, 323, 324, 326, 327, 328, 329, 330, 331, 332, 333, 334, 335, 339, 341, *342*, 343, 346, 347, 349, 350, 353, 354, 355, 358
barber, 160, 291
barograph, 123
base, 86
bath, 141, 158, 159, 282, 392
Bayon, J., 46, 148
Beaufort Scale, 121
Becker, *Frau*, 341
Becker, Konrad, 45, 47, *72*, 285, 292, *322*, 324, 334, 341, *342*, 389
Been, Ragnar Gotfrei, 52
Belgrano, General Manuel, 174
Ben Nevis, 35
Berg, Carl E., 46, 47, 53, 262, 263, 265, 282, 302, 303
Berntsen, Captain Søren, 271, 295
Binnie, Edward Beveridge, 293
birthday, 60, 134, 172, 176, *324*, 393
blizzard, 123, 138, 145, 150, 223, 258
Blue Whale, 67
Blue-eyed Shag, 85, *112*, 115, 155, *157*, 158, *159*, 162, 174, 183, 242, 280, 415
boat excursion, 280
Boneo, Alejandro Martín Guerra : *see* Guerra Boneo, Alejandro Martín
Boracchia, Alejandro J. B., 44, 46, 47, 164
Borge Bay, 52
Bosch, María Dolores [Lola], 417
Bransfield Shag : *see* Blue-eyed Shag
bridge (game), 321
British Antarctic Survey, 9, 22, 52, 270

British government, 62, 64, 66, 175, 272, 293, 380
British Legation, 20, 36
Brown Skua, 115, *119*, 238
Brown, Robert Neal Rudmose, 45, 175
Bruce, William Speirs, 20, 31, 33, 45
Bruhns, Ernst, 47, 55, 56, 87, 90, 92, 103, 105, 121, 126, 127, 128, 130, 133, 135, 137, 141, 159, 163, 171, 179, 188, 201, 204, 207, 208, 209, 210, 211, 212, 213, 214, 215, 216, 217, 218, 221, 236, 237, 238, 241, 247, 248, 250, 255, 256, *257*, 259, 261, 262, 265, 271, 273, 278, 347, 350, 383, 385
Buenos Aires, 20, 35, 36, 38, 40, 55, 87, 100, 170, 173, 253, 260, 266, 282, 321, 332, 334, 341, 352, 363, 367, 389, 421
Bullhead Notothenia, 134
Callorhinus ursinus, 192, *368*
camera, 177, 184, 231, 277, 343, 366
candle light, 247
Capdevila, Ricardo, 10, 22
Cape Petrel, 114, 199
Carcelles, Alberto, 297, 298
Casa Moneta, 9, 21, 23, *41*, 42, *57*, *124*, *150*, *216*, *284*, *307*, *405*, 412, *414*, 422
Casariego, Miss, 340
Casariego, Pedro Martín, 28, 44, 47, *72*, 284, 291, 321, 323, 326, 328, 339, *342*, 345, 355, 363, 364
castor oil, 161, 163, 391
cats, 302
Celestino y Monti, Félix, 47, 48, *193*, 373, 387, *388*, 393, 396, *398*, 399
cemetery, 22, 43, 52, 56, 58, 65, 86, 94, 127, 144, 148, 152, 164, 186, *222*, 223, 224, *225*, 275, 326, 327, 381, 402, *405*
Chakí, Alfonso, 47, 48, 347, 350
champagne, 60, 174, 248, 316
Chaquí, Alfonso: *see* Chakí, Alfonso
Charcot, Jean, 33
chess, 160, 213, 273, 276, 277, 278, 321
Chichilo, *376*
Childermass, 248
Chile, 37
Chinstrap Penguin, 92, 108, *222*
Chionis alba, 115
chocolate, 59, 99, 103, 176, 179, 188, 192, 194, 237, 247
Christmas, 85, 246, 247, 295, 298, 345
church, Grytviken, 63
cigarette: *see* tobacco
ciné film, *231*, 274, 276, 277, 280, 283, 343, 364, 365, 366, 367, *368*, *369*

Cinematografía Valle, 274, *369*
Clarizza, Captain Francisco J., 372
clothes of civilization, 171
clothing, 59, 60, 123, 161, 170, 207, 241, 264, 285, 294
coal, 58, 60, 72, 73, 77, 98, 102, 103, 117, *122*, 123, 128, 161, 245, 266, 308, 401
cocaine and adrenaline, 397
coins, 172
Collasius, Bruno: *see* Kollasius, Bruno
Collembola, 170
Comisión de Coordinación Geográfica Antártida Argentina, 422
Comisión Nacional de Cultura, 422
Comisión Nacional del Antártico, 421
Compañía Argentina de Pesca, 40, 62, 260, 286, 292, 352, 416
Congress of Tucuman, 9 July, 172, 174, 342
Cook, Puerto, 377
Copeland Observatory, 35
Copeland, Ralph, 35
copepods, 243
Coronation Island, 31, 94, 95, 178, 179, 269, 415
Cumberland Bay, 62, 282, 291, 356, 415
Customs Officer, 271, 272, 295, 298, 351
Cuthbertson, William Alexander, *39*, 45
Daption capense, 114
dark-room, 90
Davis, Point, 188, 301, 415
Davis, Walter Gould, 21, 36, 415
death, 53, 58, 65, 223, 361, 383, *see also* cemetery
Decauville railway, 102, 412
Deception Island, 27, 58, 299
decoration and award, 28
Delta Island, 415
Desimone, Alberto, 287, 333
Devoto, Rómulo, 47, 48, *125*, 373, 389, 390, *391*, *392*, 393, 397, 399, 401
Diebel, Otto, 45, 52, *222*, 223, *225*, 416
Diebel, Point, 96, 177, 201, 226, 279, *372*, 416
dip circle, 127
Dirección de Meteorologia, Geofísica e Hidrología, 35, 421
Discovery, 32
Discovery Expeditions, 64, 66
Discovery House, 64
Dobrovolski, Vladimir, 47, 263
dogs, 160, 286, 330, 341, 343, 374, 375, *376*, 386, 387, 389
Dominican Gull, 115
Don Ernesto, 67, 294, 347, 421
Dove, 30

dream, 111, 172, 205, 237, 256, 337
Drygalski, Erich von, 32
Dumont d'Urville, Jules Sebastian César, 31
dust jacket, 24, 305, 421
dwelling, 42, 86
eau-de-cologne, 78, 256
Ediciones Peuser, 422
electric light, 174, 275, 308, 309, 331
electric torch, 130, 144
electrical interference, 320, 401
Elephant Seal, 30, 194
Eliesen, Jon, 45, 52, 223, *225*
Entre los Hielos de las Islas Orcadas, 367, *368*
epileptic seizure, 394
Equinox, 183
Esbensen, Captain Wictor, 66, 282
escabeche, 136
Escalante, Wenceslao, 37
Escobar, Fortunato A., 47, 52, 53, *225*, 346, 350, 370, 371, 372, 383, 384, 385, 386, 401
Eskimo film, 365
Euphausia superba, 227
Ezeiza, 27
Factory Cove, 268, 416
fainá, 399
Fairweather, Captain Alexander, 32
Falkland Islands, 20, 33, 62, 64, 298
Falkland Islands Dependencies, 26, 65, 271, 293, 299, 352
Fállico, Luis, 44, 47, 284, 291, 292, 309, *310*, 321, 323, 325, 326, 327, 328, 329, 330, 339, 341, *342*, 344, 363
familiar form, 88, 142, 263, 273, 279, 391
Fernández, Manuel García, 37
Figueroa, Javier, 22
Filer, John Roger, 52
fire on base, 371
fish, 132, 133, 386
flag pole, 87, *95*, 100, 135, 137, 248, 260, 267
Flaherty, Robert Joseph, 365
flammables store, 277
flensing knives, 71
flensing plan, 63
Flight of the Chimney Pipes, 139
flour, 59, 106
Foca, 281, 421
food store, 42, 59, 103, 114, 117, 155, 329, 389
fresh bread, 60
fresh provisions, 162
fresh water, 89, 104
frostbite, 166
fruit and vegetables, 162
Fuerzas Aéreas de Argentina, 22
Fur Seal, 30, 194
gangrene, 167, 168, 169
García, Admiral Manuel Domecq, 286

gastric-enteritis, 396
Gauss, 32
Gentoo Penguin, 92, 108, *116*
Giant Petrel, 114, *182*, 183, 189, 194, 197, 199
Globus company, 296
Gourlay Peninsula, 52
Governor, 64, 293, 352
gramophone, 89, 103, 136, 139, 160, 344, 401
growlers, 84
Grytviken, 21, 26, 32, 40, 62, *68*, *72*, 77, 269, 282, 286, 290, 294, 347, 352, 354, 355, 416
Guardia Nacional, 40, 58, 60, 72, 74
Guerra Boneo, Alejandro Martín, 44, 47, 273, 275, 276, 277, 278, 279, 321
Haggard, William H. B., 36
Harpon, 282
harpoon, 64, 70, 270
Hartree, Cape, 178, 416
Headland, Robert, 21
Hess, Otto, 47, 263
Historic Site and Monument, 35
hoarfrost, 139, *141*
Hodson, Governor Arnold Weinholt, 352
Holy Innocents' Day, 248
Hormanstorfer, Alejandro Christian, 51, 53
Hoxmark, Thorleif Edgar, 46, 65, 148, 149
hurricane lantern, 60, 123, 124, 144, 149
huskies: *see* dogs
hydrographic survey, 401
Hydrurga leptonyx, 194
hygrograph, 124
hypothermia, 166
Iberguren, Istmo: *see* Isthmus, The
ice fishing, 132, *135*, 386
Ice stalactites, 221
iceberg, 41, 61, 62, 72, 79, 80, 81, 95, 96, 129, 179, 184, 254, 255, 300, 357, 358, 359, 360, 362, 378, 380, 381
ice-blink, 81, 120, 129
ice-foot, 165, 187, 304
ice-pick, 104
ignis fatuus, 327
Ilex paraguyensis, 374
Inaccessible Islands, 31, 416
insect, 170, 243, *see also* moth
Instituto Antártico Argentino, 21, 22, 422
International Whaling Commission, 422
iodine, tincture, 163
Irrgang, Horst E., 47, 263, 265, 282
Isthmus, The, 95, 138, 153, 200, 226, 233, 301, 382, 416
Jacobsen, Fritjof, 282
Jacobsen, Jacob, 52
James Monroe, 30

Jaramillo, Miguel Ángel, 44, 47, 61, *72*, *83*, 283, 287, 288, 290, 291, 301, 302, 303, *304*, 309, *310*, 317, 321, 322, 323, 324, 326, 328, 339, *342*, 344, 345, *348*, 349, 351, 352, 374
Jaramillo, Mrs, 340
Jessie Bay, 149, 153, 280, 388, 416
Johanssen, Captain Johan A., 282
Johnson, John, 52
Kahl, Carlos, 47, *318*, 373, 386, *391*, 397, 399, 401
Karl, 260, 421
Kelp Gull, 115
kerosene: *see* paraffin
Kew magnetometer, 127
King Edward Cove, 21, 40, *61*, 292, 416
King Edward Point, 40, *64*, *65*, 347, 354, 416
Kollasius, Bruno, 46, 47, 48, 87, 99, 164, 165
Kopelmann, Wilhelm, 46, 164, 165, 167, 168
krill, 227
L'Astrolabe, 31
La Boca, 390
La Monja Glacier, 96, 147, 149, 416
La Monja Peak, 417
La Prensa, 26, 421
La Quiaca Observatory, 56, 142
La República, 333
La Zelée, 31
Lancing, 295, 296, 297, 300
Landívar, Gervasio, 22
Landívar, Maria, 22, 24
Larsen, Captain Carl Anton, 21, 32, 63
Larsen, Captain Karl A., 77, 78, 80, 81, 85, 87, 99, 264, 266, 269, 270
Larus dominicanus, 115
latrine, 42, 74, 161
laudanum, 163, 394, 395, 396
Laurie Island, 9, 21, 24, *34*, 40, 94, 95, 153, 178, 193, 224, 280, 286, 299, 312, 385, 387, 400, *408*, 417
Le Français, 33
Leopard Seal, 194
Leptonychotes weddellii, 115
Lewthwaite Strait, 31, 94, 254, 269, 417
lichens, 84, 204, 225
LIK radio, 315
lime juice, 59, 91, 103, 107
Linnæan names, 23
lobscouse, 75
Lola, Mount, 95, *98*, 149, *150*, 151, 152, 153, 154, 185, 219, 223, 417
Lola, Point, 53, 86, 95, 102, 261, 350, 381, 388, 417
LOT Olivos, 332, 335, 336, 346
LRT radio, 312
Mabel, Cape, 153, 280, 417

Mackenzie Peninsula, 130, 304, 417
Macronectes giganteus, 114
MacVitie, Cape, 416
magazines, 180, 264
Magistrate, British, 64, 292, 293, 295
magnetic observations, 33, 35, 43, 126, 207, 218, 345, 384
magnetic storm, 320
Malvinas, Islas: *see* Falkland Islands
Manchón Austral, 85, 95, 260, 381, 417
March of San Lorenzo, 136, 174, 335
Martin, Point, 177, *178*, 196, 197, 199, 200, 228, 237, 251, 301, 309, *310*, 417
Martin, William, *39*
maté, 374, *392*
meat, 27, 75, 92, 107, 117, 204, 236
medical books, 395
medical examination, 56, 163
medical kit, 60, 162, 163, 370, 383, 397
Melville Highlands, 153, 417
meteorological observations, 20, 34, 121, 124, 128, 144, 145, 148, 164, 173, 207, 213, 238, 327, 344, 366
Mey, Carlos, 22, 44
micro-organisms, 243
Mihura, Emilio, 286
milanesas, 107, 183, 204, 232, 234, 236
Ministerio de Relaciones Exteriores y Culto, 421
Ministro Ezcurra, 377
Mirounga leonina, 30
Moneta, Cape, 417, 422
Moneta, Carlos Juan, 50
Moneta, Dora, 22
Moneta, José Manuel, *2*, 22, 29, 47, *125*, *257*, *342*, *388*, 421
Moneta, Julián, 22
Moneta, Manuel, 336, 421
Moneta, Martha, 22
Montero, Jorge Páez: *see* Páez Montero, Jorge
Montevideo, 282, 352
Monti, Félix Celestino y: *see* Celestino y Monti, Félix
Moreno, Francisco Pascasio, 37
morphine, 163, 396, 397
Morse code, 278, 309, 310, 311, 314, 320, 331, 342, 350
Mossman Hills, *91*, 155, 158, *372*, 418
Mossman Peninsula, 147, 156, 177, 178, 226, 242, 300, 418
Mossman, Mount, 95, 147, 223, 275, 306, 418
Mossman, Robert Cockburn, 35, 38, *39*, 44, 45
moth, 169, 170
Müller, Edmund, 47, 282
Murdoch, Cape, 96, 130, 184, 189, 200, 296, 297, 418

Museo de la Armada, Tigre, 22
Museo Marítimo, Ushuaia, 22, 368, 412
Museo Nacional de Ciencias Naturales, 297
Nanook of the North, 365
National Commission for Culture, 26
National Lottery, 324, 345
National Museum of Scotland, 20
nationalities, 44
Naval Arsenal, 372
Naval Communications Service, 283, 284
Naval Military Hospital, 370
nematodes, 170
nephritis, 370
New Year, 248
newspaper, 26, 160, 287, 302, 333, 336, 421
Nivea, Mount, 95, 418
Nordenskjöld, Otto, 32
Norway, 60, 62, 66, 86, 175, 269, 295, 298, 349, 352
Notothenia coriiceps, 134
Oceanites oceanicus, 115
oceanographic laboratory, 64
Odegaard, Georg, 58
Odontesthes bonariensis, 134
Olsen, Captain Hans, 294
Olson, Aksel, 52
Omond House, 34, *39*, *93*, 147
Omond, Robert Trail, 34
onions, 59, 75, 103, 109
Orcadas, 164, 247, 286, 299, 312, 317, 333, 347, 369, *412*, 418
Orkney Olympics, 325
Orlovski, Yakob, 47, 282, 373, *391*
Orwell, 268, 269, 270, 271
Orwell II, 295, 297, 298, 299, *300*, 328, 347, 349, 350, 351
pack ice, 34, 38, 81, 82, 120, 127, 137, *151*, 156, 184, 186, 219, 242, 269, 299, 381
Páez Montero, Jorge, 47, *191*, *193*, 373, 387, *391*, 393, 397, 399
Pagodroma nivea, 114
painting and cleaning, 89, 248, 250, 347
Palazzi, Rubén, 22
Palmer, Captain Nathaniel Brown, 30
Pampa, 297, 375
pampero, 288
pantry, 59, 161, 240, 305
paraffin, 60, 77, 104, 118, 144, 161, 214, 237, 247, 275, 305
Park Wardens, 51
Parlour, 42
pasta, 59
Pejerrey, 134
penguin: *see* Adélie, Chinstrap, or Gentoo Penguin
penguin eggs, 92, 93, 107, 184, *202*, 207, 234, 236, *240*, 242, 345
penguin meat, 109, 131, 201, 236

peri-Antarctic islands, 9, 24, 32, 419
peso, Argentine, 172, 323, 345, 370
Peuser, Jacobo, 422
Phalacrocorax atriceps, 115, 415
Phalacrocorax bransfieldensis, 115
phantom light, 325, 328, 329, *331*
philosophy, 61, 181, 205, 259, 265
Pilar, 283, 367
Pintado Petrel, 114
Piper, Georg, 46, 47, 164, 273, 277, 279, 280, *281*
Pirie Peninsula, 153, 418
Pirie, James Hunter Harvey, *39*, 45
Plagge, Robert, 46, 47, 55, 56, 90, 102, 105, 107, 108, 121, 126, 130, 132, 133, 136, 137, 142, 145, 148, 155, 158, 160, 171, 174, 179, 180, 201, 203, 204, 207, 208, 212, 213, 221, 226, 227, 228, 230, 232, 233, 234, 237, 238, 240, 247, 249, 250, 251, 254, 260, 261, 262,264, 271, 272
Plate, Enrique G., 286, 317, 336
Poch, Captain Ramón Augustín, 273
poker, 321, 322, *325*
Post Office, 37, 269, 270, 287, 350
postage stamps, 37
Post-Master, 37
potatoes, 59, 75, 87, 91, 103, 109, 232, 236
powder magazine, 270
Powell Island, 94, 300, 418
Powell, Captain George, 30
provisions, 56, 59, 72, 99, 102, *111*, 155, 245, 265, 285, 329, 401, *402*
Pseudoboeckella, 243
psychrometer, 123, 344
puchero criollo, 59
Puerto Belgrano, 55
Puerto Militar, 55
Pygoscelis adeliae, 108
Pygoscelis antarctica, 92
Pygoscelis papua, 92
pylon, *95*, *150*, 275, 285, 293, 305, *307*
quadruped, 387
radio broadcast, 335, 337, 340, 342
radio call-sign, 293, 303, 312, 380
radio communication, 19, 25, 42, 43, 64, 167, 247, 252, 272, 274, 276, *277*, 284, 293, 303, *310*, 315, 317, 320, 331, 349, 370, 386
Radio Operator, 52, 284, 285, 297, 298, 305, 346, 370, 373, 380, 381, 384, 386, 401
Rae, Point, 96, 418
Ramsay, Allan George, 45, 52, 223, *225*, *418*
Ramsay, Mount, *91*, 95, *150*, 242, 326, *337*, 381, 400, *418*
República Argentina, 172
Reynoso, José, 44, 46, 47
rice, 59, 109, 236

Río Negro, 296, 297
Robertson, Cape, 153, 281, 419
Robertson, Captain Thomas, 45, 419
Roca, President General Julio Argentino, 36
Rodríguez, Carlos V., 47, 375
Rodríguez, Víctor, 375
Rosita, 72, 77, 79, 100, 260, 421
Ross, Alastair, *39*, 45
Royal Geographical Society, 22, 32
Royal Scottish Geographical Society, 21
Ruberoid, 42, 249, 250, 305
Rumbolds, William Charles, 271, 272, 295, 351
Russ Point, 417
Saddle Island, 31, 33, 85, 95, 137, 153, 185, 219, 226, 254, 260, 268, 269, 300, 381, 419
San Isidro Club, 421
San Martín, General José de, 136
Sandefjord Peaks, 31, 419
Saunders, Alfred, 64
Scandinavia, 53, 66, 75, 76, 90, 148, 262
Schjær, Sigurd, 46, 400
Schwarz, José, 297
Scotia, 20, 33, *34*, 38, 45, 223, 419
Scotia Bay, *34*, 95, 96, 102, 108, 113, 130, 132, *150*, 156, 164, 177, 188, 189, 220, 226, 244, 250, 251, 254, 279, 296, 300, 301, 302, 310, *372*, 415, 419
Scott Polar Research Institute, 9, 20, 22, 24, 34, 39, 64
Scott, Captain Robert Falcon, 32
Scottish National Antarctic Expedition, 35, *39*, 45, 276
sea temperature, *83*, 164
seal: *see* Fur, Elephant, Leopard, or Weddell Seal
seal blubber, 102, 245
seal meat, 204, 235
sealers, 30, 40, 63, 66, 416
Servicio Meteorólogico Nacional, 20, 421, *see also* Argentine Meteorological Service
Shackleton, Sir Ernest Henry, 65, 235, 352
Shag Rock, 280, 419
shaven head, *292*
Sheathbill, 115, *118*, 194
sheep, 387
Sheila Cove, 419
shipwrecked crew, 328
Signy Island, 52, 224, 268, 270, 295, 299, *300*, 419
ski, 60, 89, 148, 150, 151, 152, 154, 176, 177, 184, 188, 192, 195, 196, 198, 200, 205, 226, 228, 233, 238, 279, 303, 343, 388
Skilton, Kathleen, 21
slop chest, 294
Smith, Captain William, 30
Smith, William, 38, *39*, 45

Snow Petrel, 114
soap, 142, 207, 208, 209, 211, 214, 216, 218
soda siphon, 322
Solstice, 182
Sørlle, Petter, 419
Sosa, Elías Reynoso, 44, 46
Soto, Walter, 50, 53
South Cape, 178, 419
South Georgia, 21, 25, 32, 40, 58, 61, 62, 66, 67, 73, 77, 78, 114, 179, 208, 244, 260, 269, 272, 281, 282, 285, 288, 291, 294, 295, 297, 298, 302, 313, 346, 347, 349, 351, *354*, 415, 419
South Orkney Islands, *6*, 9, 31, 33, 38, 40, 54, 65, 73, 84, 113, 142, 175, 253, 273, 274, 295, 329, 335, 370, 399, 401, *406*, 415, 420, 422
South Pole Correspondent, 287
South Shetland Islands, 30, 58, 299, 422
Southern Ocean, 25, 33, 38, 61, 66, 67, 72, 77, 133, 244, 286, 298, 299, 363, 372, 377, 419, 420
special meals, 134, 392
Stanley, 26, 33, 293
Stercorarius antarcticus, 115
Sterna vittata, 115
Stevenson screen, 35, 123, *126*, 145, 219, 223
Stuxberg, Anton, 46, 148, 149
suit, immaculately tailored, 170
Svend Foyn, 58
Szmula, Edgar Hans Ulrich Carl, 38, *39*, 45
Talamón, Gastón O., 44, 45
Tapia, Augusto, 44, 46, 164, 165, 166, 167, 168, 273
Tavella, Daniel Néstor, 53
Tero, 92
terrestrial invertebrates, 170
territorial waters, 175
thermograph, 119, 124
thermometer, 122, 124, 127, 130, 164, 219, 237, 344
Thom, Captain Ingvar, 271
Thorsen, Captain Gotfred, 288, 355, 358, 360, 361
Thorstensen, Henrik, 52
tide-crack, 187
Tierra del Fuego, 40, 55, 62, 421
Tijuca, 40, 285, 286, *289*, 290, *292*, *294*, 346, 347, 351, 352, *353*, *354*, 356, 357, 360, 378
Tijuca brygge, 294
tobacco, 28, 60, 99, 110, 172, 181, 256, 278, 294, 339, 345, 386
Tolnay, Albert de, 47, 273, 278
Tønsbergs Hvalfangeri, 268, 295, 416
tourist visit, 375
truco, 401
United Nations, 422

Uruguay, 38, 55, 85, 244, 372, 420, 421
Uruguay Cove, *83*, 86, *91*, 95, *98*, 99, 104, 135, 137, 140, *150*, *151*, 164, 186, 200, 219, 220, 226, 227, 241, 242, 244, 248, 251, 254, 260, 275, 281, 301, 304, 326, 328, *337*, 349, *372*, 380, 381, 388, 412, 420
Ushuaia, 27, 40, 309, 315, 316, 333, 346, 377
Vago, Captain Ricardo, 58
Vairo, Carlos, 22
Valentiner, Hugo, 46, 47, 54, 55, 56, 74, 79, 80, 85, 86, 87, 90, 94, 100, *101*, 103, 104, 105, 107, 111, 114, 115, 119, 121, 122, 126, 127, 128, 130, 137, 140, 144, 148, 151, 155, 159, 160, 162, 163, 164, 166, 167, 171, 172, 174, 179, 183, 184, 199, 201, 202, 203, 207, 212, 215, 216, 217, 220, 221, 234, 236, 237, 239, 243, 244, 245,249, 251, 252, 253, 256, 260, 261, 263, 264, 265, 268, 269, 270, 271
Valette, Luciano Honorio, 38, *39*, 44, 45, 54, 202, *240*, 417
Valle, Federico, 274
Vanellis chilensis, 92
Variometer Hut, 93, *97*, 127, 128, 144, 145, 146, 186, 214, 242, 252, 327, 330, *336*
vegetation, 62, 84, 179, 204, 205, 225
Vírgenes, Cabo, 313, 346
Vivot, Ana Elena, 421
Washington Strait, 94, 153, 155, 156, 179, 200, 226, 254, 269, 281, 300, 304, 420
water fleas, 243
weapon, 107, 114, 135, 155, 157, 158, 248, 379
Webster, Captain William, 30
Weddell Seal, 115, 183, 188, 189, 190, *191*, 193, 194, 195, *196*, 232, 237
Weddell, Captain James, 31
whale-catcher, 40, 58, *61*, 63, 67, 70, 73, 81, 83, 86, 90, 99, 102, 163, 208, 260, 264, 267, 269, 274, 281, 282, 290, 294, 295, 296, 300, 303, 328, 346, 350, 357, 378, 380
whaling industry, 26, 32, 62, 66, 175, 268, 272, 298
Wiggin, George Otis, 55
will-o'the-wisp, 327
Wilson's Storm Petrel, 115
Wilton, David Walter, 45, *182*
wine, 91, 134, 137
Wiström, Harald, 45, 46, 52, 224, *225*
Witt, Alfredo C., 44, 47, 263, 264, 265, 267, 282, 302
women, 78, 142, 173, 180, 181, 258, 291, 319, 340, 341, 345, 375

Zeiger, Otto, 47, 55, 91, 103, 105, 107, 109, *111*, 117, 134, 142, 159, 184, 204, 209, 210, 213, 235, 237, 246, *257*, 261, 263, 265, 272